WOOD ADHESIVES

CHEMISTRY AND TECHNOLOGY

WOOD ADHESIVES

CHEMISTRY AND TECHNOLOGY

edited by

A. Pizzi

Chemistry Division
National Timber Research Institute
Council for Scientific and Industrial Research
Pretoria, South Africa

MARCEL DEKKER, INC. New York and Basel

Library of Congress Cataloging in Publication Data
Main entry under title:

Wood adhesives.

Includes bibliographies and indexes.
1. Adhesives. 2. Wood--Bonding. I. Pizzi, A. (Antonio), [date].
TP968.W66 1983 668'.3 83-20870
ISBN 0-8247-1579-9

MARCEL DEKKER, INC.
270 Madison Avenue, New York, New York 10016

Current printing (last digit):
10 9 8 7 6 5 4 3 2

PRINTED IN THE UNITED STATES OF AMERICA

Preface

Sticking wood together is a daily experience in the timber processing industry. Thus, the field of adhesives for wood is an important one and has significant technical and economical implications. Many excellent adhesives and adhesion books already exist, some of which describe adhesive resins that, among other applications, also can be used for wood. However, books collecting in a single volume the more technically and economically important adhesives for wood bonding and their preparation and formulation as well as techniques and hints on their application are quite scarce. It is hoped that this volume will help to fill such a gap. Only the main adhesives in use today and those adhesives that show considerable promise for the near and not so near future are discussed. In this respect, I believe it is the first time that the new categories of tannin-based, lignin-based, and diisocyanate wood adhesives are collected in the same book and thoroughly described. These adhesives, some of which are already in the initial stages of extensive industrial ultilization, have already generated and will continue to generate considerable interest. For this reason they have been treated as extensively as more established wood adhesives such as phenolic and aminoplastic resins. A few adhesives have been excluded. I particularly refer to casein and soya bean adhesives; these were extensively used in the past, but their use has been drastically reduced by the introduction of the cheaper and more effective synthetic resins during the early thirties. They were excluded because it was felt that in a world in which protein food is already quite scarce and is likely to become even more so, employing even a small fraction of the available protein material for use in adhesives is bound to be, sooner or later, phased out.

It is a pleasant duty on completing a volume of this nature to acknowledge openly the help willingly given by friends, colleagues, companies, and institutions. My first thanks go to Dr. Maurits Dekker of Marcel Dekker, Inc., New York, for proposing this book and for prompting me to finish it. My sincere thanks also go to all the staff of Marcel Dekker, Inc., who helped during many stages of the work and without whom this volume would have not seen the light. Second, I would like to thank all the authors of the various chapters for their patience and for finding the time, in between their other numerous tasks, to contribute to this volume. Third, I wish to thank my friends and colleagues at the National Timber Research Institute, at other research and standard institutions, and in the adhesive and timber industries both in South Africa and abroad, for their help. Finally, last but not least, I would like to thank both the Council for Scientific and Industrial Research, Pretoria, South Africa and my family for permission and encouragements, although of a different nature, and at different levels, to compile this volume.

A. PIZZI

Contributors

JOHN M. DINWOODIE Princes Risborough Laboratory, Building Research Establishment, Princes Risborough, Buckinghamshire, England

K. C. FRISCH Polymer Institute, University of Detroit, Detroit, Michigan

TERENCE MARTIN GOULDING General Chemical Corporation, Johannesburg, South Africa

HORST H. NIMZ* Polymer Institute, University of Karlsruhe, Federal Republic of Germany

A. PIZZI Chemistry Division, National Timber Research Institute, Council for Scientific and Industrial Research, Pretoria, South Africa

L. P. RUMAO† Polymer Institute, University of Detroit, Detroit, Michigan

*Present affiliation: Institute of Wood Chemistry and General Technology of Wood, Federal Research Center for Forestry and Forestry Products, Hamburg, Federal Republic of Germany.

†Present affiliation: Ford Motor Co., Dearborn, Michigan.

Contents

1

Properties and Performance of Wood Adhesives

JOHN M. DINWOODIE / *Building Research Establishment, Princes Risborough, England*

I. INTRODUCTION

Although it is a relatively simple task to discuss the properties of an adhesive in terms of either its chemical composition or its physical and mechanical characteristics, it is much more difficult to define what is meant by performance. First, we must seek to qualify it by asking "Under what conditions?" and "How much time is involved?"; second, even with such qualifications, we are often not clear as to how to specify and quantify what we think we mean by performance; and third, and perhaps most important, is the prediction, together with the associated degree of assurance of that prediction, of performance from observed behavior under short-term tests.

In this chapter the wide range of adhesives currently in use in forming wood to wood bonds is discussed, together with their basic chemical composition and their physical and mechanical properties. The quality of the adhesive is treated separately from the performance of the bond, and considerable emphasis is placed on describing various test methods aimed at simulating long-term weathering and the degree of assurance in the use of these tests for predictive purposes. Finally, examples of the performance of bonded wood products under a wide range of conditions are presented and discussed.

The need to bond together pieces of wood has been recognized from the earliest times of recorded history, when mud and animal manure were used as early adhesives. Throughout the ages a succession of adhesives of natural origin have been used, including animal glues, casein, and starch. All these glues lacked durability when used in hazard situations, although they performed admirably under dry, cool conditions. Criticism of their performance in the past is a reflection of their misuse under unsuitable conditions.

The present century has seen the replacement of these natural adhesives in all but a few instances by a galaxy of synthetic resins. Although the majority of wood-to-wood bonds are formed using the formaldehyde-based thermosetting resins developed between 1930 and 1945, there is an increasing awareness at the present time of the advantages to be obtained from thermosetting resins, such as isocyanate and cross-linked polyvinyl acetate, or from the ever-increasing range of thermoplastic resins and elastomeric adhesives.

The principal adhesives currently in use in the formation of wood-to-wood bonds in a wide range of applications are listed in Table 1.1. This table is to a large extent based on that in BS 5442, 1979: *Classification of Adhesives for Construction: Part 3—Adhesives for Use with Wood,* and divides the range of adhesives into classes based on their performance under different environmental conditions.

Table 1.1. Adhesives for Bonding Wood to Wood

	Thermoset Resins									Elastomers					Thermoplastics			Other
	Urea-formaldehyde	Melamine/urea-formaldehyde	Phenol-formaldehyde	Resorcinol/phenol-formaldehyde	Tannin-phenol-formaldehyde	Tannin-formaldehyde	Sulfite liquor	Diisocyanate	Cross-linked polyvinyl acetate	Polychloroprene (contact)	Polychloroprene (gap)	Polyacrylonitrile-styrene	Styrene-butadiene rubber	Neoprene	Ethylene vinyl acetate	Polychloroprene latex	Polyvinyl acetate emulsion	Cement
Standard Abbreviation	UF	MUF	PF	RF	TPF	TF	SL	MDI	PVA				SBR				PVA	
Laminated beams Prefabricated trusses	4	4	1	1	1	1												
Boat building		2	1	1														
Window frames Exterior doors		2	1	1	1	1			2									
Joinery on site	3	2							2		3	3	3	3			3	
Interior doors	3								2								3	
Furniture	3	2									4				3		3	
Wood veneering	3	2							2	3					3	3	3	
Waferboard manufacture		1	1															
Particleboard manufacture	3	2	1	1	1	1	1	1-3										1
Plywood manufacture	3	2	1	1	1	1												
MDF manufacture	3																	
Wood wool cement slab manufacture																		1

1, Suitable for full exposure to weather or water immersion; 2, suitable for exposure intermediate between 1 and 3; 3, suitable for service inside heated and ventilated buildings and in other enviornments where equilibrium moisture content of softwood does not exceed 18% and bond line remains below 50°C; 4, could be suitable in special cases, but is not used in general practice.

Source: Adapted from British Standards Institution, BS 5442, 1979.

II. PROPERTIES OF WOOD ADHESIVES

A. Urea-Formaldehyde Resins

Developed in the early 1930s urea-formaldehyde (UF) resins have seen wide application throughout the world in bonding solid wood, plywood, and particularly, particleboard of different types. In many countries, including the United Kingdom, urea-formaldehyde resin is used almost exclusively for particleboard manufacture, although in other countries, including Norway, Sweden, and the Federal Republic of Germany, probably less than 75% of the particleboard manufactured is UF bonded. Use of UF resins in plywood manufacture is limited to the USSR and the Southeast Asian countries.

The resin is primarily based on urea and formaldehyde, the former being produced commercially by the reaction of carbon dioxide and ammonia at a temperature of 135-200°C and pressure of 70-230 atm, while the latter constituent is manufactured by the oxidation of methanol prepared commercially from carbon monoxide and hydrogen, or from petroleum.

The formaldehyde and urea in a molar ratio varying from 1.2:1 to 2.0:1 are heated under neutral or slightly alkaline conditions (pH 7-8) to produce mono-, di-, tri-, and tetramethylol ureas. Because of problems associated with formaldehyde emission both in the manufacture and subsequent use in service of the boards, it is now common practice to use resins with a molar ratio toward the lower end of this range. In this first stage of resin production no condensation reaction takes place and the prepolymers that are formed have no adhesive properties. In the second stage the solution of methylol ureas is refluxed and made slightly acidic (pH 4-6) to induce the condensation reaction to occur. Long-chain molecules of urea-formaldehyde are formed which become cross-linked by the formation of methylene bridges. The solution is neutralized to arrest the reaction, cooled, and then evaporated under vacuum to a liquid adhesive of about 65% solid content, or alternatively it can be spray-dried to give a resin powder. Considerable controversy exists over the actual configuration of the urea-formaldehyde polymer molecule in the uncured resin and linear, branched, and cyclic polymeric forms have been recorded, frequently within the same batch of resin. The reactive end groups of all these polymeric forms are predominantly methylol groups [1-3].

Precondensed neutral UF resins are hardened or cured by inducing a sufficient reduction in the pH; this is usually achieved by the action of a catalyst (hardener). Where curing is being effected at elevated temperatures, ammonium chloride or ammonium sulfate is used as the hardener. To control the amount of free formaldehyde in the urea resin solution, and in turn the amount of acid and hence the reaction rate, a buffer such as ammonia, urea, or hexamethylene tetramine is added; this is the system adopted in plywood and particleboard manufacture.

Where room-temperature curing is carried out, organic acids such as citric, formic, or tartaric acid are employed as hardeners, the acid usually being applied to one of the surfaces and the UF resin to the other surface; this system is used in the manufacture of glulam beams and window joinery.

During the curing process two sets of condensation reactions take place to form a rigid cross-linked three-dimensional network of macromolecules, the exact structure of which is not fully understood. The first reaction is the linking together of adjacent polymers through the amide nitrogen of each polymer by a methylene bridge which originates from the formaldehyde: the second reaction is between the methylol end groups of adjacent polymers to form an ether bridge. Curing, therefore, results in an initial increase in resin viscosity followed by gelation and finally by complete cross-linking to give a rigid thermoset resin.

Although the basic process as outlined above is fairly straightforward, resin manufacture is much more complex and has become quite sophisticated with the production of adhesives to suit a particular process or plant; thus the conditions of manufacture can be changed to modify the product of the condensation reactions, and it is also fairly common practice nowadays to incorporate additional urea to mop up the excess of free formaldehyde. Recent research has indicated that the incorporation of scavengers such as sodium disulfite or sodium dithionite can significantly reduce the release of formaldehyde [4].

One of the more important aspects of resin formulation is the molar ratio of the formaldehyde to urea. This, together with the nature of the prepolymers formed in the first stage of resin production, plays the major role in controlling the intermediate products formed during the curing process, and consequently on the performance of the hardened resin. Resins with low F/U ratios (1.2:1-1.6:1) have a longer pot life (i.e., longer gelation time) and a lower free formaldehyde content, but have a higher viscosity, lower water resistance (durability), lower strength and stiffness, and a slower curing rate than resins with a high F/U ratio (1.8:1-2.0:1) [5-7]. Until fairly recently, resins with a ratio within the range 1.4:1-1.6:1 were commonly employed in particleboard manufacture, but with the ever-increasing restrictions imposed by many countries on the permissible level of free formaldehyde given off by the boards, resins with a ratio as low as 1.18:1 are now being used. Much work is in progress to ensure that the strength, stiffness, and durability of these low-ratio resins do not fall, and this appears to be possible by the addition of various urea derivatives, melamine, or acid scavengers [8].

One of the major disadvantages of UF adhesives is their susceptibility to hydrolytic degradation in the presence of moisture and/or acids, especially at moderate to elevated temperatures; thus in cold water the rate of breakdown of the resin structure is extremely slow, but at temperatures above 40°C resin deterioration accelerates, and above 60°C the process is very rapid.

B. Melamine-Urea-Formaldehyde Resins

It is possible to produce a thermosetting synthetic resin by condensation of melamine with formaldehyde in a manner similar to that using urea. The basic reaction was discovered in the late 1930s, but it was not until the late 1950s that melamine resins were used in particleboard manufacture. In 1980 the first waferboard factory using melamine-urea-formaldehyde (MUF) resin was commissioned in Finland.

Melamine-formaldehyde is a most attractive adhesive on the grounds of its resistance to moisture in the cured state and the use of shorter press times than those necessary for phenol-formaldehyde resins. However, the adhesive is expensive, about 3.5 times the price of UF and about 20-25% more expensive than PF resin when based on 100% resin solids. Consequently, the usual practice is to dilute the melamine with urea, or from the opposite viewpoint, to fortify the urea with melamine. To achieve compliance with the French V313 cyclic test, which is now internationally regarded as an accelerated test which provides a good estimate of external weathering, the relative proportions of melamine to urea must be better than about 40:60.

The combination of melamine and urea in the resin can be brought about either as an admixture of melamine-formaldehyde resin and urea-formaldehyde resin, or by the production of a co-condensate where both melamine and urea are mixed and reacted together with formaldehyde. A whole range of MUF adhesives are now available, many with particular additives or chemical modifications [9].

C. Phenol-Formaldehyde Resins

Although the possibility of using phenol-formaldehyde (PF) resins as an adhesive was recognized as long ago as the end of the nineteenth century, it was not until the 1930s that liquid phenolics were introduced commercially. At the present time it is the most commonly used adhesive in both plywood and waferboard production and for the manufacture of glulam beams and wooden boat construction. The amount of PF-bonded particleboard is small, the principal producer being the Federal Republic of Germany.

PF resins can be prepared by two different methods. The first involves the reaction of phenol, generally produced by synthesis from benzene, with excess formaldehyde in the presence of an alkali catalyst. Mole ratios of formaldehyde to phenol of 1.8:1-2.2:1 are employed and the type of PF resin produced is usually referred to as "resol"; these resins are usually stabilized by incorporating excess alkali. Curing to produce a rigid thermoset resin is activated by increasing temperature and takes place under the action of the alkali present in the solution [2,3].

The second production process for PF involves the reaction of excess phenol with formaldehyde in the presence of an acid catalyst; mole ratios of 0.8:1-1:1 are used to produce a hard resin known as "novolac." This is ground down into a fine powder to which is added about 15% of hexamethylenetetramine. Under the action of heat, as in a press, this melts and breaks down to form both ammonia, which acts as a catalyst, and formaldehyde, which reacts with the now liquefied novalac to produce a thermosetting resin. Such resins are much more stable than the resols but must be kept dry [2,3].

In plywood, particleboard, and glulam production, resol-type PF resins are normally used, purchased either as a liquid or as a spray-dried powder which has to be mixed with water. Most waferboard plants, however, use novalac resins in powdered form.

The bond strength of PF resins is high and deterioration at elevated temperatures in the presence of moisture is slight compared with UF. Resin costs are high and current prices are about 2.8 times those of UF resin based on 100% resin solids.

D. Resorcinol-Phenol-Formaldehyde Resins

Resorcinol-formaldehyde (RF) resins have been in use since 1943, their main application being in the manufacture of glulam, boat construction, and general assembly work to be installed in hazardous situations.

Resorcinol is a phenolic substance, but it does possess much higher reactivity than phenol. Generally, RF adhesives set quickly at temperatures between 5 and 70°C, but care has to be exercised to ensure adequate wetting of both surfaces of the joint.

Resorcinol is now very expensive to produce and is seldom used by itself; rather, it is mixed in equal parts with phenol, or else a co-condensate of resorcinol and phenol is used. Formaldehyde, usually in the form of paraformaldehyde, is generally added immediately before application.

E. Tannin-Formaldehyde Resins

The continual and progressive increase in cost of petroleum-based products has added new impetus to the replacement of the synthetic phenol constituent of a PF resin by a natural polyphenol such as tannin. Although most timbers contain some tannin both in the bark and wood, the content is usually very low; however, a few species, such as mimosa wattle (*Acacia mollissima*) and quebracho (*Schinopsis* sp.) and to a lesser extent radiata pine (*Pinus radiata*), have considerable quantities of tannin present in the bark, and, for the first two species named, in the wood as well [3].

Although the possibility of utilizing the phenolic nature of natural tannins in producing formaldehyde-based adhesives was recognized nearly 30 years ago, those unfortified resins produced until fairly recent times had poor cohesive strength and low resistance to moisture, probably as a result of very limited cross-linking in the structure; poor adhesion properties due in part to the high inherent viscosity of the resin; and short pot life due to high reactivity of the tannin.

The last decade has seen the successful outcome of much research in this field with a marked reduction, and in some cases the elimination, of these problems. Increase in both strength and resistance to moisture can be effected by the addition of small amounts of resorcinol-, phenol- or urea-formaldehyde, which induces a much higher degree of cross-linking, while the use of placticizers reduces the incidence of stress concentrations in the glue line. Pot life can be extended by lowering the initial viscosity achieved by reducing the pH, or by adding hardeners such as hexamethylene tetramine, which also accelerates curing times [10]. Recent research has indicated that increased cross-linking can be achieved using metallic ion catalysts in place of the additions of phenolic- and aminoplastic formaldehyde, and that resorcinol can now be generated in situ from the tannin skeletal structure [11]. Even more recent laboratory work has indicated that diisocyanate can be successfully used as a pine tannin fortifier in the ratio 30:70 for particleboard resins [12].

Many of these developments have led to the commercial production of resins of low viscosity and with a reactivity at least equal to that of UF resins. In South Africa hot-setting powder modified tannin resins from wattle bark are available separately for plywood and particleboard manufacture, while a cold-setting liquid resorcinol tannin resin has been produced for use in glulam, finger jointing, and boat construction. In New Zealand tannin resins have been developed from *P. radiata* bark and are used unfortified in particleboard production and fortified with phenol formaldehyde in the manufacture of plywood [13].

F. Sulfite Liquor Resins

Another resin with a natural base which has been used commercially in the production of particleboard is the waste liquor from sulfite pulp mills. The Danish Pederson process, which was patented in the United Kingdom in 1965 following application in 1962 (patent no. 993,693) was developed commercially in the late 1960s and two mills were constructed, one in Denmark, the other in Finland.

About 12% waste liquor at 50% solids content plus a catalyst was sprayed onto the wood chips and the mat pressed for 1.5 min per millimeter of board thickness at 180°C, a long press time being necessary to achieve partial polymerization. After the boards emerged from the hot press, they were autoclaved at 200°C and a pressure of 10 Pa for at least 60 min to complete the polymerization process. The boards were dark brown in color due to the autoclaving, and of only moderate strength, especially in relation to their density, but had good dimensional stability and performed very well under conditions of high humidity [1,3].

Unfortunately, the process proved uneconomical, due basically to the long press time, and production was discontinued several years ago. There is renewed interest in the use of sulfite liquor as an adhesive because of its low cost and its high performance as a polymerized resin. Much research has been carried out to improve the rate of the polymerization process, and it has been found that some improvement in curing time can be achieved by the addition of small amounts of acid-curing PF resins—about 2-4% resin to 10% lignin—but unfortunately, high pressing temperatures are still required [14]. A greatly enhanced rate of curing has been achieved by acidification of lignosulfonate salt using 4-6% sulfuric acid. This produces a board of low pH, thereby necessitating some posttreatment [15].

A completely different approach has been to separate the high molecular weight content by ultrafiltration and to use these polymers as a coreactant with low molecular weight PF resin. The adhesive contains about 40-70% lignin and results in a strong, water-resistant bond after a markedly reduced cure time [16]. A commercial resin based on this research has recently been marketed.

G. Isocyanate Resins

The basic chemistry of organic isocyanates was developed in Germany in the late 1930s, and adhesives based on isocyantes were first used in in the mid-1940s. Although there was some interest in the commercial application of this type of adhesive, the combination of cost, technical problems in production, and health hazards prevented its use. Recent technical developments, on the one hand, and the demand for high-performance particleboard and a total absence of free formaldehyde emission, on the other hand, have led to the commercial production of isocyanate binders particularly for particleboard manufacture. Interest centers on the low molecular weight diisocyanates, specifically dipenylmethanediisocyanate, known technically as MDI but popularly as isocyanate. These diisocyanates are produced by the reaction of primary amines and phosgene [17].

The adhesive properties of isocyanate are based primarily on the reactivity of the NCO groups, and covalent bonds taking the form of urethane bridges are formed with the hydroxyl groups of the cellulose of the wood. Thus the wood is chemically bonded to the adhesive, resulting in an extremely strong bond which is resistant to moisture and dilute acids [17,18]. It is possible therefore to reduce the proportion by weight of the resin to below 5% without appreciable loss in board strength. The effect of the extremely high initial cost of the resin is therefore reduced considerably at this level of application, and further savings in cost are effected by reductions of some 10-15% in the volume of wood used to produce a board of similar strength performance to a formaldehyde-based board, and in the use of shorter press times [18]. Furnishes with a higher moisture content can be used and the total cost of using isocyanate is little different from that using formaldehyde-based binders.

One of the prime disadvantages of isocyanates in the past has been its ability to bond to metal surfaces during pressing. Until recent times this has restricted the use of this binder to only the core layer, but recent developments in release agents and in self-releasing isocyanates have made possible its use throughout the board [18]: performance of the board with regard to moisture resistance is very dependent on the amount of isocyanate used.

Because of the high cost of MDI much interest has recently been focused on the possibility of reacting it either with lignin pulping residues in approximately equal proportions, or with smaller amounts of certain polyols to produce a urethane-type adhesive.

H. Cross-Linked Polyvinyl Acetate Resins

The incorporation of a catalyst in specially formulated polyvinyl acetate (PVA)* resins induces a high degree of cross-linking in the glue system as it dries out, thereby conferring a considerable water resistance to the system. Durability as high as that of resorcinol has been claimed, although certainly not proven. Maximum strength and water resistance are not achieved until up to 14 days after application. It would appear that the bond strength of these adhesives is influenced predominantly by the amount of grafting of the vinyl acetate monomer on the polyvinyl alcohol at the time of emulsion synthesis [19].

Unfortunately, these adhesives are incapable of carrying a substained load and considerable creep can occur in joints over several weeks or months even at ambient conditions. Consequently, these cross-linked or catalyzed PVA adhesives are restricted in their application to non-structural uses, such as external joinery and kitchen furniture.

Most cross-linked PVAs are manufactured as two-can systems, although a few are available as a one-part system. Although the one-can system can match the durability and strength of the two-can adhesives, generally speaking the performance of the latter is superior to the former. The gap-filling capabilities of both PVA systems is considerably lower than that of the formaldehyde-based resins used for assembly gluing.

---

*Polyvinyl acetate is abbreviated PVAc in the United Kindom.

I. Elastomeric Adhesives

Elastomeric adhesives are based on natural or synthetic rubbers together with appropriate fillers and organic solvents. Many of them also contain resins, usually phenol, to improve their powers of adhesion. In recent years these elastomeric adhesives have found increasing favor in on-site semistructural applications such as the fixing of floor and wall panels to wood framing in both the United Kingdom and United States. Adhesives of this type are likely to find even wider use in the future in building and construction because of their increased tolerance to surface irregularities due to gap-filling attributes and to their ease of application on site by a caulking gun, and since their use eliminates the need for nails or other mechanical fixings in the majority of instances. Contact-type elastomeric adhesives are also available for use in veneering. A selection of the more important elastomeric adhesives is presented in Table 1.1. In general terms most of the adhesives based on neoprene are stronger and creep less than those based on reclaimed rubber or nitrile rubber. Quite frequently, maximum strength is not developed until up to 14 days after application.

J. Thermoplastic Adhesives, Including Polyvinyl Acetate

The most frequently used thermoplastic adhesive is an uncatalyzed polyvinyl acetate emulsion: developed in the early 1950s this adhesive has found widespread use in furniture making, veneering, and interior joinery. Its very poor resistance to moisture and marked tendency to creep preclude its use both externally and structurally. Brands vary considerably in performance, although it remains one of the cheaper adhesives. The other thermoplastic adhesives listed in Table 1.1 have similar performance to that of PVA.

K. Cement

Cement has been used as a binder for wood particles for many decades, either in the production of fairly dense sawdust-bound blocks used in some countries for building internal walls of houses, or in the fabrication of wood wool slabs, which, by possessing high durability and low thermal conductivity properties, have found widespread use as a decking material on flat roofs in many parts of Europe.

Early attempts to produce a cement-bound wood particleboard met with little success since the boards produced were not only very heavy but also possessed low strength, expecially resistance to impact. It was not until 1962 in the United States and 1966 in Switzerland that a dense particleboard with good mechanical performance was achieved. A pilot plant was set up in Switzerland by Durisol AG in 1969 and further development work by this company led to the production of a smooth, fine-grained, cement-enriched surface characteristic of present-day boards. The first full-scale commercial plant was erected near Zurich in 1974, and since then further plants have been built in Germany, Hungary, Finland, Japan, Malaysia, Mexico, and Russia [20].

As in all wood-cement composites, care must be taken in the selection of timber species to avoid cement poisoning. This inhibition to setting of the cement is due to the type and amounts of sugars, hemicelluloses, and phenolic compounds present in most timbers. Curing, as with the resin-bonded boards, must be effected under pressure. Since curing of cement is a very slow process, it is necessary to keep stacks of boards clamped for periods of 6-8 hr in chambers heated to 70-80°C; this slows production and increases the cost of the board [20,21].

Although the use of a cement binder results in a dense board (about 150% by weight of a resin-bonded board), exceedingly high dimensional stability is attained with portland cement-bonded boards. Durability is excellent and the board is almost incombustible with high resistance to spread of flame; modulus of rupture is generally lower and internal bond strength similar to that of resin-bonded boards, while stiffness is higher [20].

III. PERFORMANCE OF WOOD ADHESIVES

The performance or behavior of an adhesive system is dependent on a wide range of variables, some of which are related to the substrate, such as surface smoothness, pH, presence of extractives, and amount of debris present, and others which are related to the environment, such as the level and rate of change in both temperature and relative humidity. In attempting to evaluate the properties and performance of adhesives per se, it is imperative that those variables relating to the substrate be held constant. On the other hand, in the evaluation of the performance in service of a wood-bonded product, the effects of the substrate must be taken into the assessment.

There is clearly, then, a distinction to be made between the evaluation of performance of one adhesive against another, as will be discussed in this section, and the evaluation of performance of a particular resin in association with wood veneer, flakes, or chips. While the latter has considerable significance in terms of actual service, the importance of the former lies primarily in the process of quality control in resin production and in the development of new resins or modifications of existing resins. There are two basic approaches to the assessment of the performance of adhesives when these are considered as independent entities; the first approach utilizes thin films of the adhesive, while the second adopts the use of standard substrates and testing conditions.

A. Adhesive Films

It is frequently desirable when evaluating adhesive performance to remove any possible interaction of the substrate. To this end, recourse must be made to the production, and subsequent weathering and testing of isolated thin films of adhesive. Such a method is particularly applicable to the assessment of the effect of modifications to the chemical composition of adhesives.

Difficulties can arise in the production of uniformly thin, completely cured reproducible films, although successful techniques for film casting have been reported for UF resins [7], polyvinyl acetate [19], and elastomeric adhesives [22]. Samples removed from these films are assessed for strength and elasticity, usually under tensile stressing.

B. Standard Substrates

It is certainly more common to evaluate the performance of an adhesive when this is applied to some substrate, thereby facilitating the application of stress to the adhesive as well as dispensing with the rather delicate operation of producing isolated films. The use of substrates, however, does introduce another variable into the assessment, but the significance of this can be reduced by standardizing on the type of substrate used not only in terms of species of wood, but also in respect to the size and shape of the specimen, primarily its thickness. Especially where strong resins are being evaluated, it is imperative to ensure that dense hardwoods of even texture and good quality are adopted, and it is common practice in many countries to use beech, maple, or birch. Moisture content and surface smoothness of the substrates must also be kept constant.

Evaluation of adhesive quality can be assessed in terms of either the relative amount of destruction of the adhesive and wood following insertion of a knife or chisel into the glue line to produce failure in a cleavage mode, or the stress to cause failure generally in shear under mechanical loading. Each method has its own attributes and weaknessess. Thus the extent of wood failure is difficult to quantify, but the assessment can be carried out anywhere and no testing machine is required. The method, however, is seldom applied in the evaluation of adhesives, although, as will be discussed later, it is used for the evaluation of bond performance in plywood.

Mechanical testing, in contrast to the above, yields a quantitative numerical value devoid of subjective appraisal and it is this method which is commonly adopted for characterizing adhesive quality. Care must be excerised to ensure that standardized procedures are adopted, especially with regard to rate of loading. Essentially, there are just two modes of mechanically stressing the glue bond: in shear parallel to the plane of the glue line, and in tension perpendicular to that plane. While the range of standard tests methods developed for glue bond

Table 1.2. Selection of Principal Test Methods Used in the Assessment of Adhesive Quality

Test specimen	Standard	Timber substrate	Size (mm)
Single lap (close contact)	BS 1204, Pt. 2	Beech	115 × 25 × 3.2
Single lap	–	Maple	150 × 20 × 16
Single lap (gap filling)	BS 1204, Pt. 1	Beech	150 × 25 × 3.2
Single lap	DIN 68602	Beech	150 × 20 × 5
Double lap	BS 3544	Beech	114 × 25 × 3
Block shear	BS 4169	–	50 × 50 × 19
Block shear (close contact)	ASTM D905	Maple	50 × 50 × 18
Block shear (gap filling)	ASTM D905	Maple	50 × 50 × 18
Block shear (modified)	–	Maple	>25 × 25 × 7.5
Block shear	CSA 0112.0	–	100 × 50 × 50
Block shear	–	Redwood	50 × 25 × 12.5
Plywood shear	BS 1203	Beech veneer	150 × 25
Plywood shear	ASTM D906	Yellow birch veneer	75 × 25
Tension perpendicular	ASTM D897	–	30 × 22 × length
Cleavage	ASTM D143	–	85 × 18 × width

[a]See Figure 1.15.
[b]See Figure 1.19.

evaluation relate to either one or the other of these two modes, in practice, due to imhomogeneity of the tests pieces, component stresses of both modes are always operating although each to differing extents. The most commonly adopted techniques are those that subject the glue line principally to shear stresses which can be generated by either tensile stressing, using double or single lap joints, or by compression stressing using small blocks. There are only a few test methods available for determining the strength of adhesives in tension perpendicular or cleavage.

Adhesive thickness (mm)	Area of adhesive (mm)	Mode of stressing	Adhesives studied	Reference
0	25 × 25	Tension[a]	RF, PF, UF, MUF	BSI 1979
0.2-0.8	50 × 20	Tensile creep	Elastomers	[23]
1.3	25 × 25	Tension[a]	RF, PF, UF MUF	BSI 1979
0	20 × 10	Tension	PVA, PF, UF	DIN
0	25 × 12	Edge bending[a] edge bending creep[b]	PVA	BSI 1962
0	50 × 38	Compression	RF, PF, UF	BSI
0	50 × 43	Compression	PF, UF, MUF Cat PVA	ASTM
0.74	50 × 43	Compression	PF, UF, MUF Cat PVA	ASTM
–	25 × 25	Compression	PF, UF, MUF Cat PVA	[24]
–	–	Compression	PF, UF	CSA
–	50 × 25	Compression[a]	PVA	[25]
0	25 × 25	Tension	PF, UF, MUF	BSI 1979
0	25 × 25	Tension	PF, UF, MUF	ASTM
–	20 × length	Tensile perpendicular	PF, UF, MUF	ASTM
–	75 × width	Cleavage	PF, UF, MUF	ASTM

Some of the more common tests methods are presented in Table 1.2, but it should be appreciated that this list is far from being comprehensive (see also Ref. 26) and is intended to give the reader some feeling for the variation in sample size that occurs internationally. Most of these methods relate to short-term tests, but a few are concerned with long-term testing in order to assess the resistance to creep, an important parameter of structural adhesives. Selection of the appropriate test procedure will be governed to some extent by the use to which the adhesive will eventually be put. Thus a resin destined for subsequent use for plywood-type construction should be assessed by one of the tensile shear tests utilizing thin veneers, such as BS 1203 or ASTM D906.

Adhesive quality is usually quantified in terms of the mean failing load required to rupture the test specimens, but since failing load or stress is strongly influenced by rate of loading, caution must be exercised in the comparison of data from different sources to ensure that loading conditions are similar. It is certainly common practice in the United Kingdom and certain other countries to subject the adhesives to some form of pretreatment prior to strength evaluation. Although some people identify such methods as constituting some form of accelerated aging with a view to predicting durability, this was certainly not the original intention when the methods were derived. Thus the 72-hr boil test listed in BS 1203 was designed specifically to separate PF resins from all other formaldehyde-based adhesives. Similarly, requirements of 3 hr at 67°C, 3 hr boil, and 6 hr boil in BS 1204 were set to separate the different adhesives types rather than to provide an index of prediction of durability, although it is true to say that interpretations of this type are made.

IV. PERFORMANCE OF BONDED WOOD PRODUCTS

Whereas in Sec. III interest was focused on the adhesive itself, this section is devoted to a study of the performance of the interaction between the adhesive and the substrate, that is, to the actual bond. In the assessment of bond performance there are two distinct aspects. The first relates to the quality of the bond as formed initially, and the second to the durability of that bond under specified conditions.

A. Bond Quality

As in the evaluation of adhesive quality, the assessment of bond quality can be quantified in terms of the relative amount of destruction of the adhesive and wood following insertion into the glue line of a knife or chisel, or in terms of the stress necessary to cause failure under some type of mechanical loading. The knife test has been widely used in the past in both the United Kingdom (BS 1455) and overseas for determining bond quality of plywood, although its use at present is somewhat limited. Bond quality is assessed visually from the relative amounts of wood and glue exposed on the fracture face. It has been shown that differences in assessing quality of bond between operators are remarkably small, and that the reproducibility of both the knifing operation and the assessment of a single glue line by one operator is surprisingly high [27].

Mechanical stressing of plywood samples is generally carried out in a shearing mode, using samples identical to those plywood shear samples used in the evaluation of adhesive quality as listed in Table 1.2 (e.g., BS 1203, ASTM D906). Bond quality is usually quantified in terms of a mean failing load or stress required to rupture the specimen. However, an alternative approach used in certain parts of the world, as for example, in the case of the North American plywood standards, is to quantify bond performance in terms of the amount of wood failure that has occurred during the mechanical test.

Advocates exist for each mode of assessment and a number of attempts have been made to derive a relationship between them. These investigations have tended to indicate that while wood failure and shear strength are certainly not independent, their interrelationship is indeed complex. Recently, however, it has been demonstrated using a wide range of timber species and types of adhesives that by expressing bond strength as a percentage value of the shear strength of the bonded wood sample divided by the shear strength of the wood, a significant intrinsic correlation between wood failure and shear strength was obtained [28].

Unlike the use of a shearing stress to evaluate the bond strength of plywood, mechanical stressing of particleboard is carried out by the application of a tensile load perpendicular to the plane of the board. Alternatively, block shear tests have been advocated primarily because of the ease of operation, although it has been demonstrated by both other workers and ourselves that the weakest plane is weathered particle-board does not necessarily occur in the center of the board, the region assessed by the block shear test, and consequently the application of a tensile force perpendicular to the plane of the board provides a more accurate evaluation of bond strength.

Quite misrepresentative data on bond performance can be obtained by markedly increasing the amount of resin present. An illustration of this is presented in Figure 1.1, where it is shown that an increase in resin content of a UF-bonded particle board from 8% to 12% resulted in such a marked increase in dimensional stability at high relative humidity that its response was now superior to that of a PF board with lower resin content and quite unrepresentative of a UF bond.

One further variable to be taken into account in the assessment of comparative bond performance is the time that the material takes to cure. Whereas this can be as short as a few minutes as in the case of formaldehyde-based resins used in the manufacture of plywood or particleboard, it can be as long as several months with both cement and the elastomeric adhesives. Figure 1.2 illustrates the wide variation in the development of maximum shear strength obtained in testing at the Princes Risborough Laboratory. In joints, especially close contact ones, glued together with seven different elastomeric adhesives, with some adhesives maximum stress was obtained after 3 months, but in the majority of elastomers this was not achieved even after 12 months [29]. Consequently, in the case of slow-curing adhesives, true comparisons of performance can be made only after long periods of time.

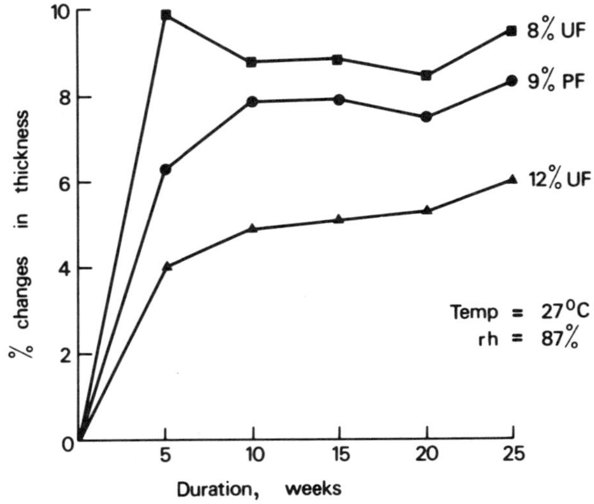

Fig. 1.1. Effect of continuous exposure to high humidity of various particleboards.

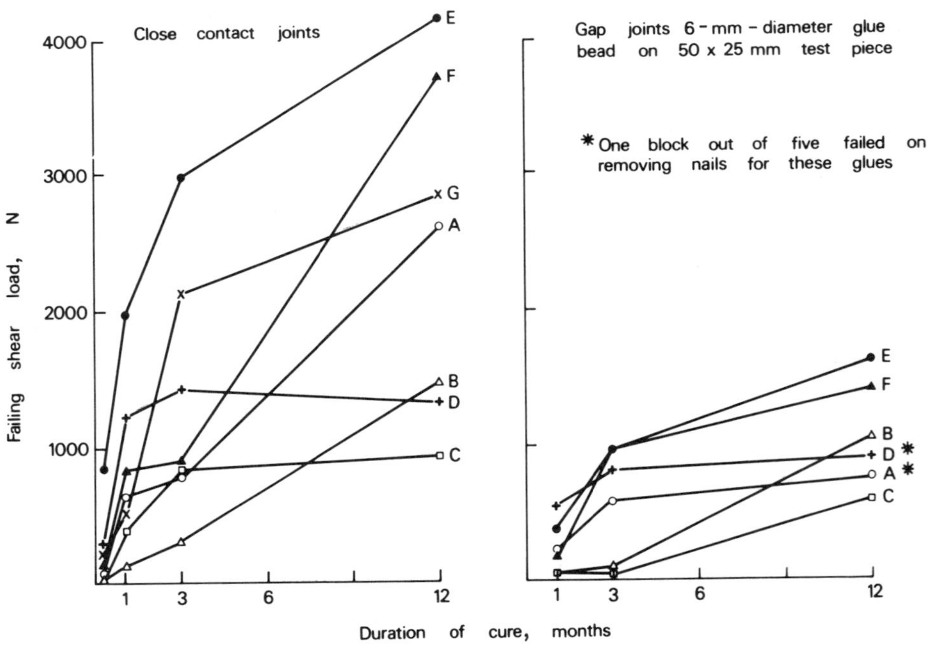

Fig. 1.2. Failing shear load for block shear test of close contact and gap joints bonded by seven different elastomeric adhesives: A and C, reclaimed rubbers; B, nitrile rubber; D, E, G, neoprene; F, styrene-butadiene rubber.

B. Bond Durability

Ideally, the durability of an adhesive should be assessed under long-term service conditions, but since this approach is impractical, recourse must be made to ways of artificially aging the adhesive. Over the years a whole host of accelerated aging procedures have been adopted, most of them unfortunately being purely empirical in nature and representing nothing more than some form of torture test. However, they do appear to be capable of placing adhesives in some form of ranking order of performance, although it is often impossible to establish good correlations between performance as indicated by these tests and actual performance in service.

The principal factors inducing bond deterioration are heat, moisture, and stress, and the accelerated aging methods developed incorporate one or more of these factors acting under under either a continous or a varying mode. The tests can be subdivided conveniently into three distinct groups: single exposure tests, cyclic exposure tests, and mechanical tests. A selection of the first two groups is presented in Table 1.3 (see also Ref. 26).

1. Single Exposure Tests

Many of the tests involved soaking, heating, or boiling test samples in water for a few hours, or in subjecting samples to either high temperatures or high humidities for long periods of time. Cold-water soak methods are used on the assessment of both PVA glues and PF resins, in the case of the latter for the determination of the presence of undercure of the resin. Following these accelerated aging methods, the samples are then assessed for bond quality according to the methods outlined in the preceding section. Generally, in the case of evaluation of shear strength, the samples are tested wet, in which case it is usual to wet up the control or unaged samples by some vacuum-pressure-soak technique such as a 30-min soak in water at 21°C under a vacuum of 0.085 N/mm^2 followed by a 30-min soak under a pressure of 0.42 N/mm^2. Performance is assessed in terms of load to failure, or as the ratio of the strength after exposure to strength before aging, or to visible changes in the glue line (e.g., delamination).

A somewhat different approach to the simulation of external conditions has been the adoption of thermal softening techniques for both thermosetting and elastomeric adhesives [32]. The method involves measuring the compressibility of a small column of adhesive powder under constant load during heating and good correlations have been achieved between softening temperature and both natural and accelerated weathering.

It should be appreciated, however, that no single condition of exposure can adequately simulate long-term natural exposure with fluctuating environmental conditions. At best this type of test will assess

Table 1.3. Accelerated Aging Techniques

Standard	Popular name	Used on adhesives	Test method
		A. Single continuous exposure methods	
Boil			
BS 1203		PF plywood	72 hr boil; joint strength >0.9 kN
BS 1203		MF, MUF plywood	3 hr boil; joint strength >0.45 kN
CAN 0188		PF particleboard and waferboard	2 hr boil; sample testes wet
DIN 68761	V100	PF particleboard	2 hr boil; sample tests wet
DIN 68705	V100	PF plywood	6 hr boil
Hot Water			
BS 1203		UF	3 hr at 67°C; joint strength >0.9 kN
Ref. 30		All types	700 hr at 70°C
Cold water			
BS 1203		All plywood	16-24 hr at 15°C; joint strength > 1.1 kN
DIN 68705	V20	UF plywood	24 hr at 20°C
CSA 0112.6		PF, RF, UF	48 hr at 21-24°C
BS 5669		UF particleboard	1 hr at 20°C
Ref. 25		PVA	48 hr at 20°C; reconditioned
High temperature, dry			
Ref. 30			200 days at 100°C
Refs. 21, 31		PF	70 days at 100, 120, 140°C; tested wet
High humdity			
BS 3544		PVA noncatalyzed	7 days at 90% RH, 25°C
Ref. 25		PVA catalyzed	7, 14, 28 days at 90% RH, 25°C

Thermal softening Ref. 32		PF, RF, MF urethane	Temperature noted at which softening under load occurs
B. Cyclic methods			
CSA 0121		PF, MUF, UF	4 hr boil + 20 hr drying at 60°C + 4 hr boil
CSA 0151		PF, MUF, UF	4 hr boil + 20 hr drying at 60°C + 4 hr boil
DIN 53254		All	4 hr boil + 16 hr dry air at 60°C, <20% RH + 4 hr boil + 2 hr cold soak
NFB 51-263	V313	MUF particleboard	3 × { 72 hr water soak at 20°C + 24 hr in air at -12°C + 72 hr in air at 70°C }
ASTM D3434-75	ABT	PF, RF plywood	10 min boil + n cycles × { 4 min drying in air at 23°C + 57 min in air at 107°C + 10 min boil }
ASTM D1037-78		PF plywood	6 × { 1 hr water soak at 49°C + 3 hr steam spraying at 93°C + 20 hr in air at -12°C + 3 hr in air at 100°C + 3 hr steam spraying at 93°C + 3 hr in air at 100°C }
FPL, Madison	VPSD	PF, RF plywood	10 × { ½ hr water soak at 21°C under vacuum of 735 mm mercury + ½-1 hr water soak under pressure of 0.42 N/mm^2 + 23 hr drying at 82°C }

Table 1.3. (Continued)

Standard	Popular name	Used on adhesives	Test method
		B. Cyclic methods (continued)	
American Plywood Association	VSD (D6)	PF, RF plywood	6 × { ½ hr water soaked at 65°C under vacuum of 370 mm mercury + ½ hr water soaked at 65°C at atmospheric pressure + 6 hr drying in air at 82°C }
FPL, Madison, Wis.	BD		½ hr water soaked under vacuum of 735 mm Hg + ½ hr soak under pressure of 0.42 N/mm^2 + n cycles × { 10 min boiling water + 3¾ hr drying in air at 107°C }
Ref. 33		PF, MUF, UF particleboard	50 × { 7 days at 30% RH, 25°C + 7 days at 90% RH, 25°C }
BS 4169		RF, PF, MUF, UF	Soak in water at 18-27°C + 1½ hr soaking under vacuum of 0.067 N/mm^2 + 1½ hr soaking under pressure of 0.5 N/mm^2 + 1½ hr water soak as above + 1½ hr pressure soak as above + 88 hr drying at 27°C and 30% RH
BAM Xenotest		Plywood and particleboard	12-week cyclic exposure to UV and IR radiation, spraying, drying and cold storage; see Ref. 34 for full programs

bond durability under one single component occurring in natural weathering. Nevertheless, such tests are useful in comparing the relative performance of different adhesive formulations and for purposes of quality control.

2. *Cyclic Exposure Tests*

The second group of tests (group B in Table 1.3) are those in which samples are exposed to alternating temperature and/or moisture levels in an attempt to overcome the limitations of the single-factor continuous exposure tests and simulate fluctuating environmental conditions. It will be noted from Table 1.3 that these cyclic tests are frequently of very long duration, with some of the tests involving part of the cycle under freezing conditions, even though the effect of freezing moist panels is not fully understood. One of the methods also includes exposure to ultraviolet (UV) light as part of a long cycle devised to obtain a high degree of simulation to natural weathering [34]. It is difficult to see why the effect of UV is not confined to the surface layers, unless there is an indirect heating effect. Recent work by the West Coast Adhesives Manufacturers' Association questions the significance of the freezing cycle in ASTM D-1037, while the French continue to uphold their decision to include a freezing cycle in the V313 test for particleboard.

The variability in the composition of these methods is conspicuously apparent from the descriptions of group B in Table 1.3. Various attempts have been made to compare different methods, as for example, a recent comparison of the "boil-dry" and "vacuum-pressure soak-dry" (VPSD) techniques [35]. The boil-dry degradation rates were shown to be much more uniform among all the materials tested than the VPSD, probably as a result of the absence of any effect of resin cure as a source of variability in the boil-dry test.

While both the continuous and cyclic test methods are useful for the purposes of initial acceptance, quality control, and comparison of different formulations, none of them provides much quantitative information on the prediction of long-term performance. To overcome this limitation, a chemically oriented rate process-to-durability evaluation has been adopted by the Forest Products Laboratory (FPL), Madison, Wisconsin [30,31,36]. From a series of curves showing residual strength after exposure to different temperatures, the half-life is calculated and plotted on a log scale against the reciprocal of exposure temperature in kelvin to give an Arrhenius plot. Extrapolation of the line is used to forecast durability.

3. *Mechanical Tests*

There is a growing body of opinion on the inadequacy of a single aging test, even a cyclic one, to effectively simulate long-term bond performance and the desirability of running parallel tests primarily of a mechanical nature (see, e.g., Refs 34, and 37). Such mechanical tests fall effectively into two groups: the first, the application of fracture mechanics on specially shaped specimens and performed in a short period of time, and the second, the assessment of rate of creep or time to failure, which can be achieved only by long-term study.

Certainly, one of the more exciting recent developments for evaluating and predicting the performance of adhesives is the use of a tapered double-cantilever beam fracture test coupled with the application of the theory of fracture mechanics [38,39]. Glue lines are subjected to cleavage rather than shear stresses as in the majority of bond evaluation strength tests recorded in Sect. IV.A. The test provides both crack initiation and crack arrest values; these, in turn, can be used to follow the development of brittleness in a resin under different environmental conditions, thus providing a most effective and sensitive tool for evaluating changes in bond performance and imparting a high degree of confidence in the prediction of future performance.

Investigations of creep, whether for the purpose of investigating the increase in deflection with time and subsequent recovery following load removal, or the time elapsed prior to rupture, have usually been performed on composite products stressed in bending under either third- or quarter-point loading; in a few investigations creep has been studied in tension either in a wood composite, or exceptionally for cast glue films. Some of these investigations are described in a later section; however, not all of these have been concerned with modeling and predicting future creep.

There have been two approaches to the modeling of creep behavior and the use of these models for prediction of further deflection. The first is to fit a regression line of some form to the available experimental data [37,40-42] but without any theoretical justification of the model equation, and the second is to derive from first principles a rheological model to describe creep behavior but without any experimental data to substantiate it [43,44]. Very few attempts have been made to coordinate the theoretically based rheological models with experimental data [45–47]. References 46 and 47 are restricted to creep in timber for short periods of time, while Ref. 45 relates to particleboard loaded for up to 2 years.

Use has been made of a model of the form

$$y = a + b \log t$$

to predict creep deflection after 27 years from data for 30 days [48]. However, it is clear from the graphs that the data do not follow a straight line even up to 30 days. Predictions made using this model must be viewed with caution, and errors on the low side should be expected.

The equation

$$\log Y = a + b \log t$$

has also been used [42] to describe creep data and to predict deflection from test data from 10^2 hr (4 days) to 10^4 hr (400 days) with a claimed accuracy of 3-4%. The published data depart markedly from a straight line, the data curve being concave upward, and it is difficult to see how a claimed accuracy as high as this can be justified. Use of this model on other experimental data resulted in a predicted deflection some 23% too low [45].

The use of equations with additional terms, for example,

$$Y = a + b \log t + c(\log t)^2$$

have at best yielded only limited improvement in accuracy of prediction. On the other hand, considerable success in accurate prediction has been achieved using rheological models of the form

$$Y = \beta_1 + \beta_2[1 - \exp(1 - \beta_3 t)]$$

and

$$Y = \beta_1 + \beta_2[1 - \exp(1 - \beta_3 t)] + \beta_4 t$$

where the first and second terms in each equation represent, respectively, the elastic and viscoelastic portions of a time-deformation curve; in the four-element model, the extra term represents viscous flow, all of which are represented in analog form by springs and dashpots [45].

For each of the nine specimens under test for periods up to 24 months, the data were curtailed over successively longer time periods and at each stage the three- and four-element models were fitted to the data. Figure 1.3 illustrates the speed of covergence of the predicted values from the two models to the actual deflection of one sample when the models are fitted using shorter lengths of test data. The four-element model converges considerably faster, coming within 10% of the actual end deflection on prediction from 5-month data, and 3% on prediction from 12-month data.

More investigations of this type are required to ensure a high degree of assurance that long-term performance can be accurately predicted from relatively short-term tests. This is especially relevant in view of the increasing body of opinion in favor of complementing accelerated weathering tests with some form of prediction of long-term mechanical behavior.

4. *Correlation of Accelerated Tests and Natural Weathering*

Various attempts have been made over the decades to correlate results on bond durability obtained with various accelerated testing techniques

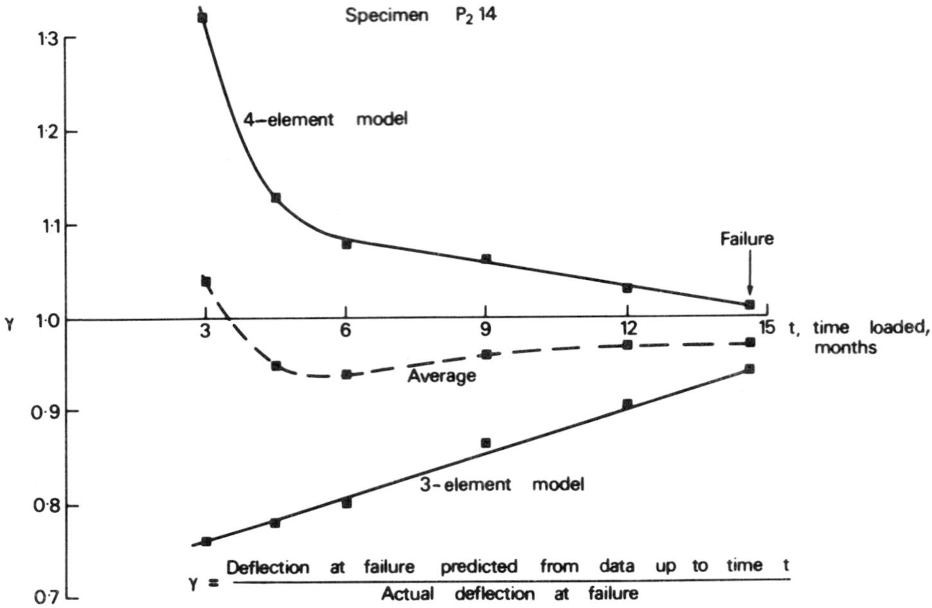

Fig. 1.3. Model convergence when predicting from various time points for particleboard under third-point loading.

and results achieved under natural weathering. On the whole, the degree of correlation has frequently been low, especially so with single continuous exposure, thereby emphasizing comments made earlier on the poor simulation of natural exposure achieved by these tests. Comments on the unsatisfactory relationship between the V100 accelerated test method and outdoor weathering have recently been published [34] and the trend has been toward the use of more varied cyclic tests. Thus using a weatherometer fitted with a UV light source (the Xenotest) [34], very good correlations have been obtained and it has been shown that for phenol-formaldehyde resins, 12-week exposure in the weatherometer has an equivalent weathering effect on bond durability to 3 years of natural exposure.

More rapid degrade takes place in the automatic boil test (ASTM 3434), where it is claimed from weathering trials that 40 cycles of the ABT method taking 2 days to perform have an equivalent effect to 1 year of natural weathering [49]. It has also been reported that when ASTM D1037 was used on matt-formed chipboard, six cycles of the aging procedure produced effects equivalent to 50 years of natural aging.

Work at Princes Risborough Laboratory on the weathering of particleboard has indicated that the highest correlations between results following 7 years of exposure to natural weathering and accelerated techniques were achieved using the V313 cyclic test (see Table 1.3). Actual correlation coefficients were considerably higher, with the exception of thickness swelling, than those obtained using the V100 method (see Table 1.4). Good correlations were also obtained using the V70 test and weatherdrum.

Work on catalyzed PVA adhesives at Princes Risborough Laboratory clearly showed a lack of any correlation between natural exposure for 12 months and boiling in water for 6 hr. On the other hand, good correlations were achieved for strength between external exposure and a 48-hr soaking in cold water (Fig. 1.4) [25]. As reported in Sec. IV.B.3, a good correlation was achieved between the thermal softening temperature of both thermoset and elastomeric adhesives and both natural exposure and two accelerated test methods [32].

Various attempts over the years have been made to compare the relative degree of representation of natural weathering by various accelerated aging methods (see e.g., Ref. 50). Perhaps the most comprehensive investigation of the effect of accelerated aging techniques was carried out some 15 years ago, and the conclusions drawn for the comparison of 11 bond degrading systems with the effects of natural weathering still hold true [51]. There is no single method which forms a satisfactory method of predicting long-term performance under all service conditions, and it was thought that treating bond degradation as chemical reactions with changes in energy levels possibly held the greatest hope of improvement in prediction of

Table 1.4. Correlation Coefficients in Performance Between Accelerated Aging and Natural Exposure for Four Commercial Particleboards

	V70		V100			
Property	Wet[a]	Dry	Wet	Dry	V313	Weatherdrums[b] (3000 hr)
MOR	0.63	0.85	-0.76	-0.33	0.96	0.94
MOE	0.61	0.82	—	0.44	0.98	0.91
Internal bond	0.91	0.98	0.19	-0.76	0.68	0.97
Thickness swelling	0.95	0.92	0.83	0.80	0.41	0.24

[a]V70 test (water soak at 70°C) now withdrawn from DIN standards (68761, Sheet 3).
[b]Weatherdrum: water spray 16 hr in every 24 hr, UV continuous.
Source: Adapted from *Proc. 15th Int. Particleboard Symp.*, Washington State University, Pullman, Wash., 1981.

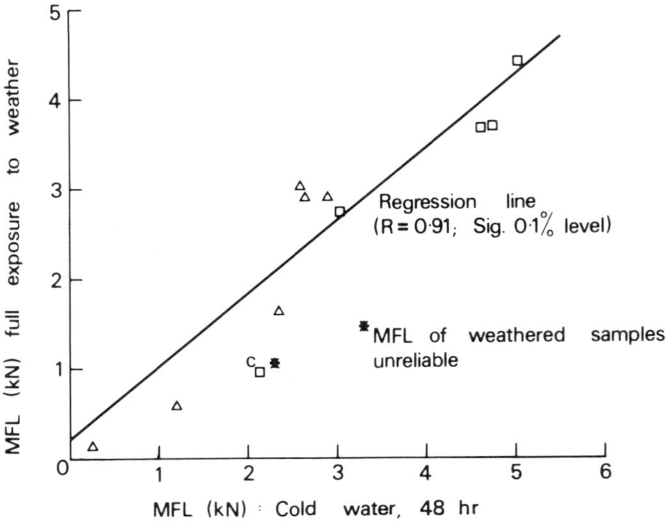

Figure 1.4. Correlation of mean failing loads (MFL) for weathered samples with cold-water immersion.

performance. The development and success of the rate-process technique [30,31,36] would certainly confirm those early conclusions and predictions.

It must be appreciated that the vast number of accelerated aging techniques are basically only torture tests, quite acceptable for the purposes of quality control, but having severe limitations for the prediction of long-term performance since they give no indication of the rate of change in performance in service. Only those methods that embrace a large number of cycles give some indication of rate of change in performance, but even the results from these tests must be accepted with considerable caution.

C. Variability in Performance

Variability in performance occurs not only between the major types of adhesives but also quite extensively within any one type; comparison is made difficult by the difference in assessment of performance provided by different test methods. The variability in performance of various wood-bonded products under differing environmental and stress situations is described below. The discussion is not intended to be comprehensive in content and only a selection of the investigations on the subject are included to illustrate the general relationships.

1. Performance at High Humidity Levels

Examination of particleboard as an example of the effect of high humidity levels on adhesive performance reveals the existence of major differences in performance between boards manufactured using different adhesives. With increasing relative humidity (RH) above 30%, and consequently increasing equilibrium moisture content of the boards, both strength and stiffness fall progressively, the rate of decrease for UF-bonded boards being greater than that for MUF or PF. At humidities below 30% it appears from limited evidence that strength and stiffness may again decrease as the humidity falls to zero.

Time of exposure plays a very significant part, and at any one level of humidity both strength and stiffness decrease and board thickness increases, at first rapidly, and thereafter more slowly with time. The increase in thickness over a 25-week period of four commercial boards with different resins at 87% RH (27°C) is illustrated in Figure 1.5. It will be noted that the change in thickness of the UF board is at least 50% greater than that of the MF/UF. PF, and sulfite liquor boards, which had very similar response to high humidity.

The corresponding loss in bond strength as measured by the tensile strength perpendicular to the board for these four boards at the end

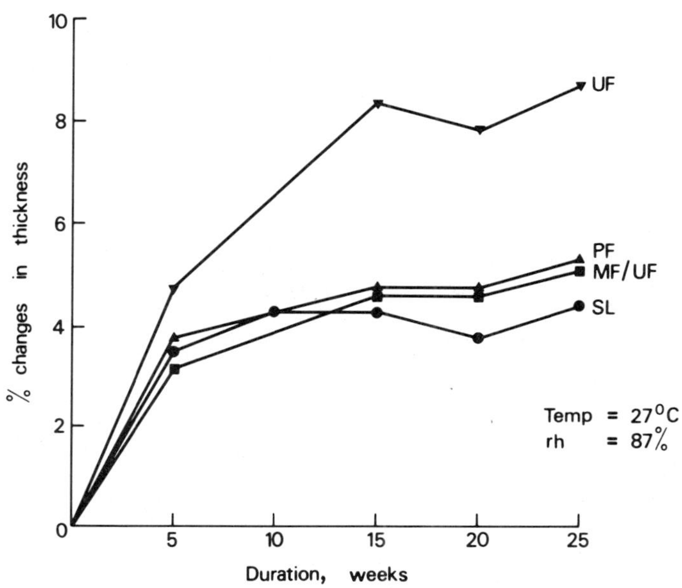

Fig. 1.5. Effect of continuous exposure to high humidity of different particleboards.

of the 25-week period is illustrated in Figure 1.6, where the strength retention of the UF board is seen to be very significantly lower than that of the other boards. Confirmatory results on the performance of adhesives in particleboard are to be found in the literature (see e.g., Ref. 52).

Significant reductions in the strength of joints made using catalyzed polyvinyl acetate adhesives generally also occur when both close contact and gap joints are stored at 90% RH for up to 28 days [25] (see Fig. 1.7). Immersion in cold water for 48 hr resulted in further significant reductions in bond strength. Confirmatory evidence of the sensitivity of PVA to the effects of moisture and water has been obtained from the tensile stressing of thin films [22].

Increasing moisture content results in considerable reductions in both the shear strength and shear modulus of various elastomeric-based construction adhesives, although their degree of sensitivity to moisture varies considerably with chemical composition. With an increase in moisture level to 20%, strength at 23°C falls by up to 70% of the original value, while with water soaking under pressure for 30 min, strength at 23°C falls to levels corresponding to 15-45% of the original values [53].

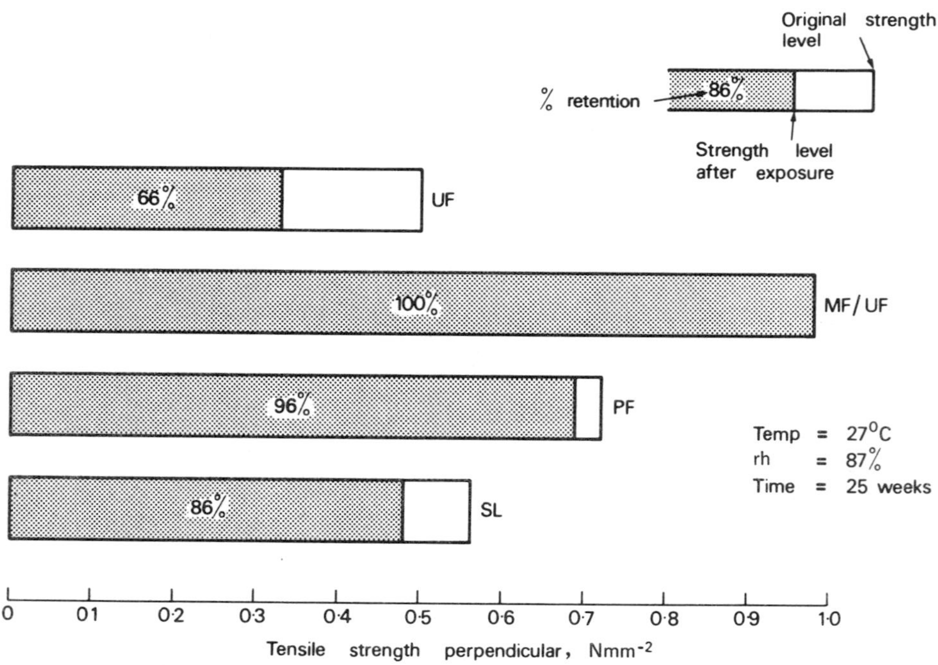

Fig. 1.6. Retention of tensile strength of particleboard following exposure to high humidity.

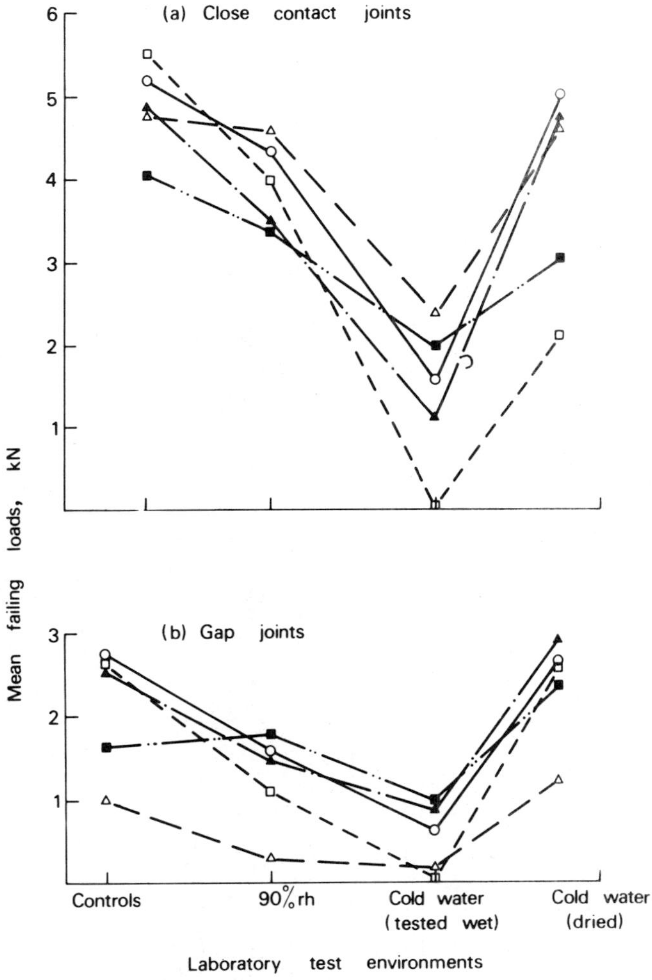

Fig. 1.7. Moisture resistance of beech slip joints bonded with five catalyzed PVA adhesives.

2. *Performance Under Alternating Humidity Levels*

Alternating levels of humidity aggravate the effect of relative humidity recorded above and reinforce comments made on the relatively inferior performance of UF resins compared with other formaldehyde-based resins. The effect of alternating humidity is more pronounced the higher the upper end of the range becomes.

It has been shown that samples of commercial UF boards subjected to 50 cycles of alternating humidity (30% to 90% RH at 25°C) in an un-

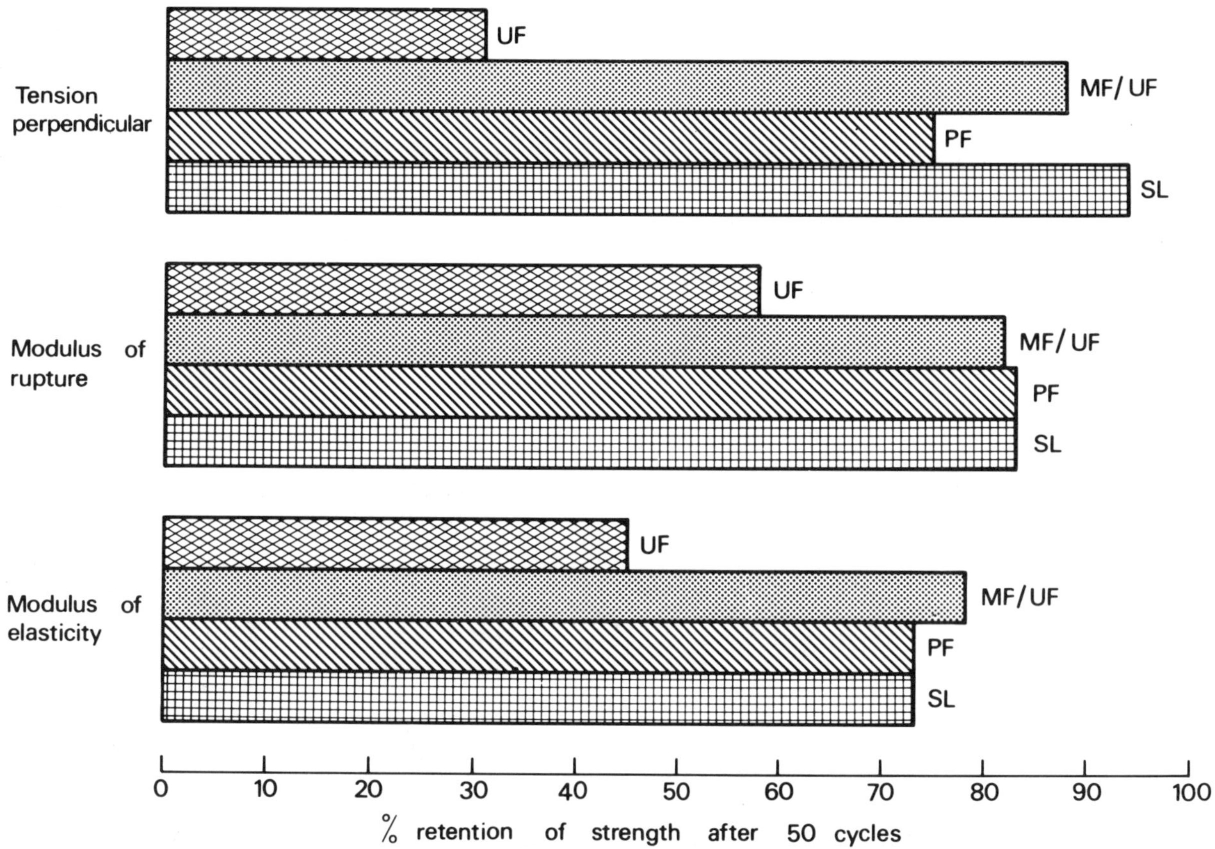

Fig. 1.8. Percentage retention of strength and stiffness of particleboard under alternating humidity.

stressed state retained only 26%, 58%, and 45% of their initial tensile perpendicular strength, modulus of rupture, and modulus of elasticity, respectively; in comparison, the retention values for MUF, PF, and sulfite liquor boards were never less than 73% and often greater than 90% (see Fig. 1.8) [33]. The reason for the poor performance of UF resins under alternating humidity is not fully understood, although recent European findings would tend to indicate that the cause is due not to chemical degradation of the resin [7], but rather to the effect of mechanical stressing set up within the glue line by the alternating swelling and shrinkage of adjacent chips [54, 55]. Rigid adhesives are affected more than flexible ones [56].

3. *Performance at Elevated Temperature*

The effect of raising the temperature of UF resin in the presence of moisture (moisture content >8%) is to initiate and subsequently accelerate the chemical deterioration of the resin, a process known as hydrolytic degradation and involving the severence of the cross-linking in the cured adhesive by water molecules. When the resins are incorporated in a board, this process leads initially to increased thickness and subsequently to a complete loss in strength. Degradation of this type does not occur with the other types of resins, some of which are represented in the results of soaking tests for particleboard illustrated in Figure 1.9. Increased thickness of the two types of UF-bonded boards compared with the others is apparent at 45°C; at 60°C swelling of the UF boards is over five times that of the other boards after 14 days of soaking and the boards have lost almost all their strength; most of these changes are irreversible on drying out.

The effect of increasing temperature in the absence of moisture on the durability of the wide range of adhesives used in bonding plywood panels of Douglas fir and yellow birch has been shown to be quite considerable [31]. Subjected to temperatures of 100, 120, and 140°C for periods up to 70 days, the rate of degradation of the panels as measured by wet shear strength was shown to be temperature dependent, following an Arrhenius relationship, which as described in Sec. IV.B.2, can be used to predict the performance of adhesives. It is interesting to note that many of the failures occurred in the wood, indicating that the rate of thermal degradation of the wood was greater than that of the adhesives, including UF.

Thermal degradation also occurs with the elastomeric-based adhesives, although its extent varies considerably depending on chemical composition. An increase in temperature from 23°C to 71°C at the same relative humidity resulted in a loss in both strength and modulus of elasticity from 40 to 100% [53]. The catalyzed polyvinyl acetates appear to possess even greater thermal sensitivity with a softening temperature of only 33°C. Such behavior stimulates concern when this type of

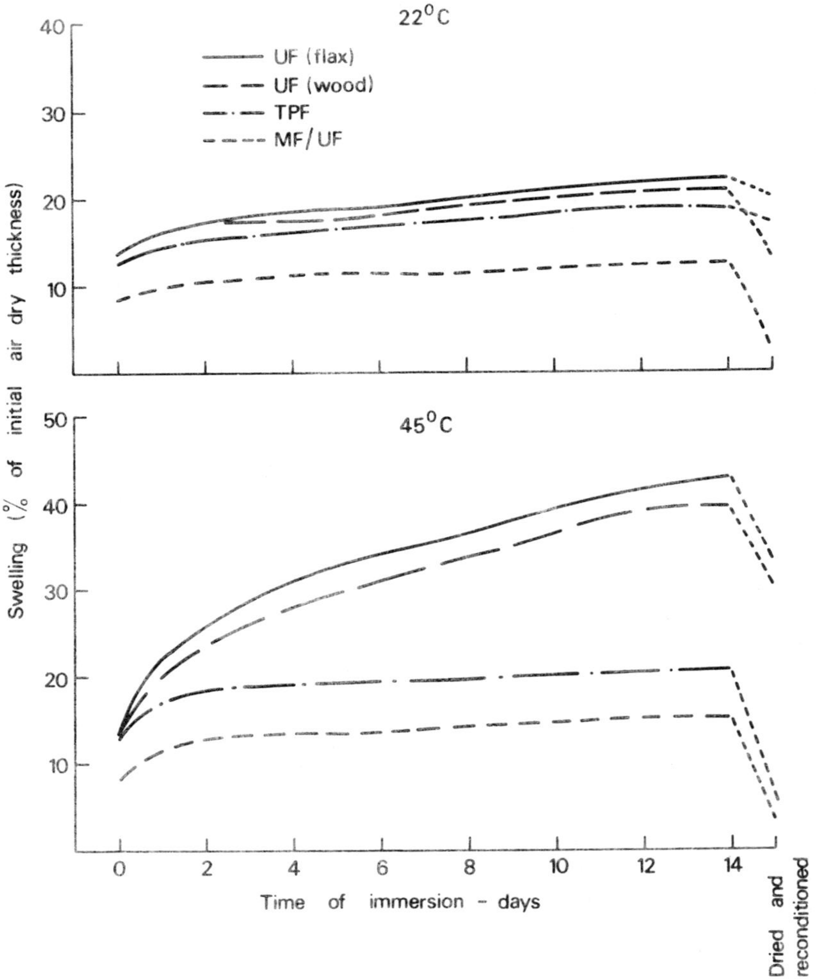

Fig. 1.9. Effect of temperature on the swelling of various particleboards.

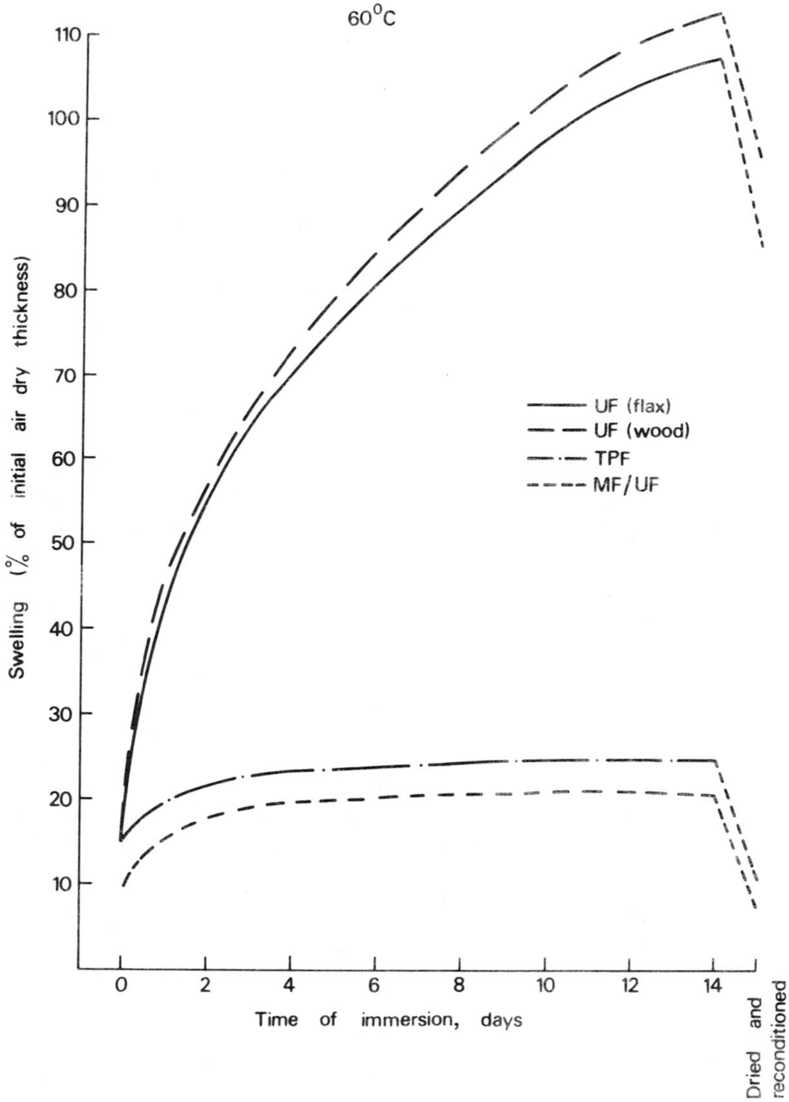

Fig. 1.9. (Continued)

adhesive is used in areas prone to the occurrence of high temperatures, as in roof spaces in summer [22].

4. *Performance at Low Temperature*

Little information is available on the effects of prolonged periods of low temperature on bond performance. However, some information is available on the effect of cyclic exposures of water-saturated Douglas fir boards laminated using a range of adhesives and subjected to temperatures of 20, 112, and -65°C. Loss in shear strength was lower for urea-formaldehyde and phenol-resorcinol-formaldehyde than for a melamine-urea-formaldehyde, phenol-formaldehyde, and cross-linked polyvinyl acetate [57].

5. *Performance Under Ambient Conditions*

Little information is available on the very long term performance of adhesives under ambient temperature and humidity. Perhaps it has been rashly assumed that degradation does not occur under these conditions. It has been demonstrated from shear strength tests on plywood panels of both Douglas fir and yellow birch stored for 4 years at 27°C and 30% RH that those panels bonded with nonacidic resins, including phenol, resorcinol, and melamine-urea, lost little if any strength during the storage period. However, panels bonded with acidic types of resins, such as urea, catalyzed PVA and an acid-catalyzed phenol resin, lost from 20 to 30% of their initial strength [31].

Even greater losses in strength are recorded for UF and PF resins, assessed by both plywood shear and lap joint tests, stored under interior conditions for 24 years. The percentage retention of initial strengths varied considerably especially among brands of any one type of adhesive. For UF and PF plywood glues the initial strength retained was 60-75% and 75% respectively, while for assembly glues the corresponding values were 35-100% and 60-90%. One sample of resorcinol-formaldehyde showed a retention in strength of 85% [58].

6. *Performance Under External Weathering*

Natural weathering of test samples has always been regarded as the most informative means of assessing the performance of adhesives that will be used under severe moisture and temperature conditions. Unfortunately, as an evaluative tool, exposure to natural weathering has distinct limitations, not the least of which are the long periods of test time involved and the poor reproducibility of the results obtained. Primarily because of the absence of any other appropriate and satisfactory method, the method continues to be widely applied.

The long-term performance of plywood glues under full exposure to

Fig. 1.10. Part of the weathering site at Princes Risborough Laboratory.

the weather has been undertaken in many laboratories throughout the world. Test work commenced at Princes Risborough Laboratory (formerly Forest Products Research Laboratory) in the early 1940s, and at regular intervals samples have been removed for bond evaluation using the knife test method described in BS 1455 (see Ref. 27). The weathering site is illustrated in Figure 1.10.

Although a number of the phenol-formladehyde glues tested in this extensive program displayed initially acceptable standards, their efficiency was not maintained during the period of exposure, which ranged up to 16 years (Fig. 1.11a). This loss in efficiency could be correlated with the amount of filler included in the resin mix: thus those phenol resins formulated for use either without added fillers, or with the addition of only very small amounts maintained a high level of bond quality throughout the period of exposure. However, as the percentage of filler in the mix increased, bond quality decreased progressively (Fig. 1.11b) [59].

Corresponding investigatons on aminoplastic resins demonstrated that unmodified urea-formaldehyde resins deteriorate rapidly with a life expectancy no greater than 3 years when fully exposed (Fig. 1.12). When UF resins were fortified with melamine or phenol or resorcinol, a much improved performance was obtained, with a service life extended to about 6 years (Fig. 1.12).

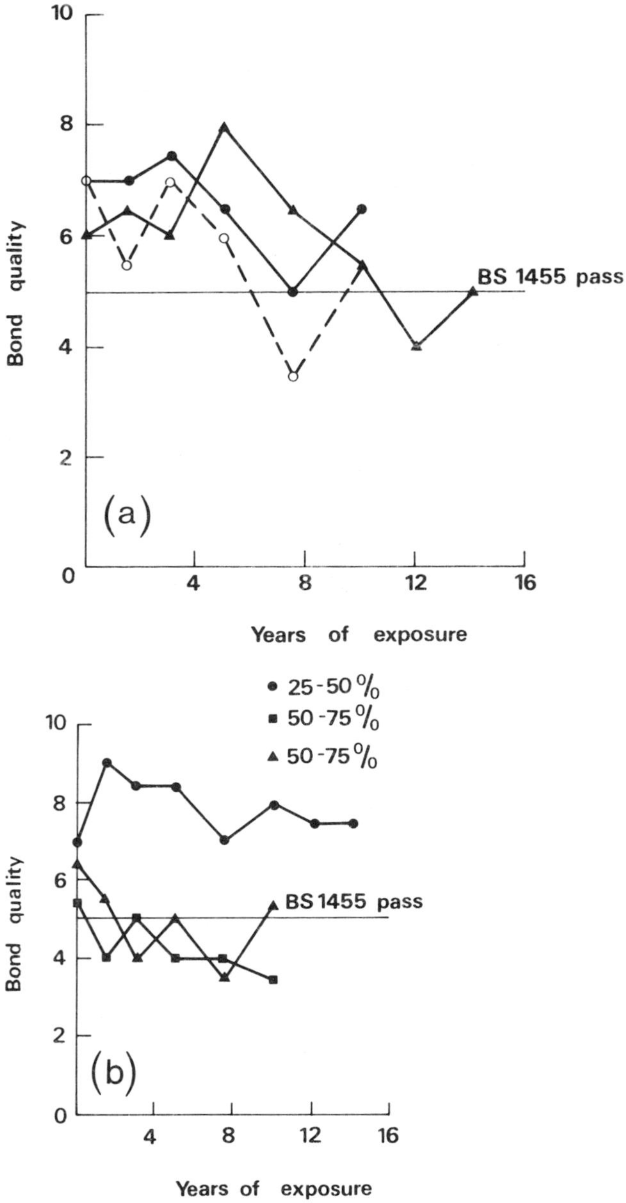

Fig. 1.11. (a) Loss in bond quality (knife test) on natural exposure of liquid applied PF resins used in manufacutre of plywood; (b) influence of filler concentration in PF resins on bond performance (knife test) under natural exposure.

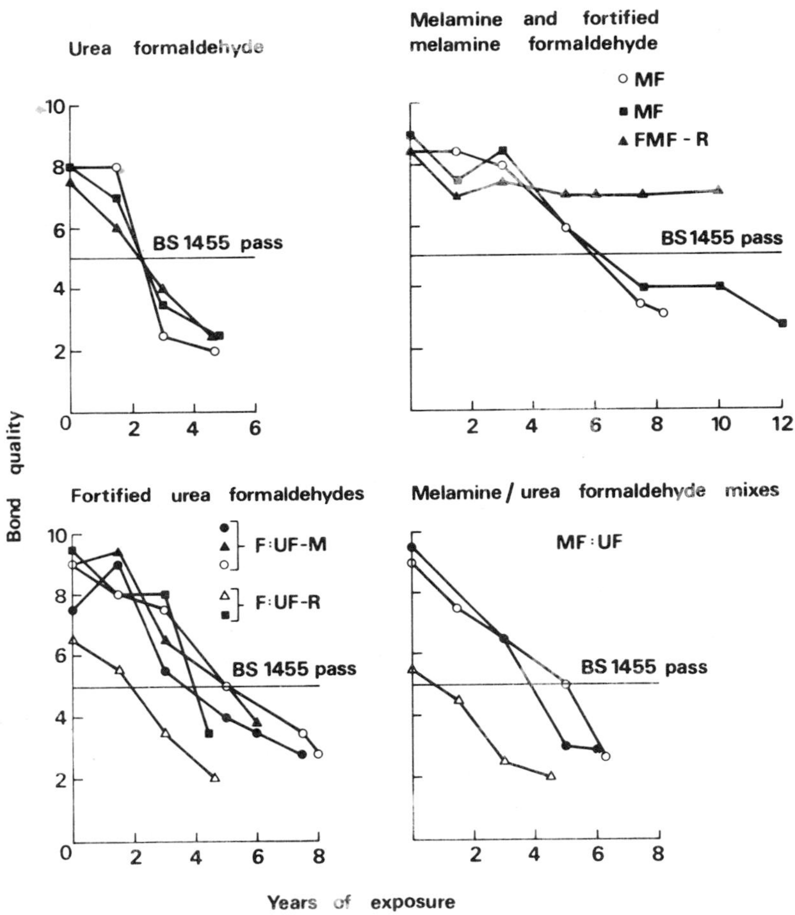

Fig. 1.12. Loss in bond quality in plywood made with amino resins exposed to natural weathering.

Comparable exposure trials have been conducted at Princes Risborough Laboratory on particleboard manufactured using different adhesives. In particleboard, bond performance is usually quantified in terms of the tensile strength perpendicular to the plane of the board or in terms of changes in board thickness. Bond performance for a wide range of adhesives extending over a period up to 7 years is illustrated in Figures 1.13 and 1.14.

All boards, with the exception of the cement-bonded boards, lost tensile strength on exposure (Fig. 1.13), but there is considerable diversity as to the rate and extent of this loss. Whereas UF boards show

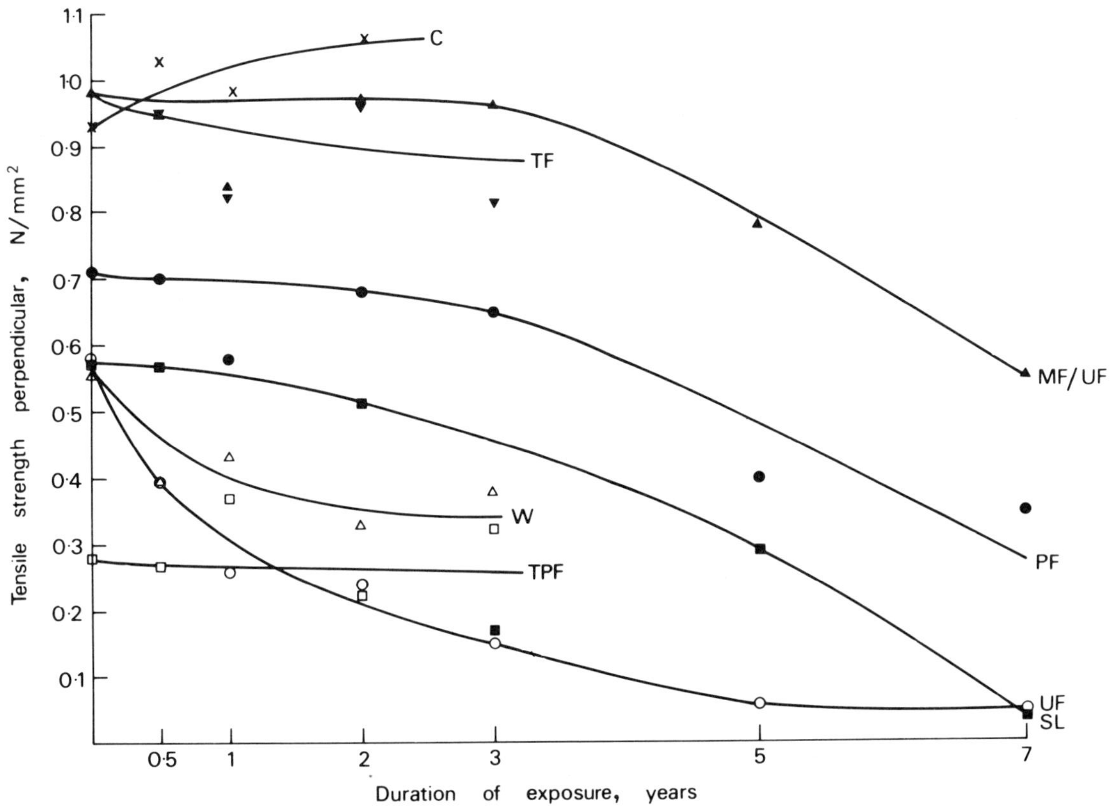

Fig. 1.13. Reduction in tensile strength perpendicular (internal bond) of various particleboards over 7 years of natural weathering. (C = cement bonded particleboard; W = waferboard; other symbols as used in text).

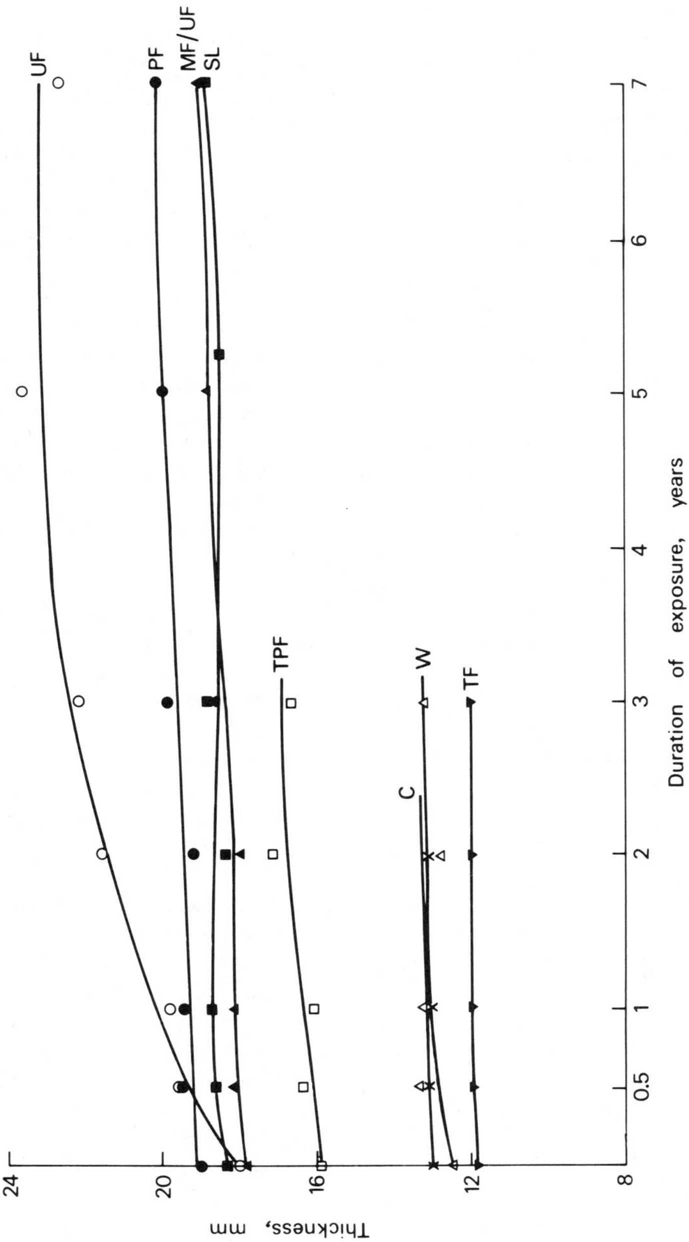

Fig. 1.14. Increase in thickness of various particleboards over 7 years of natural weathering. (C = cement bonded particleboard; W = waferboard; other symbols as used in the text).

a rapid initial loss, decrease in strength of MUF, PF, and sulfite liquor does not appear to be significant until after 3 years when the rate of decrease appears to be similar for all three binders. Although after 7 years weathering the MUF and PF boards still have acceptable strength values, they nevertheless have lost at least 40% of their initial strength, a feature that is somewhat surprising. Whereas the two tannin-formaldehyde boards show little strength loss in their 3 years of exposure, the PF-bonded waferboard displays a considerable initial reduction in strength. Cement-bonded board is the exception, with a 10% increase in both strength and stiffness after exposure for 3 years; this is probably due to prolonged curing of the cement.

The very poor performance of UF-bonded particleboard on exposure is again reflected in the marked increase in thickness of the boards that occurs (Fig. 1.14). This is about five times the dimensional change found in the MUF, PF, and sulfite liquor boards. Initial swelling of the PF-bonded waferboard was considerable, certainly greater than expected for a PF-bonded board, reflecting the very low resin contents used in waferboard manufacture. This is a very good example of the interaction of the adhesive and the wood and how the performance of the composite product bears little relationship to the performance of the same adhesive when evaluated by standard methods.

Recently published data [60] obtained in a most extensive investigation of the performance of 19 types of commercial and 6 types of laboratory particleboard under natural weathering over a period of up to 9 years confirms the relative durability of the formaldehyde-based adhesives described above, and additionally provides data on the performance of various isocyanate-bonded boards. Bond durability as measured by reduction in tensile strength perpendicular to the plane of the board appears to be lower for the isocyanate boards than for PF-bonded boards. Irreversible increase in thickness is greater for the isocyanate-bonded boards.

Little information is available on the long-term durability of elastomeric adhesives and cross-linked PVA, although it is widely acknowledged that the durability of un-cross-linked PVA and other thermoplastic adhesives is very low. Beech slips bonded together with cross-linked PVAc and exposed to the weather for 12 months showed no loss in strength in the majority of brands tested when gap joints (Fig. 1.15C) were employed: however, close-contact joints (Fig. 1.15B) appeared much more vulnerable and all glues suffered severe losses in mean strength [25]. It is probable that given longer periods of exposure, the gap joints would show similar strength losses.

7. *Performance Under Prolonged Loading*

Figure 1.16 illustrates the characteristic deformation-time relationship for a piece of timber, board, or glued joint under a sustained load and

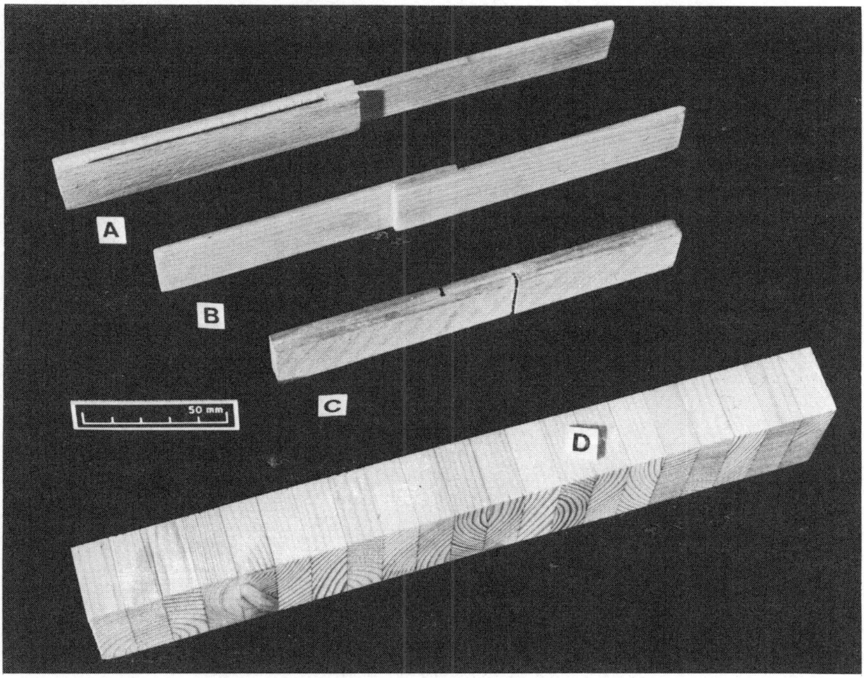

Fig. 1.15. Selection of test samples for bond evaluation: A, sustained load test (BS 3544); B, BS 1204 beech slip sample—close-contact type; C, BS 1204 beech slip gap joint; D, redwood multilaminate sample.

constant temperature and humidity where the applied stress is not great enough to initiate early failure. On the application of a load at time zero, an instantaneous and completely reversible deformation occurs which represents elastic behavior. On maintaining the load to time t_1, the deformation increases, although at an ever-decreasing rate; this increment is known as creep. On removal of the load at time t_1 an instantaneous recovery occurs which is approximately equal in magnitude to the initial elastic deformation. With time, the remaining deformation will decrease at an ever-decreasing rate until time t_2, when it

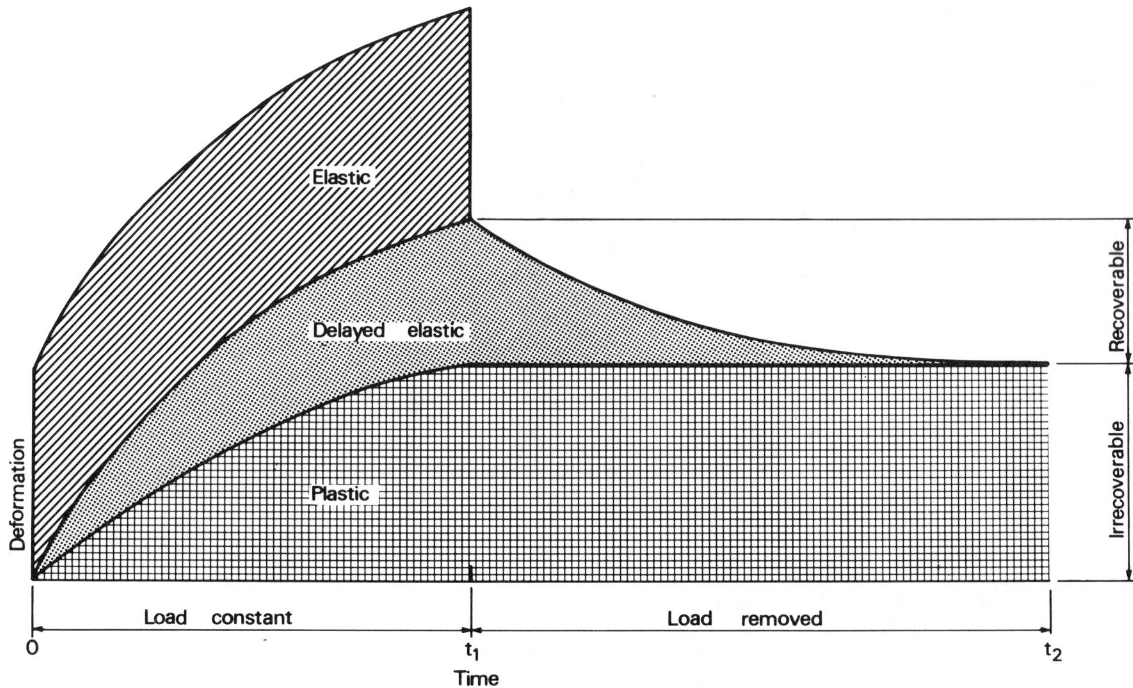

Fig. 1.16. Various elastic and plastic components of the deformation of particleboard under constant load.

becomes constant. The amount of creep that has occurred during loading can be conveniently split into a recoverable component, which displays delayed elastic behavior, and an irrecoverable component, which is due to plastic or viscous flow.

Although this overall pattern of creep and recovery is similar in timber, board materials, and glue joints, the extent of creep deflection under similar conditions is generally considerably lower in solid timber than in all systems incorporating adhesives. The amount of creep in wood-adhesive systems varies dramatically with type of adhesive used, and is also influenced considerably by moisture content, temperature, and level of stressing. For most wood-adhesive composites the relationship between the magnitude of creep deflection and stress level is linear up to about 40% of the short-term stress level; thereafter, creep increases at a rate greater than the increasing level of stressing [40,48].

Looking specifically at particleboard, it has been recorded that the amount of creep deflection at a moisture content of 15% is about 10 times higher than that occurring at a moisture content of 6% [61]. Such a significant influence of moisture content has been confirmed by many other workers, including our own investigations at the Princes Risborough Laboratory [62]. However, where the level of moisture content is varying (i.e., un-steady-state conditions), the extent of creep deflection is very much greater than that occurring under steady-state conditions [40,63].

In particleboard, high inverse correlations have been recorded between creep and stiffness [40,61], although there are no indications as to whether stiffness reflects processing or material variables; certainly both the level of binder and the ratio of binder to wood will have a marked effect on both stiffness and creep. Considerable controversy prevails as to whether or not the types of binder used in particleboard manufacture have an effect on creep behavior. Thus both the rate of creep and the relative creep at a particular period of time (i.e., either the ratio of deflection or increase in deflection at time t_1 to the initial elastic deflection at t_0) have been found to be equal in both UF- and PF-bonded particleboards by some workers [40,64,65], while other investigators have shown that creep in PF-bonded boards is significantly greater than that in UF boards under the same environmental conditions [37,48, 66,67], due, it is thought, to the greater flexibility of the cured phenolic resins [37].

Extensive investigation on creep of commercial particleboard at Princes Risborough Laboratory has so far revealed the existence of considerable variation in the magnitude of creep among boards produced from the same type of adhesive by different manufacturers [45]; this is confirmed by recent results from Germany [60]. Both

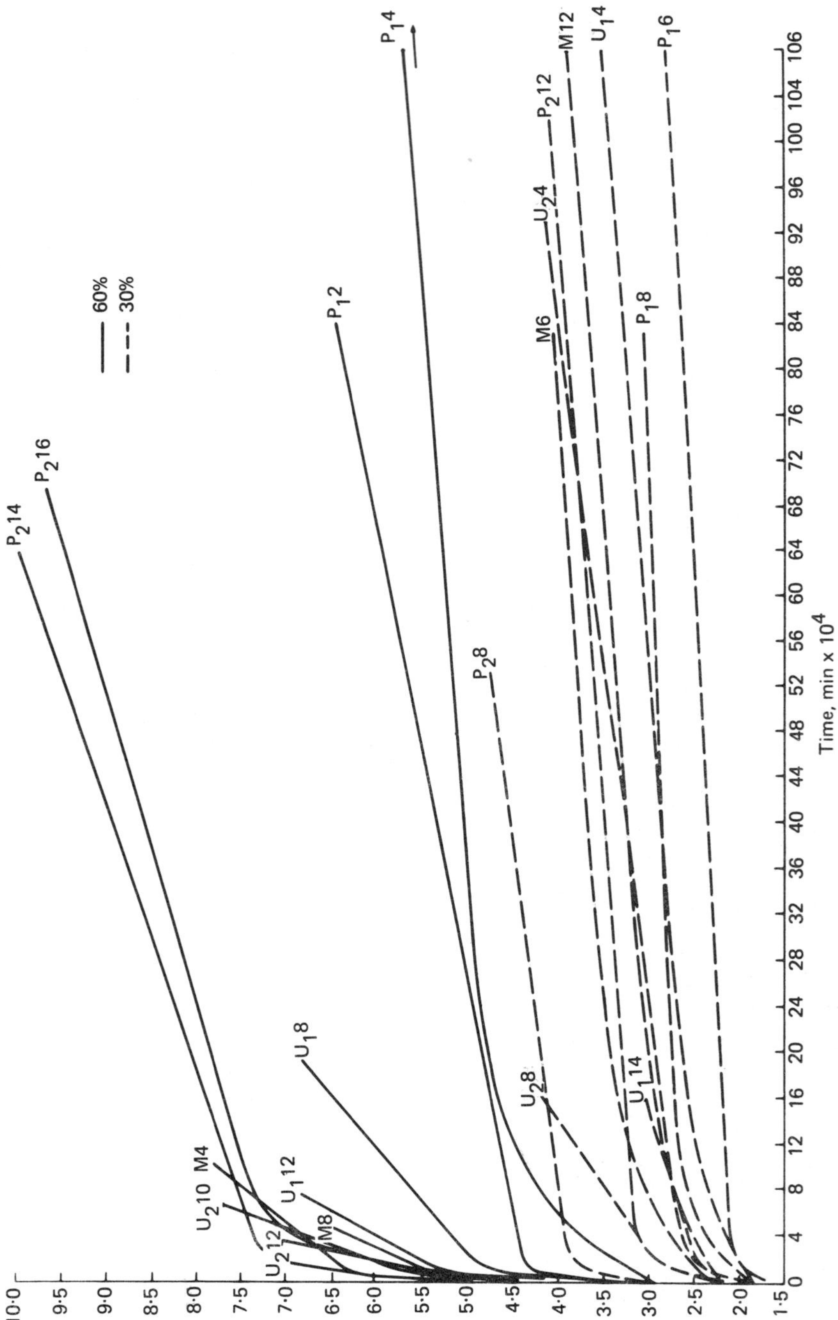

Fig. 1.17. Creep behavior of particleboard at 30% and 60% stress levels represented by four-element fitted curves.

the lowest and highest creep values obtained in the Princes Risborough Laboratory results occurred with PF-bonded boards, while the UF and MUF samples were intermediate (see Fig. 1.17). Density profiles of the two types of PF board were identical and it would appear, therefore, that such marked diversity in creep performance is due to differences in resin formulation. Thus the range in creep behavior appears to be as great among the boards of any one type of adhesive as it is between boards produced from different adhesives, thereby casting considerable doubt on the generalized statement that PF boards creep more than UF boards.

Much interest is being expressed at the present time in the use of isocyanate as an adhesive, but so far little information is available regarding the creep behavior of isocyanate-bonded particleboard, especially in comparison to boards bonded with amino- or phenol-formaldehyde resins. A considerable number of isocyanate boards from different manufacturers have been included in a second investigation on creep at Princes Risborough Laboratory, and it would appear that in a comparison of boards manufactured on the same production line taken immediately prior to and immediately following the change in resin, relative creep values are similar for both types of board, although deflections are usually greater for the isocyanate boards than for the formaldehyde-bonded boards. Although the relative creep values of the wood cement particleboard are high, actual deflection values are very low.

The rate of creep deflection for elastomeric and catalyzed PVAc adhesives is very much greater than that indicated above for the formaldehyde-based resins. As we noted above for the latter group of adhesives, the variability in performance within the group is very considerable. Similar wide variation in the rate of creep has been observed in a set of six elastomeric-based adhesives at Princes Risborough Laboratory [29]. Extension during a 10-day period varied from as much as 8 mm for adhesives based on reclaimed rubbers or nitrile rubber (A, B, and C, Fig. 1.18), to only 0.5 mm for adhesives based on neoprene (D and E, Fig. 1.18); a styrene-butadiene rubber was intermediate (F). Whereas some samples carried their load throughout the investigation (D, E, and F), others failed early in the test (A, B, and C). Confirmation of the variability in creep performance of elastomeric adhesives is recorded by the Forest Products Laboratory, Madison, Wisconsin, where the creep behavior of two ranges of elastomeric adhesives is characterized in terms of their creep modulus, shear slip, and fractional recovery [41,68].

Tensile creep studies on thin films of styrene-butadiene rubber, polyurethane, and catalyzed PVA adhesives remove any possibility of interaction with timber substrates and have confirmed; first, the marked tendency for elastomeric adhesives to creep appreciably, and second, the extemely variable creep and recovery response of differ-

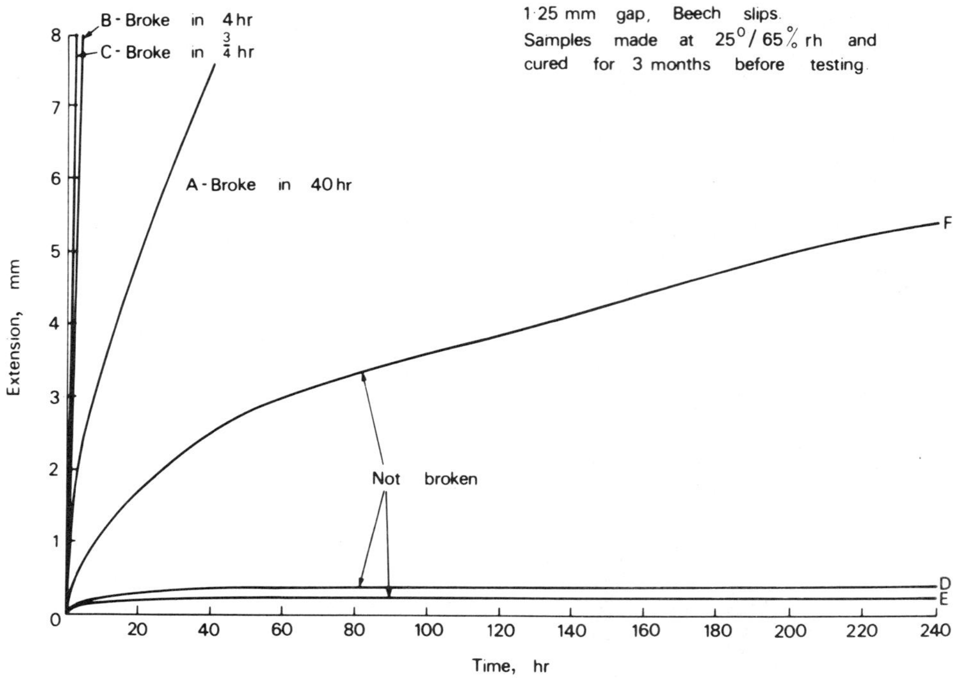

Fig. 1.18. Extension of beech slips bonded with different elastomeric adhesives under a sustained load of 50 N: A and C, reclaimed rubbers; B, nitrile rubber; D and E, neoprene; F, styrene-butadiene rubber.

Table 1.5. Sustained Load Test (BS 3544) on Catalyzed PVA Adhesives Under Various Environments

	Time to failure (days)				
				Exposure to weather	
PVA adhesive	65% RH, 25°C	90% RH, 25°C	Cold water 48 hr reconditioned	Under cover	Full exposure
1	5.5	1.4	>15	306	23.2
2	>15	2.4	>15	321	23.4
3	1.3	0.7	0.1	23	8.3
4	2.7	1.2	>15	226	23.3
5	3.8	2.6	>15	505	24.2

Source: Princes Risborough Laboratory, Building Research Establishment, Princes Risborough, England. Crown copyright.

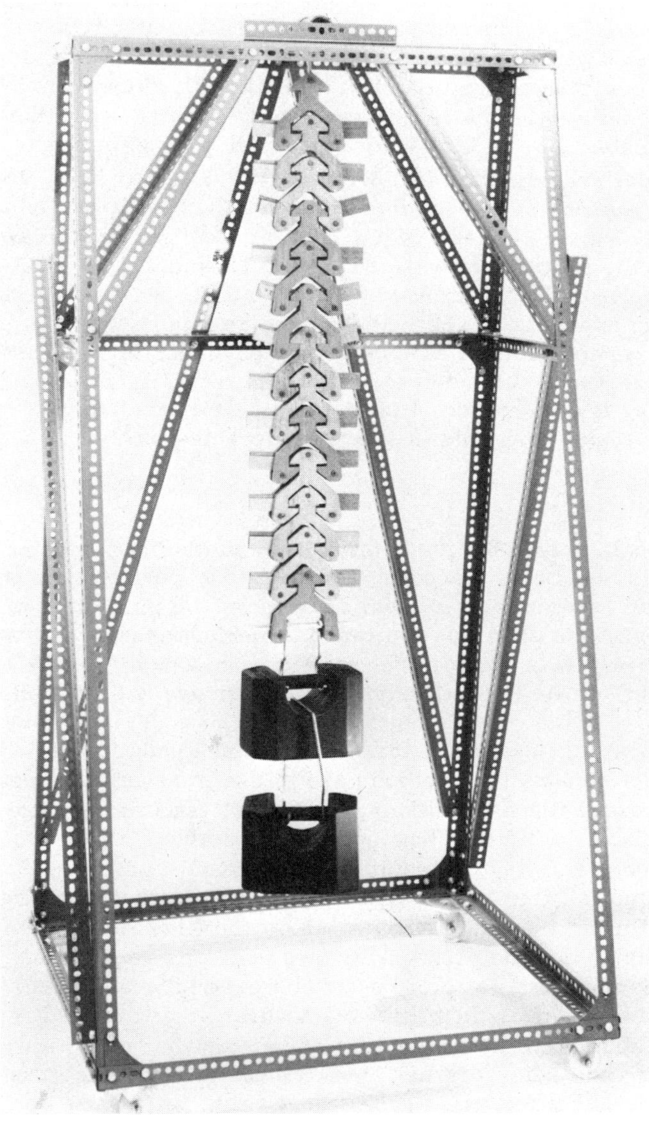

Fig. 1.19. Sustained load samples (PVA bonded) under load in jig (BS 3544).

ent adhesives [22]. The influence of various environmental factors in influencing creep, as demonstrated above for the formaldehyde-based adhesives used in particleboard, is again manifest in the case of catalyzed PVA adhesives. Results of a study at PRL of creep in five different PVA adhesives under different environmental conditions are presented in Table 1.5. A standard load of 445 N was applied according to the methods included in BS 3544 and illustrated in Figure 1.19, and four of the five adhesives failed to meet the requirements of BS 4071 when tested at 65% RH and 25°C. The rate of failure increased appreciably when the samples were tested at 90% RH and 25°C, but decreased for samples soaked in cold water for 48°C followed by reconditioning, with one exception. The longest times to failure were obtained with exposure to weather, samples under cover lasting at least 10 times longer than those subjected to full exposure [25]. Such improved performance is a reflection of ambient temperatures being consistently lower than those used in the laboratory tests.

D. Conclusions

In an age of increasing costs for both initial construction and maintance, greater emphasis is now being placed on the long-term performance of materials, including adhesives, and even more so on the prediction of long-term behavior. The need for assurance in performance is an aspect of building control which is going to become even more significant; to this end it would appear that adhesives for timber are not too well placed.

It is true that many methods exist for the artificial simulation of weathering and that various test methods are in use to assess the loss in bond strength under these conditions. However, remarkably little information is available to indicate how good these methods of predicting performance are, with the result that the level of assurance provided is really not all that high. Granted that the methods employed are extremely useful in assessing comparative performance of different adhesive formulations, but each test on its own provides very little information on the long-term performance of one particular adhesive.

The answer to the requirement to improve assurance is threefold. The first solution relates to the assessment of performance using more than one yardstick, and reference has already been made to the growing body of opinion which feels that accelerated aging tests must be complemented by tests in which the glue line is subjected to mechanical stressing under different environmental conditions.

The second approach is the adoption of more meaningful and representative test methods, and the adoption of cyclic tests, especially those encompassing low temperatures to maximize internal stress, is certainly a step in the right direction. New tests based on thermal softening to define durability, and crack propagation to assess the development of brittleness in bonds, represent marked improvements in the sensitivity of test methodology.

The third and final solution to the need for higher degrees of assurance in the prediction of performance lies in a more detailed understanding of the process of adhesion to wood substrates. Little has been done until fairly recent times, and it is gratifying to see the considerable current research interest in this field. Recent work has included studies on understanding the curing reaction of various adhesives [69], the development of strength during the curing process [70], and on the interaction at the interface between adhesive and the wood (reviewed in Ref. 71). It would appear that in the past insufficient attention has been paid to the physics and chemistry of adhesion, and too much to the comparative performance of different adhesive formulations using somewhat arbitrary test methods. Let us hope that the present trends can redress this imbalance of interest.

V. SELECTED RELEVANT STANDARDS

Standard no.	Title
BS 1203:1979	Specification for synthetic resin adhesives (phenolic and aminoplastic) for plywood
BS 1204:Part 1:1979	Synthetic resin adhesives (phenolic and aminoplastic) for wood: Pt. 1. Specification for gap-filling adhesives
BS 1204:Part 2:1979	Synthetic resin adhesives (phenolic and aminoplastic) for wood: Pt. 2. Specification for close-contact adhesives
BS 1455:1972	Specification for plywood manufactured from tropical hardwoods
BS 3544:1962	Methods of test for polyvinyl acetate adhesives for wood
BS 4169:1970	Specification for glue-laminated timber structural members.
BS 5442:Part 3:1979	Classification of adhesives for construction: Pt. 3. Adhesives for use with wood
BS 5669:1979	Specification for wood chipboard and methods of test for particleboard
ASTM D143	Standard methods of testing small clear specimens of timber
ASTM D897	Standard method of test for tensile properties of adhesive bonds
ASTM D905	Standard test method for strength properties of adhesive bonds in shear by compression loading

Standard no.	Title
ASTM D906	Standard test method for strength properties of adhesives in plywood-type construction in shear by tension loading
ASTM D1037	Standard methods of evaluating the properties of wood-based fiber and particle panel materials
ASTM D2339	Standard test method for strength properties of adhesives in two-ply wood construction in shear by tension loading
ASTM D3434	Standard recommended practice for multiple-cycle accelerated aging test (automatic boil test) for exterior wet-use wood adhesives
CSA 0112 Series M 1977	Standards for wood adhesives
CSA 0121-M1978	Douglas fir plywood
CSA 0151-M1978	Canadian softwood plywood
CSA CAN3-0188.0-M78	Standard test methods for mat-formed wood particleboards and waferboard
CSA CAN3-0.188.2-M78	Waferboard
NF B51 263	Accelerated aging test by the so-called V313 method
DIN 53254 (1980)	Testing of wood adhesives--determination of binding strength of lap joints in shear.
DIN 68602	Valuation of adhesives for joining of wood and derived timber products
DIN 68705 (1968)	Pt. 1. Plywood: definitions: general requirement, testing
	Pt. 2. Plywood: plywood for general purposes, quality condition
	Pt. 3. Plywood: veneer boards for building, quality condition
	Pt. 4. Plywood: blockboards for building purposes: quality conditions
DIN 68763 (1973)	Particleboards; flat pressed boards for building: concepts, properties, testing, and supervision

ACKNOWLEDGMENTS

All figures in this chapter are courtesy of the Princes Risborough Laboratory, Building Research Establishment. Crown copyright.

REFERENCES

1. A. A. Moslemi, *Particleboard,* Vol 1: *Materials,* Southern Illinois University Press, Carbondale, Ill., 1974, pp. 67-108.
2. F. F. P. Kollmann, E. W. Kuenzi, and A. J. Stamm, *Principles of Wood Science and Tchnology II: Wood-Based Materials,* Springer-Verlag, New York, 1975, pp. 1-93, 312-550.
3. T. M. Maloney, *Modern Particleboard and Dry-Process Fiberboard Manufacturing,* Miller Freeman, San Francisco, 1977, pp. 367-412.
4. B. Meyer, W. E. Johns, and J. E. Woo, Formaldehyde release from sulfur-modified urea-formaldehyde resin systems, *For. Prod. J. 30*(3):24-31 (1980).
5. P. R. Steiner, Durability of urea-formaldehyde adhesives: effects of molar ratio, second urea, and filler, *For. Prod. J. 23*(12):32-38 (1973).
6. E. R. Roffael, Effect of formaldehyde content on the reactivity of urea formaldehyde resins and formaldehyde release from UF-bonded particleboard, *Holz Roh.- Werkst. 34*:385-390 (1976).
7. J. M. Dinwoodie, Causes of deterioration of UF chipboard under cyclic humidity conditions: 1. Performance of UF adhesive films, *Holzforschung 31*(2):50-55 (1977).
8. M. Higuchi, H. Shimokawa, K. Kuwazuru, and I. Skata, Studies on the improvement of urea-formaldehyde resin adhesives: Pt. 1. A possibility of water resistance improvement, *Mokuzai Gakkaishi 25*(7):496-502 (1979); Pt. 2. Effect of some urea derivatives and melamine as additives on water-resistance of a UF resin, *Mokuzai Gakkaishi 25*(10):630-635 (1979); Pt. 3. Water-resistance improvement by an acid scavenger, *Mokuzai Gakkaishi 26*(5):310-314 (1980).
9. W. Clad, and C. Schmidt-Helleran, Particleboards for the building industry: performance and suitability tests, *Proc. 11th Int. Particleboard Symp.* Washington State University, Pullman, Wash., 1977, pp. 33-61.
10. A. Pizzi, and H. O. Scharfetter, The chemistry and development of tannin-based adhesives for exterior plywood, *J. Appl. Poly. Sci. 22*:1745-1761, (1978).
11. A. Pizzi, Tannin-based adhesives: new theoretical aspects, *Int. J. Adhes. Adhes. 1*:13-16 (1980).
12. A. Pizzi, Pine tannin adhesives for particleboard, *Holz. Roh.- Werkst. 40*:293-301 (1982).
13. J. K. Woo, Bark adhesives for particleboard and plywood, 2nd Int. Symp. Adhesion and Adhesives for Structural Materials, Pullman, Wash., 1982.

14. E. Roffael, and W. Rauch, Production of particleboard using sulphite waste liquor as bonding agent: II. a new and speedy method for production of particleboard bonded with sulphite waste liquor, *Holzforschung 25*(5):149-155 (1971).
15. K. C. Shen, Spent sulphite liquor binder for exterior waferboard, *For. Prod. J. 29*(5):32-38 (1977).
16. K. G. Forss, and A. Fuhrmann, Finish plywood, particleboard, and fiberboard made with a lignin-base adhesive, *For. Prod. J. 29*(7):39-43 (1979).
17. O. Wittmann, Wood bonding with isocyanate, *Holz Roh.-Werkst. 34*:427-431 (1976).
18. H. J. Deppe, Technical progress in the gluing of particleboards with isocyanates, *Holz Roh.- Werkst. 35*:295-299 (1977).
19. K. Motohashi, B. Tomita, and H. Mizumachi, Relationship between chemical properties of the poly(vinyl acetate) emulsion adhesives and tensile strength of cross-lap joint of wood, *Holzforschung 36*:183-189 (1982).
20. J. M. Dinwoodie, Wood cement particleboard—a technical assessment, *Build. Res. Establish. Inf. Pap. 4/83*, 1983.
21. J. R. Stillinger, and I. W. Wentworth, Product, process, and economics of producing structural wood-cement panels by the Bison-Werke system, *Proc. 11th Int. Particleboard Symp.*, State University, Pullman, Wash., 1977, pp. 383-410.
22. P. R. Steiner, and G. E. Troughton, Interrelationship of bond performance and some physical properties of elastomeric construction adhesives, *For. Prod. J. 30*(1):49-54 (1980).
23. B. H. River, and R. H. Gillespie, Measurement of shear modulus and shear strength of adhesives, Report prepared for U.S. Department of Housing and Urban Development by Forest Products Laboratory, Madison, Wis., 1978.
24. M. D. Strickler, Adhesive durability: specimen designs for accelerated tests, *For. Prod. J. 18*(9):84-90 (1968).
25. J. C. Beech, The performance of some catalysed polyvinyl acetate (PVAc) wood adhesives, *Build. Res. Establish. Curr. Pap. 1/78*, 1977.
26. M. A. Millet, R. H. Gillespie, and B. H. River, Evaluating wood adhesives and adhesive bonds, Report prepared for U.S. Department of Housing and Urban Development by Forest Products Laboratory, Madison, Wis., 1977.
27. R. A. G. Knight, and L. S. Doman, Analysis of the mechanical and knife tests for glues and plywood used in British Standard Specifications, *J. Inst. Wood Sci. 4*:28-43 (1959).
28. S. Chow, and K. S. Chunsi, Adhesion strength and wood failure relationship in wood-glue bonds, *Mokuzai Gakkaishi 25*(2):125-131 (1979).

29. J. Beech, The gluing of timber components with elastomeric glues, *Build. Res. Establish. Inf. Sheet 21/76*, 1976.
30. R. H. Gillespie, Accelerated ageing of adhesives in plywood-type joints, *For. Prod. J. 15*(9):369-378 (1965).
31. R. H. Gillespie, and B. H. River, Durability of adhesives in plywood: dry-heat effects by rate-process analysis, *For. Prod. J. 25*(7):26-32, (1975).
32. S. Chow, and R. W. Caster, Relationship of adhesive softening temperature to exposure tests for bond durability, *For. Prod. J. 28*(6):38-43 (1978).
33. J. M. Dinwoodie (1979) Today's adhesives: their properties and performance, *Proc. FESYP Int. Particleboard Symp.* Hamburg, 1978, pp. 38-61.
34. H. J. Deppe, and K. Schmidt, (1979) Vergleichende Lang- und Kurzzeit-Bewitterungsprüfungen an Holzwerkstoffen, *Holz Roh.-Werkst. 37*:280-294 (1979).
35. B. H. River, R. H. Gillespie, and A. J. Baker, Accelerated ageing of phenolic-bonded hardboards and flakeboards, Forest Products Laboratory, Madison, Wis., *Res. Paper FPL 393*, 1981.
36. M. A. Millett, R. H. Gillespie, and A. J. Baker, Precision of rate-process method for predicting durability of adhesive bonds. In *Durability of Building Materials and Components*, ASTM STP 691, P. J. Serada and G. G Litvan eds., American Society for Testing Material, Philadelphia, 1980, pp. 913-923.
37. P. Gressel, Zeitstandbiegeverhalten von Holzwerkstoffen in Abhängigkeit von Klima und Belastung: i. Mitteilung: Bisherige Untersuchungen, Versuchsplan, Versuchsdurchführung, *Holtz Roh.-Werkst. 30*:259-266; ii. Messergebnisse in Abhängigkeit von der untersuchten Kriechparametern, *Holz Roh.-Werkst. 30*:347-355; iii. Diskussion der Versuchsergebnisse, *Holz Roh.-Werkst 30*:479-488 (1972).
38. R. Ebewele, B. River, and J. Koutsky, Tapered double cantilever beam fracture tests of phenolic-wood adhesive joints: Pt. I. *Wood Fiber 11*(3):197-213 (1979).
39. R. Ebewele, B. River, M. Luce, and J. Koutsky, The bonding of wood—a fracture mechanics approach, 3rd Annu. Meet. Adhesion Soc., Savannah, Ga., 1980.
40. J. Haygreen, H. Hall, K.-N. Yang, and R. Sawicki, Studies of flexural creep behaviour in particleboard under changing humidity conditions, *Wood Fiber 7*(2):74-90 (1975).
41. B. H. River, and R. H. Gillespie, Long-term load deformation behaviour and strength of elastomer-based adhesives, Report prepared for U.S. Department of Housing and Urban Development by Forest Products Laboratory, Madison, Wis., 1979.
42. A. Lundgren, Determination of creep by short term loading, *Proc. 36th Meet. Tech. Comm. FESYP*, Frankfurt, 1977.

43. A. Ylinen, Prediction of the time-dependent elastic and strength properties of wood by the aid of a general non-linear viscoelastic model, *Holz Roh.- Werkst.* 5:193-196 (1965).
44. R. Ganowicz, and K. Kwiatkowski, Die Untersuchungen des Kriechens mehrschichtiger Platten, *Holztechnologie 15*:95-99 (1974).
45. C. B. Pierce, J. M. Dinwoodie, and B. H. Paxton, Creep in chipboard: Pt. 2. The use of fitted response curves for comparative and predictive purposes, *Wood Sci. Technol. 13*: 265-282 (1979).
46. T. Szabo, and G. Ifju, Influence of stress on creep and moisture distribution in wooden beams under sorption conditions, *Wood Sci. 2*:159-167 (1970).
47. J. F. Senft, and S. K. Suddarth, An analysis of creep-inducing stress in Sitka spruce, *Wood Fiber 2*:321-327 (1971).
48. L. Eriksson, Rapport d'essais sur le fluage, *Proc. 17th Meet. Tech. Comm. FESYP*, Brussels 1967.
49. R. Caster, (1980) Correlation between exterior exposure and automatic boil test results, Wood Adhesives—Research, Application and Needs Symp. Madison, Wis., Sept. 23-25, 1980.
50. E. Back, and E. Sandstrom, Critical aspects on accelerated methods for predicting weathering resistance of wood-based panels, *Holz Roh.-Werkst. 40*:61-75 (1982).
51. P. L. Northcott, R. E. Kreibich, and R. A. Currier, First replications comparing bond-degrade-accelerating systems, *For. Prod. J. 18*(5):58-65 (1968).
52. J. D. McNatt, Properties of particleboards at various humidity conditions, Forest Products Laboratory, Madison Wis., *Rep. 225*, 1974.
53. B. H. River, Strength and shear modulus of several construction adhesives as influenced by environment and loading conditions, Report prepared for U.S. Department of Housing and Urban Development by Forest Products Laboratory, Madison, Wis., 1978.
54. W. Ginzel, Zum chemisch-hydrolytischen Abbau von Harnstoff-Formaldehydharz in Spanplatten nach dreijährige Bewitterung, *Holz Roh.- Werkst. 29*(8):301-305 (1971).
55. J. M. Dinwoodie, Characterizing the performance of particleboard in the UK, *Proc. 15th Int. Particleboard Symp.* Washington State University, Pullman, Wash., 1981.
56. R. H. Gillespie, Effect of internal stresses on bond strength of wood joints, Report prepared by U.S.Department of Housing and Urban Development by Forest Products Laboratory, Madison, Wis., 1976.
57. P. R. Steiner, and S. Chow, Low-temperature durability of common wood adhesives, *For. Prod. J. 25*(8):26-30 (1975).

58. R. A. G. Knight, The efficiency of adhesives for wood, *For. Prod. Res. Lab. Bull. 38,* 1968, p. 16. HMSO, London.
59. N. C. White, The durability of glues for plywood manufacture, *Build. Res. Establish. Curr. Pap. CP 84/75,* 1975.
60. P. Gressel, Prüfung und Beurteilung der Dauerhaftigkeit von Spanplatten-Verleimungen—Ein Beitrag zur Aufstellung klebstoffunabhängiger Prüfrichtlinien: Teil 1. Aufgabenstellung—Planung und Durchführung der Versuche—Ergebnisse, *Holz Roh.- Werkst. 38*:17-35; Teil 2. Diskussion der Ergebnisse, *Holz Roh.- und Werkst. 38*:61-71; Teil 3. Polgerungen und Vorschlag von Prüfrichtlinien, *Holz Roh.- Werkst.* 38:109-113 (1980).
61. E. L. Bryan, and A. P. Schniewind, Strength and rheological properties of particleboard affected by moisture content and sorption, *For. Prod. J. 15*(4):143-148 (1965).
62. J. M. Dinwoodie, B. H. Paxton, and C. B. Pierce, Creep in chipboard: Pt. 3. Initial assessment of the influence of moisture content and level of stressing on rate of creep and time to failure, *Wood Sci. Technol. 15*:125-144 (1981).
63. W. F. Lehmann, T. J. Ramaker, and F. Hefty, Creep characteristics of structural panels, *Proc. 9th Int. Particleboard Symp.*, Pullman, Wash., 1975, pp. 151-170.
64. A. F. Halligan, and A. P. Schniewind, Effect of moisture on physical and creep properties of particleboard, *For. Prod. J. 22*(4):41-48 (1972).
65. J. Carre, Contribution à l'étude du fluage des panneaux de particules en atmosphère sèche et humide, *Rapport d' activité*, Laboratoire Forestier de l'état, Gembloux, Belgium, 1967.
66. E. L. Bryan, (1960) Bending strength of particleboard under long-term load, *For. Prod. J. 10*(4):200-204 (1960).
67. W. Kratz, Untersuchungen über das Dauerbiegeverhalten von Holzpanplatten, *Holz Roh.- Werkst. 27*(10):380-387 (1969).
68. B. H. River, and R. H. Gillespie, Behavior of construction adhesives under long term load, *Res. Pap. FPL 400,* Forest Products Laboratory, Madison, Wis., 1981.
69. S. Chow, Thermal analysis of liquid phenol-formaldehyde resin curing, *Holzforschung 26*:229-232 (1972).
70. A. J. Bolton, and P. E. Humphrey, Measurement of the tensile strength development of urea formaldehyde resin-wood bonds during pressing at elevated temperatures, *J. Inst. Wood Sci. 7*(5):11-14 (1977).
71. J. D. Wellons, Adhesion to wood substrates. In *Wood Technology: Chemical Aspects,* Series 43, American Chemical Society, Washington, D.C., 1977, Chap. 10.

2

Aminoresin Wood Adhesives

A. PIZZI / *National Timber Research Institute, Council for Scientific and Industrial Research, Pretoria, South Africa*

I. INTRODUCTION

Aminoresins are polymeric products of aldehyde reaction with compounds carrying $—NH_2$ or —NH groups. Such groups are mainly amide groups, such as those in urea and melamine. They constitute the most important members of this class of compounds, more so than the amine groups as in the case of aniline. Formaldehyde is the main aldehyde used. Other aldehydes, such as furfural, are generally not used for wood adhesives. The advantage of aminoresin adhesives (or aminoplastic adhesives as they are often called) are their (1) initial water solubility (this renders them eminently suitable for bulk and relatively inexpensive production), (2) hardness, (3) nonflammability, (4) good thermal properties, (5) absence of color in cured polymers, and (6) easy adaptability to a variety of curing conditions.

Although many amidic and aminic compounds have been investigated for use in production of aminoresins, only urea and melamine and, in rare cases aniline, are extensively used. Thermosetting aminoplastic resins produced from urea and melamine are built up by condensation polymerization. Urea and melamine are reacted with formaldehyde, which results in the formation of additional products, such as methylol compounds. Further reaction, and the concurrent elimination of water, leads to the formation of low molecular weight condensates which are still soluble. Higher molecular weight products, which are insoluble and infusible, are obtained by further condensing the low molecular weight condensates.

Urea- and melamine-formaldehyde (UF and MF) resins have a great deal in common as regards the chemical and physical characteristics of both the cured and uncured resins. MF is superior to UF because of its superior water and heat resistance, hardness, and shorter curing time under less drastic conditions. The greatest disadvantage of these aminoplastic resins is their bond deterioration, caused by water and moisture. This is due to the hydrolysis of the aminoplastic or aminomethylenic bond, which is the same for both UF and MF resins.

The higher resistance of MF resins to water attack is due to the considerably lower solubility of melamine in water. (Melamine dissolves in hot water only, whereas urea dissolves in cold water as well.) Therefore, UF adhesives are used for interior application only; MF or melamine-urea-formaldehyde (MUF) resins can be employed successfully even for rather severe outdoor conditions. If full exterior-grade quality is needed, it is safer to use phenolic-type resins rather than aminoplastic resins.

II. CHEMISTRY OF AMINORESINS

A. Urea-Formaldehyde Condensation

The reaction between urea and formaldehyde is very complex. The combination of these two chemical compounds results in both linear and branched polymers, as well as tridimensional networks, in the cured resin. This is due to a functionality of 4 in urea (due to the presence of four replaceable hydrogen atoms), and a functionality of 2 in formaldehyde. The most important factors determining the properties of the reaction products are (1) the relative molar proportion of urea and formaldehyde, (2) the reaction temperature, and (3) the various pH values at which the condensation takes place. These factors influence the rate of increase of the molecular weight of the resin. Therefore, the characteristics of the reaction products differ considerably when lower and higher condensation stages are compared, especially solubility, viscosity, water retention, and rate of curing of the adhesive. These all depend to a large extent on molecular weights.

The reaction between urea and formaldehyde is divided into two stages. The first is the alkaline condensation to form mono-, di-, and trimethylolureas. (Tetramethylolurea has never been isolated.) The second stage is the acid condensation of the methylolureas, first to soluble and then to insoluble cross-linked resins. On the alkaline side, the reaction of urea and formaldehyde at room temperature leads to the formation of methylolureas. When condensed, they form methylene-ether links between the urea molecules. The products from urea and formaldehyde, and from mono- and dimethylolureas, are as follows:

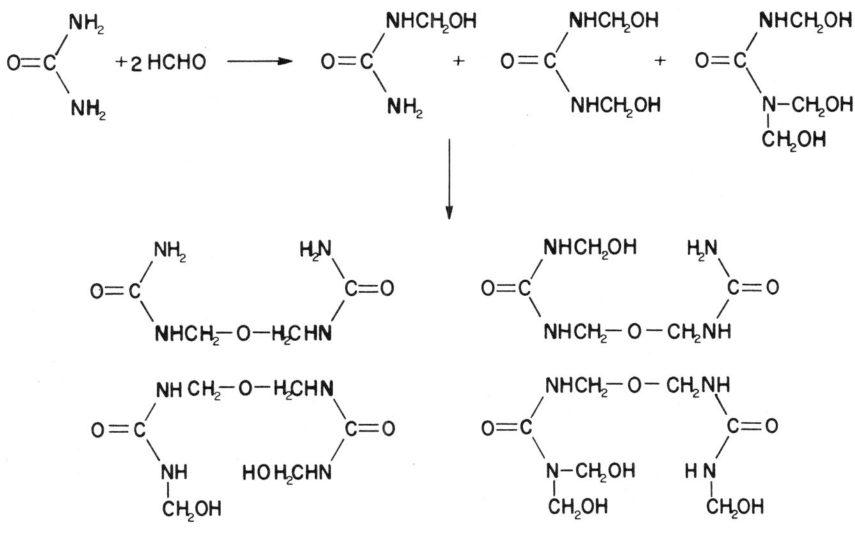

The reaction also produces cyclic derivatives: uron, monomethyloluron, and dimethyloluron.

On the acid side, the products precipitated from aqueous solutions of urea and formaldehyde, or from methylolureas, are low molecular weight methyleneureas [1].

$$H_2NCONH(CH_2NHCONH)_nH \qquad \text{I}$$

These contain methylol end groups in some cases.

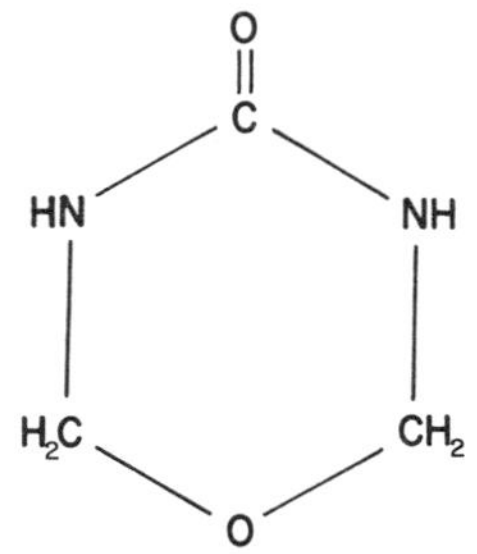

Uron

The structures of the reaction products of urea and formaldehyde under different reaction conditions are the following:

1. $H_2NCONH_2 + 0.75HCHO \xrightarrow{pH\ 1} H_2NCONH\ (CH_2NHCONH)_4H$
2. $H_2NCONH_2 + 1.0HCHO \xrightarrow{pH\ 1} H_2NCONH\ (CH_2NHCONH)_{5-6}H$
3. $H_2NCONH + 1.0HCHO \xrightarrow{pH\ 3}$ monomethylol-H_2NCONH $(CH_2NHCONH)_{5-6}H$
4. $H_2NCONH_2 + 1.0HCHO \xrightarrow{pH\ 3}$ dimethylol-H_2HCONH $(CH_2NHCONH)_{6-7}H$

Zigeuner [1] sees the acid condensation of methylenebisurea producing higher methylene ureas as a process of intermolecular cracking. In this process the methylene link of one molecule is attacked by the amide group of another.

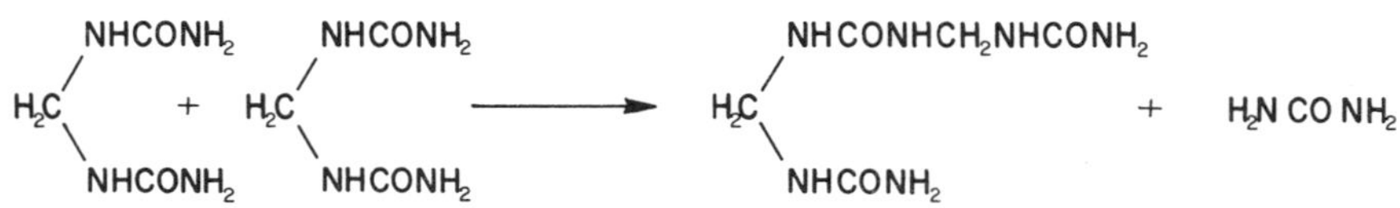

This process is probably important in the resinification step. The following conclusions can be drawn:

1. In alkaline, and probably in weak acid solution (pH 4—7), urea and formaldehyde react to form methylene-ether links between urea molecules.
2. In acid solution (pH 4), only methylene bridges are formed.
3. The molecules are linear chains in the early states of reaction.
4. The resinification may proceed by an intermolecular cracking of methylene-ether and methylene linkage.

There are facts in the chemistry of the urea-formaldehyde resins which are inconsistent with the classic theory of resin formation, that is, the formation of methylene bridges between urea molecules with solubilizing methylol groups [2]. For instance, (1) methyleneureas are already practically insoluble when three to four urea units are linked in chain. (2) The solubilizing effect of methylol groups in model compounds is negligible [2]. (3) Furthermore, the resinification process is a slightly endothermic one, whereas methyleneurea formation is strongly exothermic. Fahrenhorst [2] considers that neither the methylene bridge nor the methylol group plays an important role in the resinification reaction. He advances the theory that the reaction of urea and formaldehyde in a 1:1 molar ratio proceeds through an azomethin derivative:

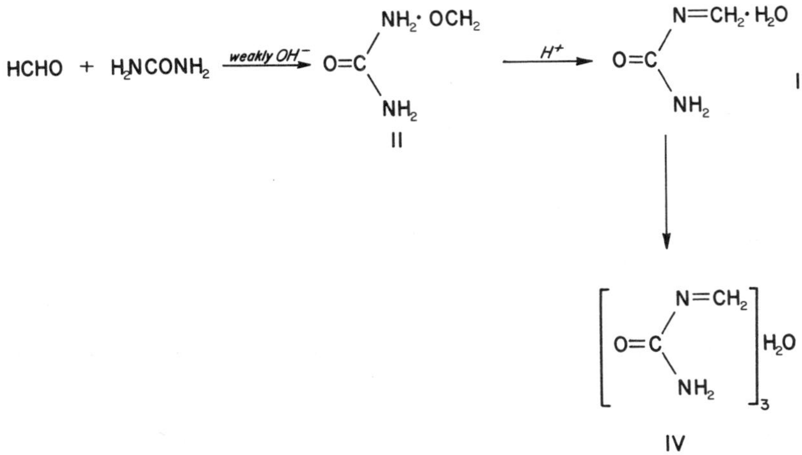

In weak alkaline solutions, the first product of the reaction is a complex (II), which is capable of rearranging itself exothermically into monomethylolurea. On acidification, the complex (II) eliminates water, resulting in unstable trimethylene urea hydrate (IV). The hydrated azomethin (III), which was identified by Fahrenhorst [2] in urea-formaldehyde resins, is regarded as characteristic of the intermediate stage of the reaction.

The electron theory provides a possible bonding mechanism between azomethin groups and either the solvent or other resin molecules. In this bonding, the π electrons of the C=N bonds and the free electron pair on the nitrogen atoms are involved. The higher pH stabilizes the degree of polymerization or association by permitting the formation of ionic complexes with water or the solvent. The lower pH causes the loosening of the water from the hydrated azomethin groups, allowing association. The resin eventually proceeds to the liophobic stage.

Indirect evidence strongly points to the existence of this mechanism as well as of the mechanism proposed by the classic theory of UF resin formation. However, no conclusive evidence of the participation of structure IV in the UF resinification process has yet been obtained. The association through azomethine-type intermediates has been mentioned to explain resin formation and to interpret the mechanism of etherification of methylol groups under acid, neutral, and alkaline conditions [2,3]. This theory opposes the classic theory.

To conclude, the increase in the molecular size of UF resins is probably caused by the following reactions, all of which occur with the simultaneous liberation of water, and in some cases of formaldehyde:

1. The formation of unsaturated azomethine groups (e.g., those from monomethylolurea). This unsaturated group is supposed to polymerize easily to polymethyleneurea:

$$nNH_2CONHCH_2OH \rightarrow NH_2CON{=}CH_{2\ n} + nH_2O$$

2. The formation of methylene bridges between methylol and amino groups of neighboring molecules:

$$RNHCH_2OH + NH_2R' \rightarrow RNHCH_2NHR' + H_2O$$

3. The formation of methylene-ether bridges by reaction of methylol groups of neighboring molecules:

$$RNHCH_2OH + HOCH_2NHR' \rightarrow RNHCH_2OCH_2NHR'$$

4. The formation of methylene bridges between two methylol groups by splitting off of water and formaldehyde:

$$RNHCH_2OH + HOCH_2NHR' \rightarrow RNHCH_2NHR' + H_2O + HCHO$$

In the first reaction, monomeric methyleneurea is formed as a result of the intramolecular loss of water. An unsaturated azomethine group is formed, followed by rapid polymerization. This gives the insoluble end product. The other reactions are condensation polymerizations in which the methylolureas are merely the building blocks of the polymers and of the insoluble end product. The polymers formed in both cases are mainly linear polymers obtained by the intermolecular splitting off of water. Under certain conditions water may also be split off intramolecularly. Cyclic compounds called urones [4] are then formed. In both cases, further splitting off of water and formaldehyde leads to the formation of hardened or cured resins.

B. Melamine-Formaldehyde Condensation

The condensation reaction of melamine (V) with formaldehyde is similar to the reaction of formaldehyde with urea. Formaldehyde first attacks the amino groups of melamine, forming methylol compounds.

Formaldehyde addition to melamine occurs more easily and completely than to urea. The amino group in melamine accepts easily up to two molecules of formaldehyde. Thus up to six molecules of formaldehyde are attached to a molecule of melamine. The methylolation step leads to a series of methylol compounds with two to six methylol groups.

Because melamine is less soluble in water than urea, the hydrophilic stage proceeds more rapidly in MF resin formation than in UF condensations. Therefore, hydrophobic intermediates of the MF condensation appear early in the reaction. Another important difference between MF

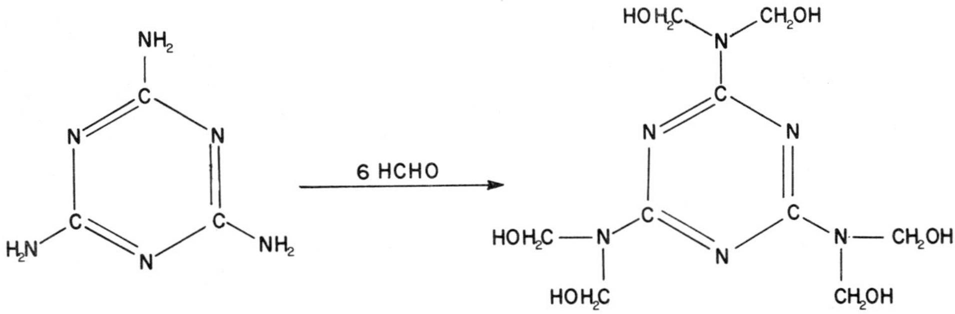

V

and UF is that the MF condensation and curing occurs not only under acid conditions, but also under neutral or even slightly alkaline conditions.

The mechanism of the further reaction of methylolmelamines to form hydrophobic intermediates is the same as for UF resins, with splitting off of water and formaldehyde. Methylene and ether bridges are formed and the molecular size of the resin rapidly increases. These intermediate condensation products constitute the large bulk of the commercial MF resins. The final curing process transforms the intermediates to the desired MF insoluble and infusible resins through the reaction of amino and methylol groups which are still available for reaction.

A simplified schematic formula of cured MF resin has been given by Koehler and Fry [5,6]. They emphasize the presence of many ether bridges besides unreacted methylol groups, and the methylene bridges. This is because in curing MF resins at temperatures of up to 100°C, no substantial amounts of formaldehyde are liberated. Only small quantities are liberated during curing up to 150°C. However, UF resins curing under the same conditions liberate a great deal of formaldehyde.

Wohnsieldler, Updegraff, and Hunt [7] have tried to correlate the best physical properties of melamine-formaldehyde resins with the degree of curing or condensation. They have found that the various properties attain their peak at different degrees of reaction. However, the best physical properties of MF resins were always associated with significant cross-linking.

C. Aniline-Formaldehyde Condensation

When aniline and formaldehyde are reacted in equal amounts, under neutral or alkaline conditions, N-methylolanilines are formed. These form rapidly N-methyleneanilines by eliminating water. If heated further, they form soluble and fusible aniline-formaldehyde resins (VI).

In the presence of acids, especially hydrochloric acid, the amino group is protected by the formation of aniline hydrochloride. In this

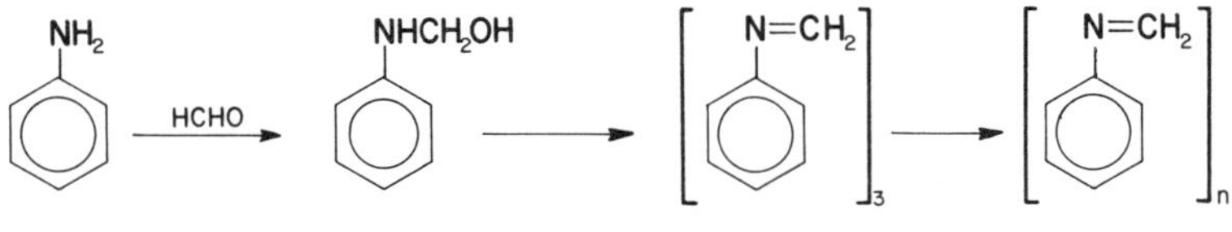

VI

case the formaldehyde attacks the free *para*position on the aromatic ring. With an aniline formaldehyde molar ratio of 1:1 the hydrochloride of the *p*-aminobenzylalcohol is formed. The free base is produced by neutralizing the solution of *p*-aminobenzylalcohol hydrochloride. Water is split off and long linear chains of brittle aniline-formaldehyde resins (VII) of low mechanical strength are formed by condensation polymerization.

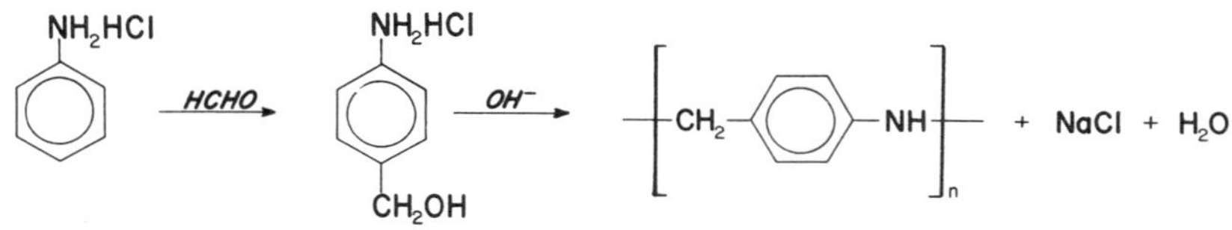

VII

By using an excess of formaldehyde over the equimolar amounts, the linear chains of this product are cross-linked, with the formation of methylene bridges. The resins produced have a three-dimensional network of good mechanical strength. In these resins both aminomethylene linkages ($-CH_2-NH-$) and methylene linkages ($-CH_2-$) between the aromatic nuclei are present [8]. This indicates that the formaldehyde is able to attack both the *para* and *ortho* positions of the aromatic rings (exactly as in phenolic resins), as well as the aromatic amino group (as in aminoplastic resins).

Amino groups are able to induce nucleophilicity in aromatic nuclei, higher than the hydroxy groups of phenols. Therefore, aniline and some of its derivatives (*m*-hydroxyaniline and phenylendiamine) are sometimes used as terminal grafted groups of linear phenolic resins to accelerate curing [9]. In these resins, the aromatic rings and the amino groups form bonds generally in a 75–80:25–20 molar proportion, respectively. This gives rise to both aminomethylene ($-CH_2-NH-$) and methylene ($-CH_2-$) linkages [9].

D. Reaction Kinetics: Urea-Formaldehyde

The kinetics of the formation and condensation of mono- and dimethylolureas and of simple urea-formaldehyde condensation products has been studied extensively. The formation of monomethylolurea in weak

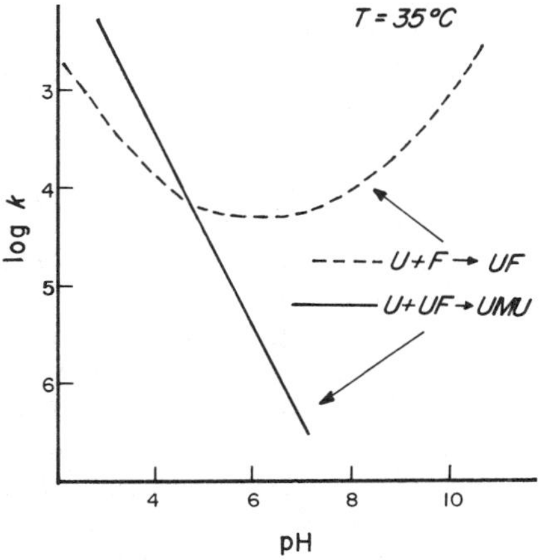

Fig. 2.1. Influence of pH on the addition and condensation reactions of urea and HCHO. U, urea; F, HCHO; M, $-CH_2-$.

acid or alkaline aqueous solutions is characterized by an initial fast phase followed by a slow bimolecular reaction [10,11]. The reaction is reversible. The formation of methylolurea is bimolecular and its dissociation monomolecular [11]. The rate of reaction varies according to the pH with a minimum rate of reaction in the pH range 5–8 for a molar ratio of 1:1 for urea/formaldehyde and a pH of ±6.5 for a 1:2 molar ratio [12] (Fig. 2.1).

The 1:2 urea/formaldehyde reaction has been proved to be three times slower than the 1:1 molar ratio reaction [13]. The dehydration of formaldehyde (present largely as methylene glycol) and the formation of the urea anion are considered to be the controlling factors [11].

The rapid initial addition reaction of urea and formaldehyde is followed by a slower condensation, which results in the formation of polymers [13]. The rate of the condensation of urea with monomethylolurea to form methylenebisurea (or UF "dimers") is also pH dependent. It decreases esponentially from a pH of 2–3 to neutral pH value. No condensation occurs at alkaline pH values. The marked influence of the pH range on the reactions rates indicates that such reactions are of the hydrogen-ion-catalyzed type.

The initial addition of formaldehyde to urea in dilute solutions (0.1 M) is reversible, and is subject to general acid and base catalysis. The forward bimolecular reaction has an activation energy of 13 kcal/mol. The reverse unimolecular reaction has an activation energy of 19 kcal/mol [11]. The proposed mechanism of the acid catalysis is that of a protonated formaldehyde carbocation with urea. The alkaline-catalyzed reaction proceeds instead through the reaction of the urea anion with formaldehyde. The subsequent reaction of monomethylolurea with formaldehyde in dilute solution, to give dimethylolurea, corresponds closely to the 1:1 monomethylolurea formation reaction in type, reaction mechanism, and activation energies [14].

It is also reversible in concentrated solutions (2–4 M) at pH 7.0, and at 35°C, the addition reactions have the same rate constants as in dilute solutions and the reactions are very similar. No trimethylolurea is detectable in the reactions of urea and formaldehyde in dilute solutions containing a 6–8 M excess of formaldehyde [15]. The rates of introduction into the urea molecule of one, two, and three methylol groups has been estimated to have the ratio 9:3:1, respectively. The formations of N,N'-dimethylolurea and of trimethylolurea are also bimolecular, and their decomposition monomolecular. The formation of N,N'-dimethylolurea from monomethylolurea is about 1.5 times that of monomethylolurea from urea. The decomposition of N,N'-dimethylolurea to monomethylolurea is three times that of monomethylolurea to urea.

The study of condensation reactions of methylolureas with urea and with themselves in dilute aqueous acid solutions is a standard type of reaction. A methylene bridge is formed by an acid-catalyzed bimolecular reaction between a methylol group and an unsubstituted amide group [16]. The following four reactions occur at decreasing rates in the order given below, where U = urea, UF = monomethylolurea, UMU = methylenebisurea and so on:

1. $U + UF \rightarrow UMU + H_2O$
2. $U + UF_2 \rightarrow FUMU + H_2O$
3. $UF + UF \rightarrow FUMU + H_2O$
4. $UF + UF_2 \rightarrow FUMUF + H_2O$

No reaction was found between two molecules of dimethylolurea under the conditions chosen. It appears that (1) the amide group in urea is more reactive than that of monomethylolurea; and (2) the methylol group in the latter is more reactive than in dimethylolurea. The results cannot be interpreted in terms of either dimethylene-ether formation between urea molecules, or dehydration of methylolureas to methyleneureas followed by polymerization [16]. However, an analysis done by the same authors [17] of the insoluble fractions of urea-formaldehyde condensates indicates that the properties of the condensates are the same as those produced by stepwise condensation to form methylene bridges between urea residues.

Methylenebisurea undergoes further condensation with formaldehyde [18] and monomethylolurea [19], behaving like urea. The capability of methylenebisurea to hydrolize to urea and methylolurea in weak acid solutions (pH 3–5) indicates the reversibility of the methylene link and its lability in weak acid moisture. It explains the slow release of formaldehyde over a long time in particleboard and other wood products manufactured with UF resins.

Equilibrium constants for urea-formaldehyde-methylolureas have been theoretically derived, and found to agree with experimental results [20]. They can be used to derive formulas, thereby quantifying these components in solution of different initial formaldehyde concentrations. Study of the addition and condensation reactions of urea and formaldehyde have led some authors [21] to conclude that the addition process is a continuous one, with the hydrogen ion acting as a negative catalyst by addition to the nitrogen atom of the urea molecule. The condensation, on the other hand, reaches a state of equilibrium. In both reactions, the hydroxyl ion can be regarded as an indirect catalyst.

A few studies have been done on the pH changes in urea-formaldehyde solutions during reactions [22–24]. The main cause of the observed pH drop was the Cannizzaro reaction, disproportioning formaldehyde to methanol and formic acid. Oxidation by atmospheric oxygen also contributed. The latter could be eliminated by working in an inert atmosphere. This, or the use as a surface layer of a solvent which will not mix with water, inhibits the formation of methylene ureas [24]. Using ammonia as neutralizing agent, an initial rise in the pH of the reaction mixture has been observed. This has been attributed to the decomposition of unstable methylolamines. The effect of methanol on the reactions has also been investigated. It slightly retards the addition reactions. Methanol hinders the formation of methylene links [25] in the condensation reactions.

E. Reaction Kinetics: Melamine-Formaldehyde

Melamine and formaldehyde react in similar way to urea and formaldehyde, although basic differences are evident in the reaction rates and mechanism. The primary products of reaction are methylolmelamines, and evidence indicates that such compounds are formed only at ambient or higher temperature, except in acid pH ranges. The reaction is reversible throughout the pH range. Its forward rate is proportional to either [melamine][HCHO], or [melamine][H^{+}CHOH] or [melamine][HCHO], according to the pH of the media.

Methylolmelamine forms "dimers" by condensation with melamine under neutral and acid conditions (70°C). This process is irreversible. The initial hydroxymethylation is very rapid. Its rate is determined by the condensation of conjugated acids of methylolmelamines with melamine. The reaction rate is proportional to $[\text{melamine}]^2[\text{HCHO}]$ [26]. Where the [mineral acid]/[melamine] ratio is 0.0 to 1.0, the early-stage hydroxymethylation of melamine is dependent on the concentration of the melamine molecule (base species) MH and its conjugated acid MH_2^+ in the following manner [27]:

$$\text{Rate} = k_{H_2O}[MH][HCHO] + k_H[MH_2^+][HCHO] + k_{MH_2^+}[MH_2^+][MH][HCHO] + k_{MH}[MH]^2[HCHO]$$

In the absence of added acid, thus when the ratio [mineral acid]/[melamine] is = 0, the rate of the reaction can be represented as:

$$\text{Rate} = k_{H_2O}[MH][HCHO] + k_{MH}[MH]^2[HCHO]$$

The condensation reaction has been studied by investigating the kinetics of the initial stage of the condensation of di- and trimethylolmelamine (MF_2 and MF_3) in the pH range 1–9. Regardless of pH, the initial rate is equal to

$\text{Rate} = k[MF_n]^2$ (with n = 2 or 3) [28]

In the presence of mineral acid, the main reaction at the early stage of the condensation is the reaction between the methylol-melamine molecule and its conjugated acid (MF_nH^+) [22]. This was found at an [acid]/[MF_n] (n = 2 or 3) ratio lower than 1.0 (pH 2.7). With an [acid]/[MF_n] ratio higher than 1.0–1.2 (pH < 2), the main condensation takes place between the conjugated acids themselves.

At equal pH values the condensation rate of trimethylolmelamine is considerably faster than that of dimethylolmelamine. This is the opposite of the rates of mono- and dimethylolurea. This means that while the nitrogen of the amino group in the case of urea is more reactive, and therefore more nucleophilic, than the nitrogen of the aminomethylol group, the opposite is true in the case of melamine. The reaction for MF_2 is mainly between the carbon of the methylol group next to the nitrogen in HM^+CH_2OH, and the nitrogen of the amino group in MCH_2OH. With regard to MF_3, the condensation is mainly between the carbon of the methylol group next to the charged nitrogen in H^+MCH_2OH, and the nitrogen of the aminomethylol group in MCH_2OH [28]. The condensation rate therefore increases with the increasing electrophilicity of the carbon of the methylol group and with increasing nucleophilicity of the nitrogen of the amino group or aminomethylol group. Therefore, in MF_3 the carbon in HM^+CH_2OH is more electrophilic than the same carbon in MF_2. On the other hand, the nitrogen of the aminomethylol group in $HMCH_2OH$ of MF_3 is less nucleophilic, and therefore less reactive, than the nitrogen of the amino group of MF_2. The effects of the carbon and nitrogen atoms are consequently opposite to each other in the MF_n condensation. Since the effect of the carbon is greater than the effect of the nitrogen on the reaction rate, the MF_3 condenses faster than MF_2. At lower pHs, the effect of the nitrogen becomes negligible and MF_3 is even faster than MF_2 in condensing to polymers.

The difference between the kinetic behavior of urea and melamine can be ascribed to the different effect of the nitrogen atom in the two compounds. With regard to the formation of methylol compounds as a result of hydroxymethylation, the functionality of melamine has been observed to be 6 against formaldehyde, but only 2.4 for urea [29]. Similarly, melamine easily reacts with formaldehyde to form MF_3. It also forms MF_6 in concentrated formaldehyde [26,30]. Urea readily forms dimethylolurea, but forms trimethylolurea with marked difficulty [29,31].

These results suggest that the nitrogen of the aminomethylol group in methylolurea is considerably less nucleophilic than the nitrogen of the amino group in urea. However, the nitrogen of the aminomethylol group in methylolmelamine is not markedly less nucleophilic than the nitrogen of the amino group in melamine. This is presumably due to the difference in basicity between urea and melamine. The same is also true of their condensation reactions, since the rate constant, hence the rate of reaction, of dimethylolurea is approximately 10^3 lower than that of monomethylolurea. The reaction mechanism is shown in some detail in the paragraph on acid and alkaline catalysis.

F. Reaction of Methylolureas in the Presence of Cellulose

Steele and Giddins [32] have analyzed the resin formed after treatment of paper sheets with solutions of methylolureas and ammonium chloride. Monomethylolurea combines in the paper in a urea/formaldehyde ratio of 1:1. It also combines with dimethylolurea in a ratio of 1:5. They concluded that in the first case, side chains of about three urea units are attached to the cellulose by ether linkages. In the second case, two urea units, on average, are linked by a methylene group and joined to the cellulose by ether links. This gives the approximate formulation VIII, where "Ce" represents the cellulose chains.

$$\mathrm{Ce{-}O{-}H_2C{-}HNCONH{-}CH_2{-}NHCONH{-}CH_2{-}O{-}Ce}$$

VIII

This type of ether link has also been suggested to explain the enhanced condensation of dimethylolurea in the presence of wood pulp [33].

G. Reaction Mechanisms: Urea-Formaldehyde

By alkaline catalysis the reaction can proceed only to the formation of monomeric methylolureas in the following manner:

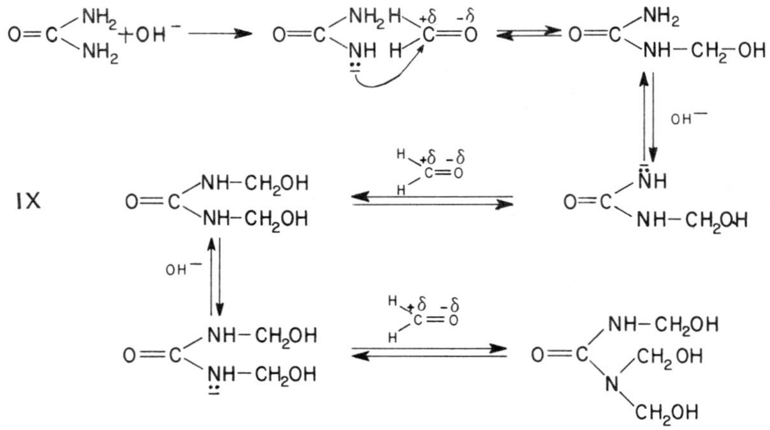

IX

The monomethylolureas formed copolymerize by acid catalysis and produce polymers and then highly branched and cured networks.

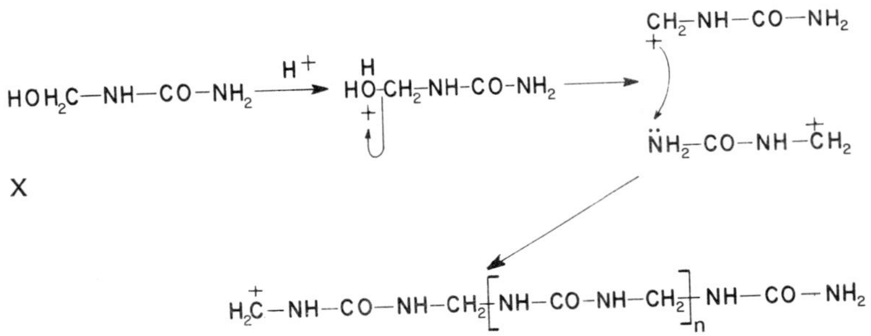

X

H. Reaction Mechanisms: Melamine-Formaldehyde

The mechanism of the initial stages of the reaction of melamine with formaldehyde, leading to the formation of methylolmelamine, is very similar to that of urea. The reaction mechanism of the acid catalyzed condensation reaction of methylolmelamine to form polymers and resins (XI and XII) (which is somewhat different from that of methylolureas) has been elucidated by Sato and Naito [28].

I. Hardening

Although much information has been published on the kinetics and synthetic pathways of the initial monomeric reactions of urea and formaldehyde, little is known of how this reaction proceeds during the condensation from higher molecular weight polymers to cured, hardened tridimensional networks. Apart from the complexity of this reaction,

The insolubility of the products in many solvents makes their characterization by conventional means difficult.

Steiner [34] developed a method by which the UF polymer's cure profile is determined by bromination, followed by an x-ray analysis of the brominated products. The bromination is carried out in CCl_4 as UF resins oxidize in aqueous bromine solutions. This is because amide groups are rapidly attached by bromine in neutral aqueous solution, and gaseous products are released according to the following reaction:

$$2NH_2\underset{\underset{O}{\|}}{C}NHR + 3Br_2 + 2H_2O \rightarrow N_2\uparrow + 2CO_2\uparrow + 2RNH_2 + 6HBr$$

Using the same method, Steiner also studied the curing profiles of three different UF resins, each one of which had a formaldehyde/urea molar ratio of 2.0, 1.4, 2.0 and methylol content of 70, 48, and 67%, respectively. The last one is a commercial UF adhesive for plywood. Figure 2.2 shows the curing curves of these resins at 60°C. All the three resins exhibit a decrease in bromine as the curing time increases. This is consistent with the decrease in methylol groups and the formation of methylene and/or methylene-ether linkages from condensation reactions which take place during curing of the resin, as the amount of bromine present was found to be roughly proportional to the amount of methylol groups in the resin before bromination [34]. The differences, although not great, are evident in the cure profiles of the three resins. However, in all cases, at least 1 hr of heating at 60°C is required before the bromine content, and therefore methylol percentage, decreases to a constant low level.

J. Analysis

Methods of formaldehyde analysis include the iodometric, sulfite [35], and mercurimetric [36,37] methods. The sulfite method measures only the formaldehyde present, whereas the iodometric method can also estimate the methylol groups. Another method is based on the partition of formaldehyde between water and isoamyl alcohol [38]. Estimation of the formaldehyde in the alcohol phase of a mixture of an aqueous solution of the resin and of isoamyl alcohol allows deduction of the amount of free formaldehyde. This procedure has the advantage that no risk of reaction between free formaldehyde and the resin components arises.

Kappelmeier [39] has suggested the use of aniline, benzylamine, and phenylethylamine as reagents, or the identification and analysis of urea in UF resins. He has provided evidence that the methylene-ether groups form a bridge between urea residues in urea-formaldehyde resins. The use of benzylamine, in particular (which yields dibenzylurea from urea derivatives), has been developed as a method of analysis. In determining the ratio of urea to formaldehyde in UF resins,

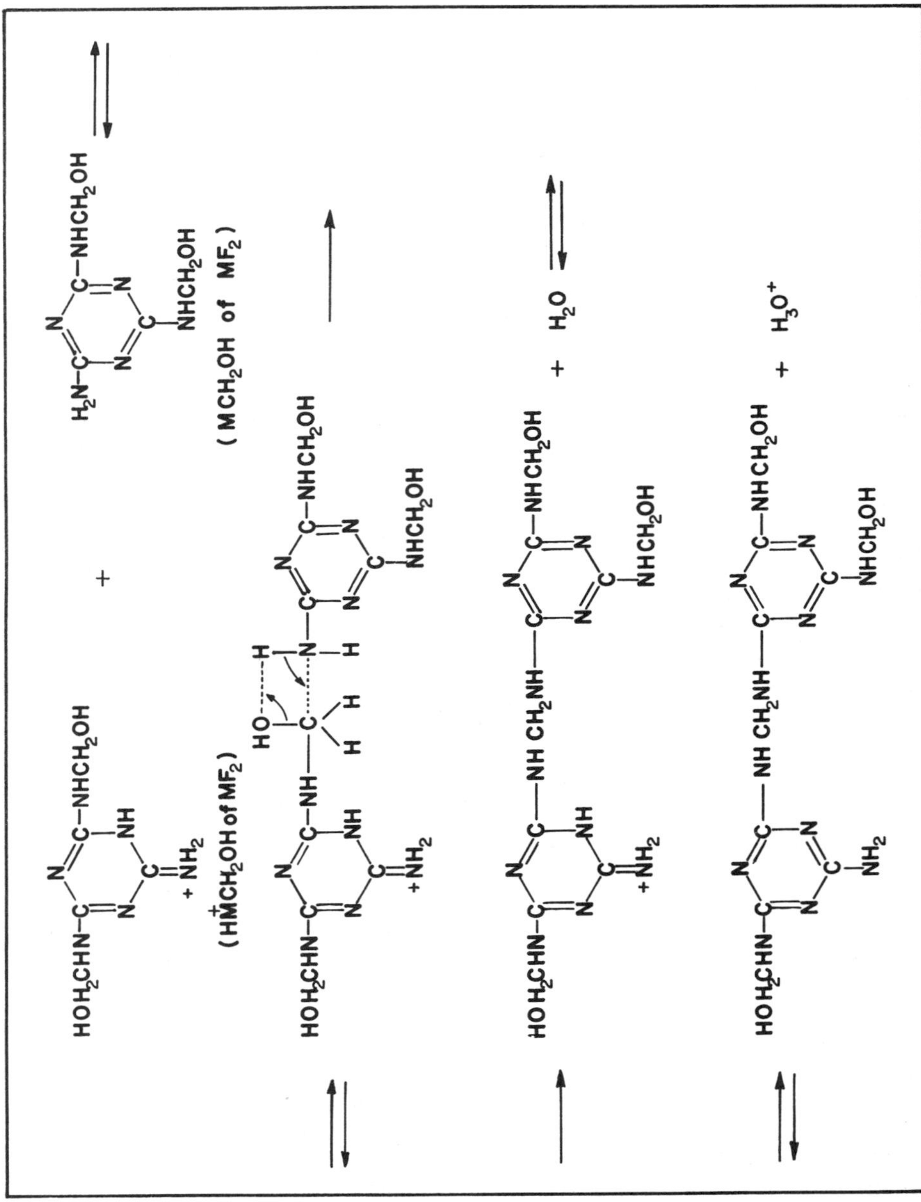

Structure XI.

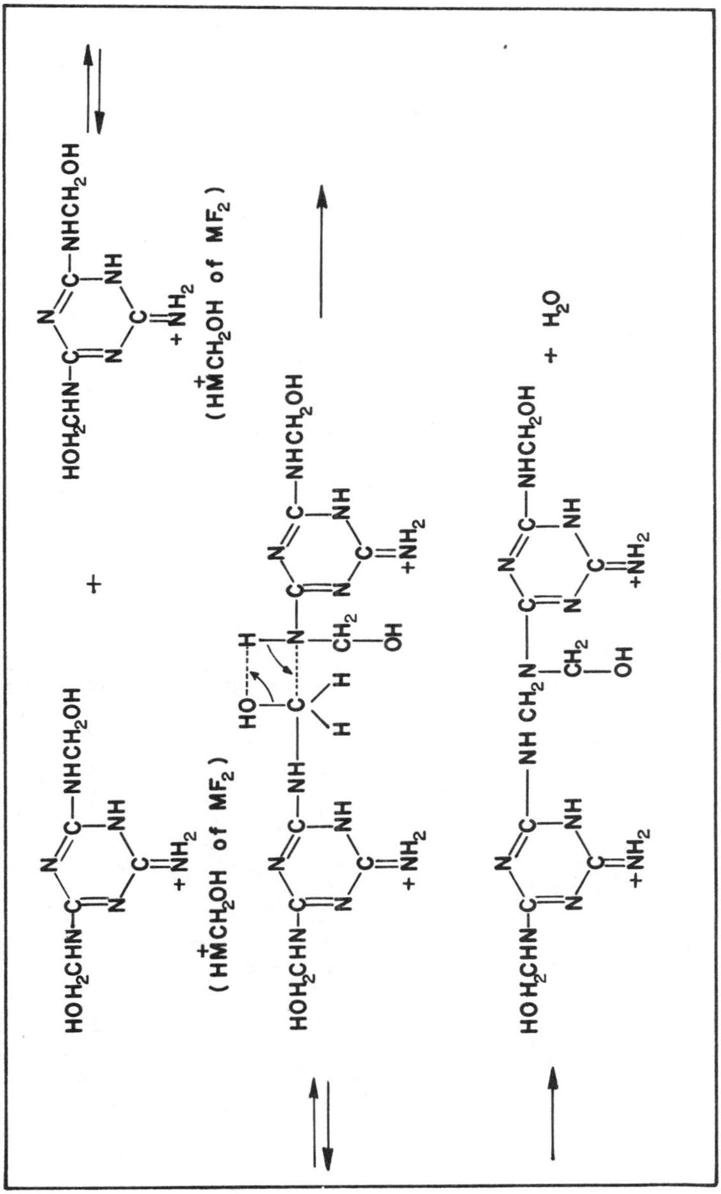

Structure XI. (Continued)

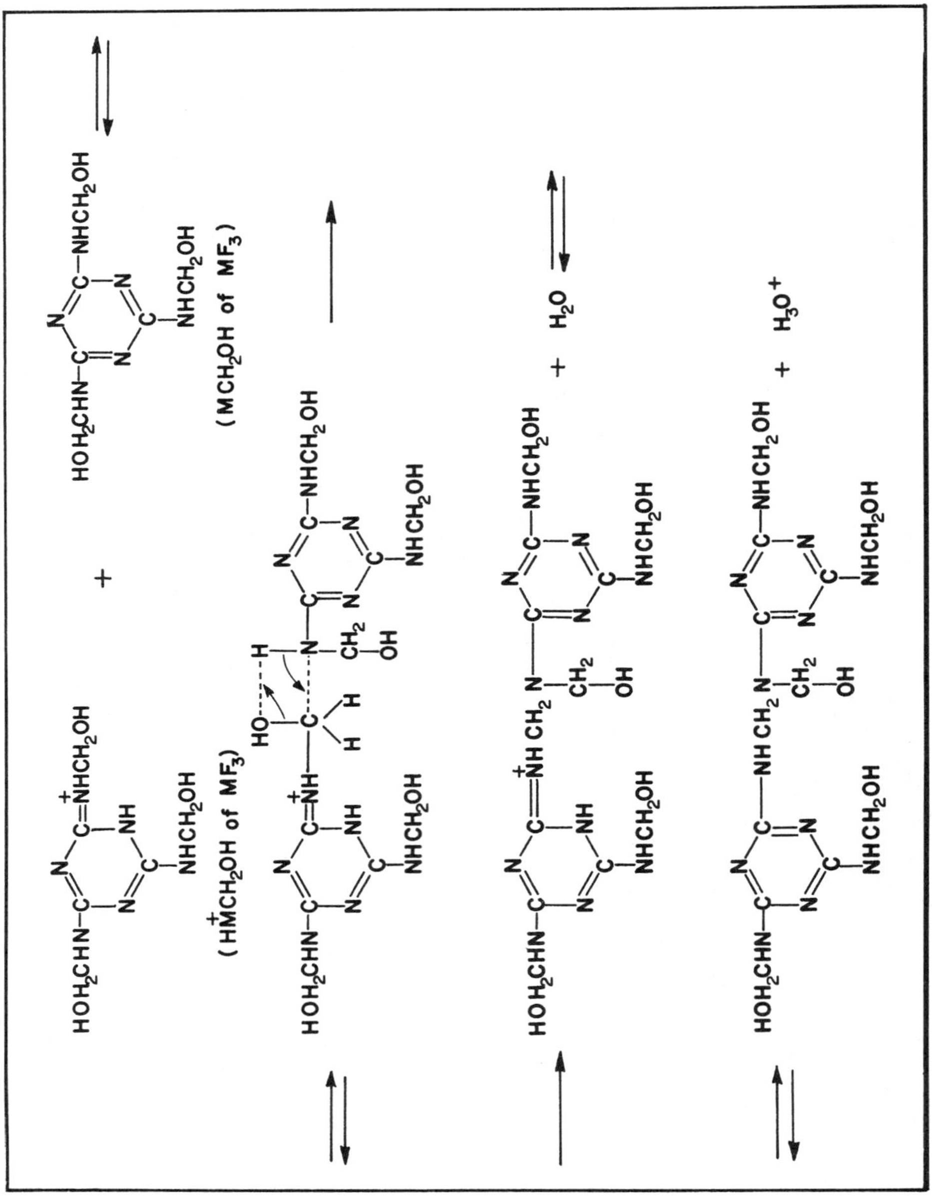

Structure XII.

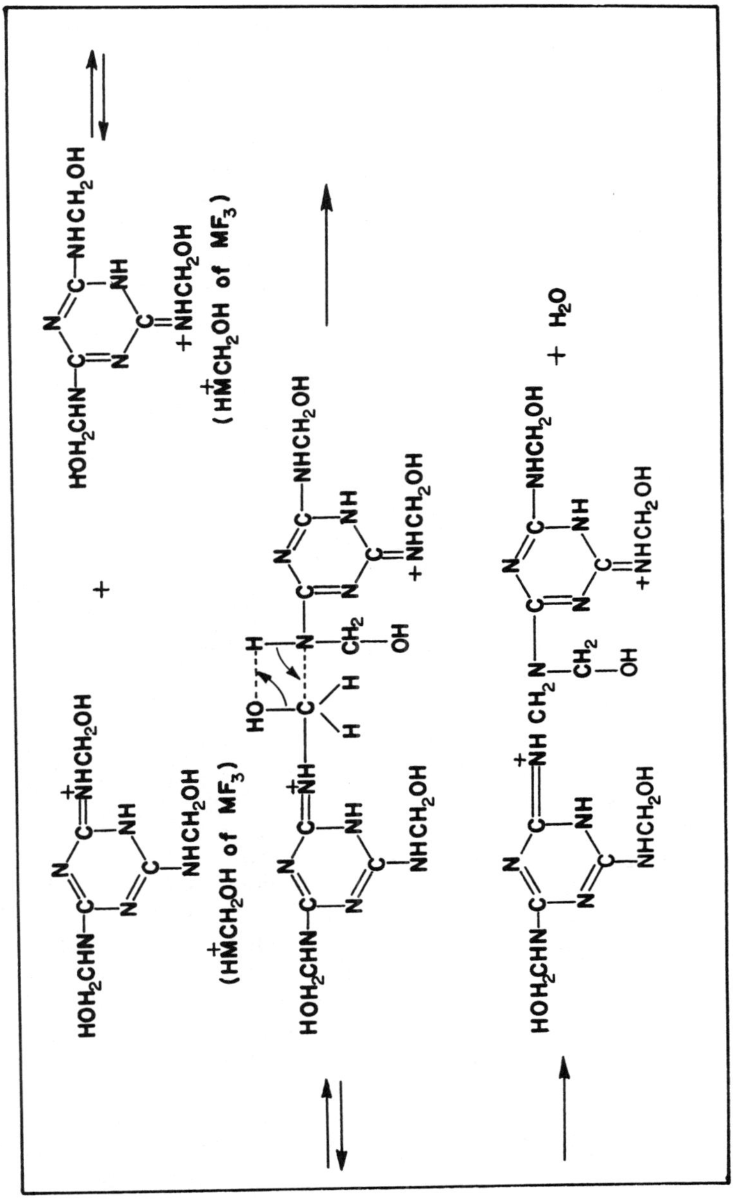

Structure XII. (Continued)

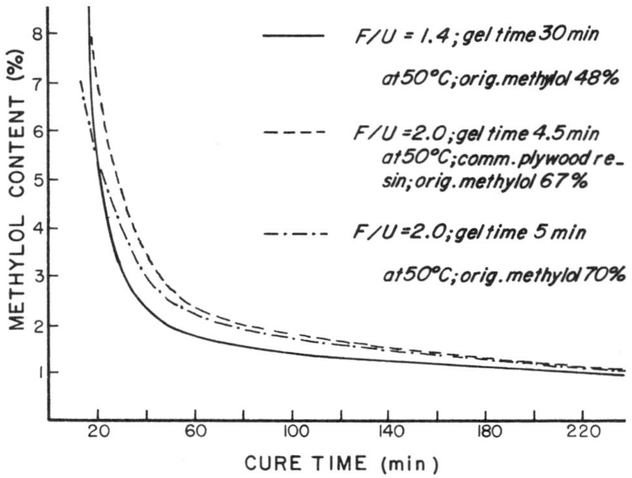

Fig. 2.2. Cure profiles for UF resins at 60°C. F, formaldehyde; U, urea; F/U, molar ratio.

the benzylamine method has been coupled with a process of formaldehyde estimation which involves depolymerization with phosphoric acid, followed by distillation into alkaline potassium cyanide solution [40].

Steiner [34] advocates the use of bromination in CCl_4 and subsequent x-ray analysis to determine available reactive methylol groups. High-resolution nuclear magnetic resonance (NMR) has also been used to analyze UF resins and trace their kinetic behavior [41]. Methylolureas, uron, and methylol ions can be detected using this method.

Molecular weight determinations of UF resins have also been carried out extensively by a variety of methods and have produced contradictory results. Vogel [42] deduced a mean molecular weight of 10,000 for a still-water-soluble resin with a 1:1 molar ratio, from osmotic pressure measurements. However, other authors [43] who have used ebulliometric methods have given results of only 800 for butylated urea resins. This appears more reasonable in comparison with other authors' evidence. Staudinger and Wagner [44] have found that thiourea-formaldehyde condensates have a low molecular weight by doing cryoscopic determinations. They explain all these anomalous findings as being due to association effects. These cause a low molecular weight resin to resemble a high molecular weight resin. They assume that association through hydroxyl groups may be responsible for this effect. The accumulated evidence of many investigations supports the idea of a low average molecular weight for UF resins.

Viscosity measurements have provided a means of investigating the shape of the polymer molecule. Vogel [42,45] fractionated methylated resins by precipitation from ethanol-water solution using methanol. He compared the viscosities of solutions of macromolecules of known shapes, and concluded that the molecules are elongated spheres. Staudinger and Wagner [46], by similarly measuring viscosity and concentration ratios, deduced that water-soluble UF resins condensates contain elongated, rather than spherical particles, although these are not of the filamentous type. Takahashi concluded [47] on the basis of viscosity measurements taken during the reaction that the resin grows linearly and then cross-links.

The analysis becomes more difficult when unknown products, particularly fully cured articles, have to be tested for UF and MF resins. Widmer [48] offers a method for the identification of UF and MF resins in technical products. This involves preparing crystalline products of urea and melamine and identifying them under the microscope. Melamine (in the form of melamine crystals) and urea (in the form of long, crystalline needles of urea dixanthate) can be seen. This method allows one to distinguish between urea and melamine even in a cured adhesive joint.

Quantitative determination of UF and MF resins is also rather difficult. Apart from Kappelmeier's method [39], another method was developed and published by Widmer [48] for the quantitative determination of urea and melamine in UF and MF condensation products. In this method, the resins are destroyed under pressure by aminolysis, leaving the melamine intact. This is then converted to melamine picrate, which is easily crystallized, and weighed. The combination of the Kappelmeier and Widmer methods makes it possible to determine quantitatively the presence of urea and melamine in intermediate condensation products and in cured UF and MF resins (even when they have been mixed). Estimations are seldom in error by more than a few percent.

Hirt, King, and Schmitt [49] have published an effective and rapid method for the detection of melamine by ultraviolet spectrophotometry. This method can be used for products containing MF. It makes use of the strong absorption of the melamine ion at 235 nm. The resin is extracted from comminuted melamine-formaldehyde samples by hydrolizing it to melamine by boiling it under reflux in 0.1 N hydrochloric acid.

Sometimes the question arises of identifying the filler in the aminoresin. This can be done at any stage of resin condensation, cured or uncured. The method takes advantage of the ability of slightly acid formaldehyde solutions to dissolve such resins, which is very easily accomplished if they are heated for a few hours in a pressure vessel at about 120°C. This method can be used to identify fillers, fibers, pigments, or any insoluble substances, even in mixed UF and MF resins. Microscopic examination of such formaldehyde-extracted fillers often discloses all the information desired as to the nature of the filler.

III. CHEMISTRY AND TECHNOLOGY OF APPLICATION OF AMINORESIN ADHESIVES FOR WOOD

A. General Principles of Manufacture and Application

It is very important in the commercial production of urea-formaldehyde (UF), melamine-formaldehyde (MF), and melamine-urea-formaldehyde (MUF) resins to be able to control the size of the molecules by the condensation reaction, since their properties change continuously as they grow larger. The most perceptible change is the increase in viscosity. Low-viscosity syrups are first formed. These change into high-viscosity syrups which are clear to turbid. These products may still be completely water soluble, but they may prove to be alcohol insoluble in later stages of the reaction.

Molecular weight may vary from a few hundred to a few thousand with a wide range of molecular size. These molecules are built up by water splitting off at random between reactive groups of neighboring molecules, thereby increasing their size. Once their solubility, viscosity, pH, concentration, and so on, have been determined, they constitute the resins available commercially. The most important factors influencing the final properties of aminoplastic resins in industrial manufacture are the purity of the reagent, the molar proportions of the materials used, the preparation process used, and the pH variation and control.

1. *Purity of Reagents*

Every commercial grade of formaldehyde contains some methanol from its manufacture in order to stabilize the formalin solution. If methanol is not present, paraformaldehyde is formed. On storage, first turbid solutions, and then white colloidal deposits are formed. These form faster in winter than in summer. Formalin solutions with a 37–41% content of formaldehyde by volume are usually stabilized with 6–12% methanol according to the storage temperature. A 30% solution (by mass) of formaldehyde needs less methanol to stabilize (i.e., from traces up to 1%).

Higher concentrations of formaldehyde require higher methanol amounts. In MF resins, no difference exists in the water resistance of the cured resin whether formaldehyde containing methanol is used or whether methanol-free formaldehyde is used. However, in UF resins, the methanol content makes a considerable difference. With methanol-free formaldehyde regular UF resins can be manufactured with normal boiling-water absorption (+2%).

With formaldehyde containing methanol (i.e., 40% formalin solution with 8–11% methanol), and identical manufacturing conditions, the UF resins obtained are much less stabile in water and have an unusually high hot-water absorption (6–10%). To avoid this problem, either unstabilized formaldehyde is used or methanol is distilled from commercial formaldehyde solutions before urea is dissolved in it.

This unusual effect is due to the formation of methylated urea-formaldehyde resins, which, on curing, remain unchanged and, because of their easy solubility in water, destroy the water resistance of the finished article. This does not occur with melamine-formaldehyde resins because the cured MF resins have a higher cross-link density and are less soluble, and therefore less water sensitive. They are also more stable at higher temperatures. Commercial UF resins are generally manufactured by using formalin solutions which contain less that 1% methanol and which are stored in heated tanks at ±60°C.

In the case of melamine resins, attention must be paid to the formation of hydrolysis products of the melamine before manufacture starts. The hydrolysis products of melamine are obtained when the amino groups of melamine are gradually replaced by hydroxyl groups. Completed hydrolysis produces cyanuric acid.

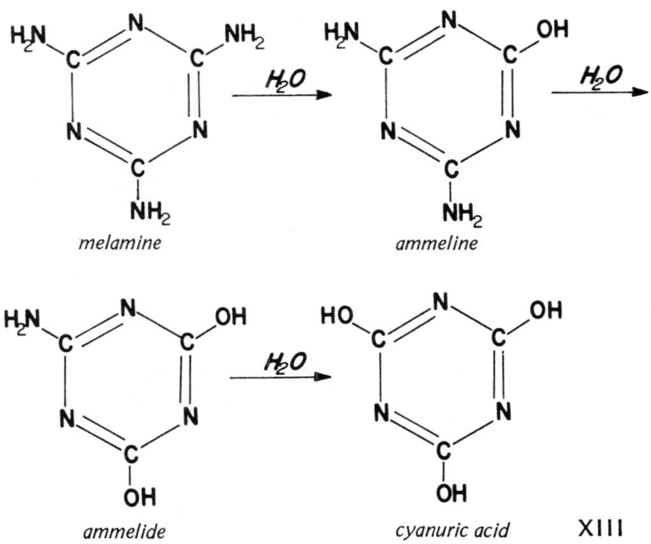

XIII

Ammeline and ammelide can be regarded as partial amides of cyanuric acid. They are acids and have no use in resin production. They are very undesirable by-products of the manufacture of melamine because of their catalytic effect in the subsequent melamine-formaldehyde resin production, due to their acidic nature. Both must be removed from crude melamine by an alkali wash and/or crystallization of the crude melamine.

2. *Proportions of Materials Used*

The molar ratio of urea to formaldehyde used in UF resin manufacture is in the range 1:1.4–1:1.6, with the formaldehyde in the form of an aqueous solution. This solution always contains some formic acid and is therefore neutralized with an alkali or ammonia before the urea is added. After the urea has been dissolved, methylol formation is induced by reacting the components at moderate temperature (+50°C) for a few hours or overnight. Free formaldehyde determination allows the monitoring of this methylolation step. A turbid solution of methylol compounds develops, which can be used as a rough and unrefined UF resin.

The most important method for the preparation of commercial UF resin adhesives is the addition of a second amount of urea during the preparation reaction. This consists of reacting urea and formaldehyde in more than equivalent proportions. Generally, an initial urea/formaldehyde molar ratio of 1:2.2 is used. Methylolation can in this case be carried out in a much shorter time, by using temperatures of up to boiling point. At about 90°C the exothermic reaction usually leads to vigorous boiling. The boiling continues without external heating and the mixture is maintained under reflux.

When the exotherm, and hence boiling, subsides (usually after 10–30 min), the methylol compounds have formed, and the reaction is completed under reflux, preferably by adding a trace of an acid to act as a catalyst until the viscosity of the solution increases. As soon as the right viscosity is reached, the resin solution is cooled to about 25–30°C. More urea (called second urea) is added to consume the excess of formaldehyde, until the molar ratio of urea to formaldehyde is in the range 1:1.4–1:1.7. After this addition of urea, the resin is left to react at 25–30°C for as long as 24 hr. The excess water is eliminated by vacuum distillation until a resin solids concentration of 64–65% is reached, and the pH adjusted to achieve suitable shelf life or storage life.

The final addition of urea can be done in one operation, or the urea may be added at suitable time intervals in smaller lots. The two methods produce slightly different resins. Some UF resins used for joinery are also produced without any final, or second, urea addition. The pH used during the condensation reaction (not the methylolation) is generally in the range 4.8–5.0.

The chemical constitution of UF resin (which can be influenced by molar ratio, type and degree of condensation, and hydrogen-bonding ability) is an important factor relating to the durability of UF adhesive wood bonds. Several authors [50–59] have studied the preparation, properties, and adhesive characteristics of UF resins in plywood. Some authors [54] report that the properties and bonding strengths of UF adhesives for plywood are primarily influenced by the degree of condensation and the molar ratio of the resin. They found that for two resins prepared under identical conditions, the one with a urea/formaldehyde molar ratio of 1:1.6 has better water resistance than the one with the ratio 1:2.0 when extended with wheat flour.

Inoue, Kawai, and Itow [51] report that bond strength increases with the degree of condensation in a resin with a molar ratio 1:1.2. They note that adhesion strength appears to be at its highest at a relatively low average molecular weight, and that it decreases with further condensation. Rice [55] shows from aging studies on commercial UF resins that higher resin viscosities which are related to higher molecular weights result in a more durable glue-wood bond.

The significance of a second addition of urea to improve adhesion strength [60] has been clearly demonstrated by Horioka et al. [50]. They conclude that dimethylolureas are the main components governing the adhesive force between glue and wood, while higher molecular weight species govern the cohesive force. Other factors, such as the degree of cure [57, 58] and the type and quantity of hardener [56], have also been reported to affect the water resistance of UF resins.

In an important article, Steiner [59] indicates that resins with higher molar ratios are more durable than those with a lower formaldehyde/urea ratio. He concludes that for samples without fillers, urea/formaldehyde molar ratios of 1:2.0 and 1:1.8 show better durability at all levels of second-urea addition than molar ratios of 1:1.6 and 1:1.4. When fillers, such as walnut shell flour, are incorporated, resins of molar ratio 1:1.6 have improved durability, and together with molar ratios of 1:2.0 and 1:1.8 have a notably slower deterioration rate than resins with a ratio 1:1.4.

In addition, an increase in the amount of second urea added to resins containing filler improves the durability of the lower molar ratio resins significantly, especially the 1:1.4 ratio resins. He finds, however, a practical limit to the amount of second urea which can be added. As the percentage of second urea is increased, the initial formaldehyde/urea molar ratio present during resin synthesis also increases [59]. The reaction times become excessively long, and eventually a point is reached where insufficient urea is present to effectively allow the condensation stage to continue. The resulting resins have low viscosities and a strong smell of free formaldehyde which limit their use as wood adhesives.

While the rate of bond deterioration is an important concept in establishing durability ranking, it does not take the initial bond quality into account. For UF resins without filler, increased second urea addition improves bond quality, especially in the low molar ratios [59]. This improvement, however, is maximized after a 20–30% addition [59]. Additional second urea tends to slightly decrease the initial percentage of wood failure for glues with filler, with the exception of resins made with a urea/formaldehyde molar ratio of 1:1.4, which show a variable behavior [59].

In all cases, the higher molar ratios resins tend to exhibit an overall better initial bond quality [59]. On cyclic wetting-drying testing, UF resins with a urea/formaldehyde molar ratio of 1:2.0 show a larger change in shear strength than resins with molar ratios of 1:1.8, with 10–20% second urea added [59]. To conclude, UF resins of urea/formaldehyde molar ratio ranges of 1:1.8–2.0 exhibit slower bond deterioration under accelerated aging conditions than resins of molar ratio ranges 1:1.4–1:1.6. Glues containing filler, made from resins with a molar ratio of 1:1.4–1.6, show improved durability with increased amounts of second urea.

Any consideration of the obvious improvements in durability resulting from an increased second urea addition must also reflect on the amount of methylol groups available. The mechanism of condensation requires sufficient methylol groups to be present before all the second urea can be incorporated in the UF polymers. It appears that a minimum amount of methylol is necessary for a satisfactory wood-glue bond. This is suggested by the following factors apparent in resin preparation when more second urea is added:

1. Resins of low formaldehyde/urea molar ratio, containing fillers, improve.
2. Only minor effects occur with higher formaldehyde/urea ratios of high methylol percentages.

This supports the concept that methylolureas are responsible for the adhesive force of the UF resin during wood-glue bond formation, while higher condensation products provide the cohesive force [50]. The methylol groups of the UF resin have been found to form ether links with cellulose [32], which explains their strong effect on the adhesion characteristics of UF resins. Therefore, the higher the amount of methylol groups, the better the adhesion of the glue to the wood.

In the manufacture of melamine-formaldehyde (MF) resins, the melamine/formaldehyde molar ratio is generally 1:2–3. If only 2 mol of formaldehyde is used, better heat stability can be achieved. As an example, 2.5 mol of formaldehyde (in the form of 40% formalin solution) is brought to pH 9.0 by adding a diluted solution of caustic alkali. One mole of melamine is added and dissolved by heating and stirring the solution. The solution is kept at about 80°C until the methylol compounds formed no longer crystallize. A small sample is then cooled in iced water, after which melamine-formaldehyde solution is ready. Products of the early condensation stages are rich in methylols of monomeric size. They perform better than the products of the medium condensation stages, which are hydrophobic in character.

3. pH Control

The control of the pH during the reactions of urea or melamine with formaldehyde is of paramount importance. Technical commercial formaldehyde always contains some formic acid, often from 0.5 to 1% upward, which is more than is allowed in this reaction. Formaldehyde can be bought according to specification with only 0.05–0.1% formic acid content, which is satisfactory. If the acid is not removed before reaction with urea or melamine, even this small amount acts as a catalyst, and the reaction is uncontrollable. In this case, the reactants heat up and form a soft white gel, which is soon transformed into a block of white, gelled, and finally hard resin that is difficult to remove from the reaction vessel, and worthless for further production.

In the condensation of urea and formaldehyde, the reaction is started at a urea/formaldehyde molar ratio of 1:2.0–2.3. The pH is held at between 7.5 and 8.5 while the solution is heated to near the boiling point. After the formation of methylolureas at ±90°C, the second urea is added at a lower temperature, and reacts with the free formaldehyde. The urea-formaldehyde solutions obtained are clear and may be further condensed to the desired viscosity. Low pH values may cause incontrollable condensation; solutions with a pH of less than 4.8–5.0 may heat up exothermically and form white compounds containing methylene bridges instead of methylol groups, which solidify into useless masses.

With melamine and formaldehyde, the optimum methylolation pH is in the range 8.5–9.0. Within this range, methylolation and further condensation to medium — and even high — molecular weight resins can be safely carried out at about 80°C. Insoluble, granular methylene compounds that may occur with UF resins are not produced in MF resins. However, if the reaction is not well controlled, it may lead to viscous, hydrophobic MF resins, with separation of an upper watery layer. These resins are rather troublesome to work with. A pH lower than 7.0 should not be used to produce melamine resins, because on warming, rapid curing sets in and the material becomes useless.

4. *Control of the Reaction*

Careful pH control during the reaction of urea or melamine with formaldehyde done in combination with repeated tests for solubility and viscosity are the methods used to control the molecular weight of the resins. The increase of molecular size can more easily be seen in UF resins than in MF resins. This is done by measuring viscosity in UF resins and by determining solubility of the respective solutions during condensation in MF resins. A MF resin solution can be evaluated by diluting a 5-ml sample at 20°C with water at 20°C. The number of milliliters of water that is necessary to add to reach permanent turbidity is a satisfactory means of estimating the condensation stage of a given resin solution.

Repeated or even continuous determination of the pH during the reaction is a recommended procedure, because in warm formaldehyde solutions which are exposed to air, formaldehyde is slowly oxidized to formic acid (Cannizzaro reaction).

$$2HCHO + O_2 \rightarrow 2HCOOH$$

The Cannizzaro reaction also occurs in alkaline formaldehyde solutions and forms sodium formate and methanol, both of which are harmful to the reaction.

$$2HCHO + NaOH \rightarrow HCOONa + CH_3OH$$

The Cannizzaro reaction can be kept within acceptable limits by keeping the solution in a closed vessel and carefully controlling the pH. Ignoring these details may cause failure.

Methylolation and subsequent resinification can be carried out by neutralizing formaldehyde with ammonia or another amine, instead of an inorganic alkali. The condensation reaction of amines and formaldehyde is slower than the neutralization reaction, and consequently formic acid is neutralized. Methylolation of urea or melamine then occurs in a neutral or slightly basic medium. Therefore, the critical stage of methylol formation is safely passed. The simultaneous condensation of amine and formaldehyde slowly proceeds until all the amine that neutralized the formic acid is consumed, and formic acid is set free again to act as an acid catalyst.

By carefully controlling the pH and viscosity, the aminoresin solution can be kept at the desired condensation stage by arresting the reaction. This is done by cooling and by bringing the pH to the level which guarantees best stability. For a UF resin, for instance, the pH of the formaldehyde solution as purchased is 3.5, and the pH after adding ammonia and urea is 8.5. The series of reactions are then as follows: formic acid is neutralized, and methylolureas are formed. At the same time, there is a slow liberation of formic acid due to the equilibrium.

$$HCOONH_4 \rightleftarrows HCOOH + NH_3$$

Parallel to this is the slow formation of hexamethylenetetramine as a result of the reaction between six molecules of formaldehyde and four molecules of ammonia. The pH drops from 8.5 to 5.5. The condensation reaction of methylolurea to UF resin occurs at pH 5.0–2.0. After the desired viscosity has been reached, the pH is raised to 7.5–8.0 by adding diluted caustic soda. The careful control of the pH makes it possible to achieve the safe manufacture of UF resins. Some manufacturers prefer this "ammonia" method since it also gives a very good resin quality.

By whatever method the aminoresin is produced, the final pH has to be such that the desired curing speed results. Gel time is also strongly dependent on molar ratio of the resin components; U/F resins with a molar ratio of 1:1.4 have much slower gel times than 1:2 resins at the same gel temperature. Factors other than the pH of the resins must be taken into account in meeting this demand. One of the most important is the final condensation stage of the aminoresin.

Control of the average molecular size of the finished resin is essential for the correct flow in plywood and particleboard applications while in the hot press prior to curing. Too low a level of condensation (i.e., low molecular weight resins) may give too much flow; the resin "runs away" from the wood or rapidly sinks into it under pressure, leaving "starved" glue lines. This can be corrected by lowering the pH by adding an acid or acid-producing substance, usually a curing agent, hardening catalyst, or simply hardener.

If a resin of too high a condensation stage (i.e., high molecular weight resins) is present, its flow under normal pressure and temperature may be too low to produce good results. This can usually be corrected by adding flow agents to it, provided that at least some flow is left in the resin. It is generally an advantage to produce resins with ample flow in the factory. Their storage life is longer and finishing can be done at any time, at short notice, to specification, particularly by adjusting the flow and speed of cure.

Many substances have been suggested as curing agents. These include the following acid products: (1) boric acid, (2) phosphoric acid, (3) acid sulfates, (4) hydrochlorides, (5) ammonium salts of phosphoric or polyphosphoric acid, (6) sodium or barium ethyl sulfate, (7) acid salts of hexamethylenetetramine, (8) phtalic anhydride, (9) phtalic acid, (10) acid resins such as poly basic acid-polyhydric alcohol, (11) oxalic acid or its ammonium salts, and many others.

However, the most widely used curing agents in the wood products industry are still ammonium chloride or ammonium sulfate. Their effect can be altered by retarding the reaction of the resin. This is done by simultaneously adding small amounts of ammonia solution (which is eliminated during hot curing) to lengthen the pot life of the glue mix. Latent catalysts, which produce acid only on heating, may also be used, such as dimethyloxalate and other easily hydrolizable esters; or alogenated substances such as 0.1–0.2% of bromohydrocinnamic acid, dibromosuccinic acid [61], 1,3-dichloro-2-propanol, and so on.

Resins that have lost part of their flow during manufacture or storage must be corrected by the addition of a flow agent. The simplest means is often the addition of water sprayed on the compound and mixed in well. If a resin is still capable of flowing, this procedure produces a resin with properties that are still acceptable. In cases where moisture content control is critical, it may be necessary to allow a little more time for "heating" to let the added moisture escape. However, if the flow is very low, and large quantities of water must be used to bring the flow back to normal, this method is not recommended. The large amount of water would cause longer "breathing" times to be necessary due to excessive volatile components, and excessive shrinkage may take place, causing excessive stress on the glue lines.

The best way to correct flow in these cases is to mix the resin with large amounts of an equal resin of the same quality which has a higher flow. Any proportion may be used to bring the flow back to normal. If increased flow is desired, 0.5–2.0% of spray-dried UF or MF resin can be added to function as a flow agent. Methylol compounds, such as dimethylolurea, also increase flow, but they increase the water released during reaction more than spray-dried resins. Lubricating agents, such as calcium stearate, are also able to give a fair degree of flow increase.

5. *Fillers*

Many inorganic and organic fillers have been tested, but the most suitable are wood, wheat and corn flours, and cellulose fillers, which give a translucent, milky appearance to the resin. Bleached sulfite cellulose is also well suited to this purpose. Cellulose fillers not only have the advantage of being a lighter color than wood flour, but because of their fibrous nature, they impart to the resin higher mechanical strength and a surprising translucency, which differentiate them very clearly from wood-flour-filled aminoresins. This is desirable for the manufacture of decorative plywood in which thin decorative veneers are used. A chemical interaction appears to take place between the cellulose, or edible flours, and the aminoresins, to form ether bridges.

It is not clear why cellulose-filled UF resins are more translucent than similarly filled MF resins. Wood, wheat and corn flours, and cellulose fillers (as opposed to inorganic nonabsorbent fillers) offer the advantage of absorbing the water produced during the curing of the aminoresin, thus rendering this water relatively harmless to the resin. Some of the improvements that can be achieved by adding absorbent fillers to aminoresins are (1) decreased shrinkage of the resin during curing, (2) the resultant improved property of gap filling, (3) decreased cracking of the cured resin, and so on.

B. Formulaire

An introduction to the typical resin synthesis of a few UF resins used for wood products and in industrial application is given below. It is aimed at explaining laboratory and industrial procedures used for the production of UF resins.

1. Resin 1: Approximate urea/formaldehyde molar ratio 1:1.8. Preparation useful for any other molar ratios higher than 1:1.2–1.4. First, 100 parts by mass of 46–47% formalin solution (obtained by distilling methanol off a ±41% commercial formalin solution to a level of less that 1%), and 51 parts of urea are placed in a reaction vessel furnished with high-torque mechanical stirrer, thermometer, and condenser. The solution is adjusted to a pH 8.6 with a 45–50% sodium hydroxide solution. The reaction is carried out for 60 min at 75°C, then cooled to 60°C. Then a 10% HCl solution is added to bring the pH down to 5.1. The mixture is heated again at 80°C and the ensuing condensation reaction is checked by frequent viscosity measurements. At a viscosity of ±85 cp (or viscosity C on a Gardner-Holt tube) the reaction mixture is cooled and neutralized with ±50% sodium hydroxide solution. The total reaction time is approximately 150 min. Water is removed under vacuum at 40°C to give ±60% solids. The whole reaction is carried out with continuous mechanical stirring.

2. Resin 2: Approximate urea/formaldehyde molar ratio 1:1.8. Preparation useful for any other molar ratios higher than 1:1.2–1:1.4 and

with a second urea addition of 5–40% of total urea added. First, 100 parts by mass of 46% formaldehyde solution obtained as in the previous case and 46 parts of urea are reacted at 75°C, at pH 8.6 for 60 min. The solution is cooled to 60°C and adjusted to pH 5.1 with 10% HCl solution. The reaction is continued until a viscosity of ±110 cp (i.e., Gardner-Holt viscosity DE) is achieved and the pH is then altered to 8.0. Then 5.2 parts of second urea is added and the mixture is kept at 60°C while being stirred for 60 min. Excess water is removed under vacuum at 40°C to obtain a resin of 61–64% solids content.

Resins prepared according to the procedure described for resin 2 tend to have higher viscosity, a longer gel time, and a higher methylol content, at equal solids content and molar ratio, with an increase in the amount of second urea. Resins in which the molar ratio is different tend to have higher viscosity, longer gel time, and lower methylol content, with a decrease in the formaldehyde/urea molar ratio (i.e., passing from 1.8 to 1.4). The effects of the change of molar ratio are more marked than the effect caused by the change in the percentage of second urea. Temperature during the reaction can be controlled by vacuum.

3. Resin 3: Urea/formaldehyde molar ratio final range 1:1.4–1.6. First, 100 parts by mass of 46% formaldehyde solution obtained as in for resins 1 and 2, and 39 parts of urea are reacted at ±90°C with pH 8.5 for 10–30 min. The solution is then cooled to 25–30°C and the pH adjusted to 4.8–5.1. From 15 to 22 parts of second urea is added, for a molar ratio variable from 1:1.6 to 1:1.4, respectively, and the reaction is then carried out at 25–30°C for 24 hr. Water is subsequently eliminated by vacuum distillation, after adjusting the pH to 7.8–8.2, until the UF resin produced has a concentration of 64–65% solids content. This preparation is particularly suited to industrial manufacture.

4. Resin 2: UF resin used for the preparation of resorcinol-urea-formaldehyde cold-setting adhesives for laminated beams, finger joints, and thermosetting adhesives for plywood. Recipes for this type of UF resins are reported in Chapter 3.

C. Plywood and Particleboard Adhesives

The introduction of UF resins in the wood products industry has caused a revolution in gluing techniques, especially in plywood manufacture. The manufacture of plywood has been carried out for hundreds of years by gluing thin veneers with protein-based natural glues. However, plywood is not very stable in the presence of moisture once it has been joined with such adhesives.

The UF adhesives for plywood generally contain less than 2 mol of formaldehyde per mole of urea and most of them are condensed to a slightly viscous, hydrophilic stage, and are quite soluble in water. The degree of polymerization, and hence the viscosity under comparable measure conditions, of UF resins for plywood is generally higher than those of UF resins for particleboard.

The application of urea-formaldehyde resins for gluing purposes is based on the excellent control which can be exercised on the condensation reaction by varying the pH, a procedure easily applicable to a production scale. A small amount of an acid as hardener is added at ambient temperature. This produces no visible change at first, or possibly for a few hours; finally thickening sets in, and the resins gels and hardens into an insoluble material. While the adhesive is still in liquid form, it can be spread on the wood surfaces, which have to be glued and joined under pressure. These have to be cured either at room temperature for a few days, or at elevated temperature for a few minutes. Solutions of ammonium salts, generally ammonium chloride or ammonium sulfate, or mixtures of ammonium chloride with urea, are generally used as resin hardeners. Often, ammonia solution is added, to lengthen the usable life of the glue mix.

The driving force in the use of these salts as hardeners is their capacity of releasing acid, which decreases the pH of the resin and thereby accelerates curing. The speed of the reaction between the ammonium salt and formaldehyde (or ammonia and formaldehyde when this is present) also determines, together with the amount of heat supplied, the rate of acid release, and therefore the rate of curing:

$$4NH_4Cl + 6HCHO \rightarrow 4HCl + \underset{\text{Hexamethylenetetramine}}{(CH_2)_6N_4} + 6H_2O$$

Ammonium chloride is a better hardener than hydrochloric acid, as the latter produces weaker joints. The effect of a fixed amount of ammonium

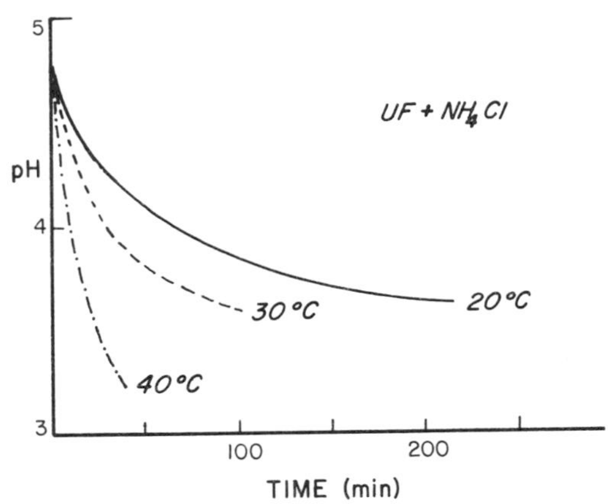

Fig. 2.3. pH change of UF resin with ammonium chloride hardener as a function of temperature and time.

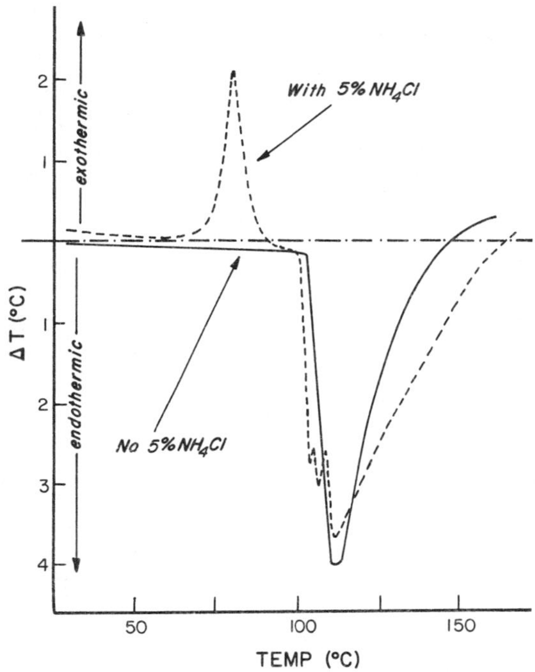

Fig. 2.4. Differential thermal analysis thermogram of UF resin (U/F = 3:3) with 5% NH_4Cl and with no NH_4Cl.

chloride on the pH change and on the rate of resin curing, as a function of time and temperature, is shown in Figure 2.3.

To understand the difference between an ammonium chloride-catalyzed UF resin and a resin without ammonium chloride, Chow and Steiner carried out a differential thermal analysis of a UF resin with a molar ratio of U/F of 1:3, both with and without adding 5% ammonium chloride. The results are shown in Figure 2.4. This shows that resin curing is accompanied by an exothermic peak at a characteristic temperature which reflects on the resin performance. The exothermic peak appears only when an ammonium salt is added. The salt presence expedites resin cure considerably. The peaks observed in both cases are due to second urea addition.

Hydraulic presses with multiple openings are usually used for the production of plywood or flat veneer work. They can operate at pressures of 10–16 kg/cm^2, but generally operate in the range 12–14 kg/cm^2, according to the wood species, to avoid destruction of the porous structure of the wood.

The temperature used is usually in the range 120–160°C according to the type and moisture content of the veneers. It is chosen according to its capacity for the fastest pressing time, and its ability to produce a good joint without blisters. Different pressing conditions are used in different countries and the resin must be manufactured keeping the differences in the conditions of application in mind. There is quite a difference, for instance, between a UF adhesive and glue mix, which is capable of giving good joints at 5–8% moisture content of the veneer and pressing temperature of 125°C, and a UF resin and glue mix usable at a veneer moisture content of 0–1% and pressing temperatures of 140–160°C. The former needs better flow characteristics and faster curing under standard measured conditions than the latter if optimum pressing times and production schedules are to be maintained. Lower temperatures lengthen the curing time of the resins considerably, but have the advantage that when the cured plywood sheets are taken out of the hot press, they tend to warp less on cooling or drying.

On their introduction a few decades ago, UF adhesives were supplied in the form of viscous solutions with a 50–70% solid content. A large number of users still prefer them in such a form and many produce the solutions themselves. Besides being available in liquid form, a large production of dry UF adhesives has developed, produced by spray-drying the solution to a very low water content, usually below 1%. These products have good stability, transport costs are lower, and they can even be shipped to tropical countries without deterioration.

Melamine-formaldehyde adhesives have replaced UF resins in wood gluing to only a certain extent. They produce high-quality plywood because their adhesive joints are boil-proof [62]. Considerable discussion has occurred and many investigations have been carried out on the weather resistance of MF glues. Many authors uphold the good weather resistance of the more recently developed MF resins, especially those in which small amounts of phenol have been incorporated. The more general trend, however, is to consider the wood products manufactured with these resins capable of resistance to limited weather and water exposure only, such as those used in flooring, rather than being capable of true exterior-grade weather resistance, for which phenolic adhesives are preferred.

It can be seen from Figure 2.5 that a phenolic resin-bonded board deteriorates rapidly and then stabilizes to a constant value. This indicates that the properly cured phenolic bonds are impervious to further water attack. The MF-bonded board has slower initial deterioration (at least in the case of excellent MF boards). However, it never stabilizes, and continues to deteriorate with time and further wet-dry cycles. This indicates the basic instability of the aminoplastic bond to water attack. The rate of deterioration, and therefore the bond hydrolysis, is faster as temperatures rise. Considering the insolubility of melamine in cold water, this is quite understandable.

This means that the upward slant of the MF curve in Figure 2.5 increases with temperature increases. Newer and more stable MF resins, which are manufactured with the addition of up to 10% grafted phenol,

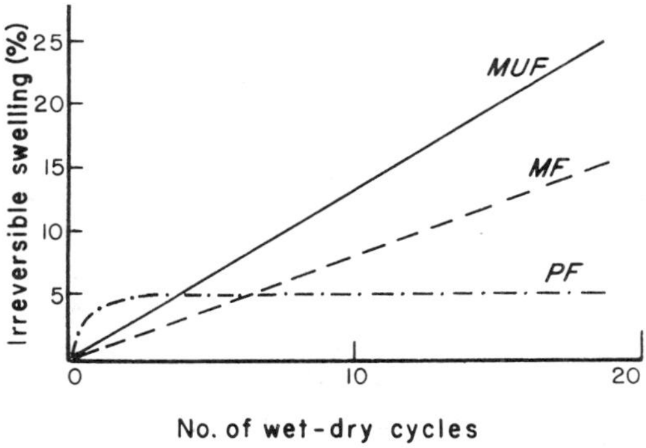

Fig. 2.5. Irreversible thickness swelling characteristics of particleboard bonded with UF, MF, and PF adhesives during a wet-dry cyclic test.

perform better in accelerated cyclic weathering tests. They are definitely more weather resistant, although a certain amount of doubt as to their long term performance still remains.

MF resin adhesives are suitable only for hot cure because they need higher temperatures to cure properly. Without heat, curing will not begin, and a glue dried on a wood surface may stay unchanged for days at ambient temperatures. On supplying heat, immediate cure and hardening takes place. The higher price of melamine-formaldehyde compared to urea-formaldehyde adhesives has also counted against them in spite of their better quality.

In order to improve the hot water resistance of UF adhesives economically, commercial formulations of mixed urea- and melamine-formaldehyde resins are used. These consists of UF resins with melamine or MF resins added. Relatively small amounts of very fine powdered melamine or melamine-formaldehyde resins (i.e., 15–20%), produce a remarkable increase in resistance to boiling water. Mixtures of this kind are very often prepared by replacing up to 50% of the UF resin with MF resin. Such melamine-urea-formaldehyde resins (MUF) give outstanding bonding.

The use of fillers with aminoresins adhesives has important economical consequences and is necessary for technical reasons, because the fillers produce "body" in the glue solution and therefore prevent joint "starvation" in porous wood. Without filler, it would be difficult to prevent part of the adhesive from flowing away or flowing into the open pores of the wood; or, in the gluing of medium to thin veneers, from flowing through them to the other side, thereby causing undesirable resin patches on the outer veneer surfaces. As a rule, 20–50% filler is used for joinery and up to 100% for plywood. The most common fillers are wheat flour, corn flour, rye flour, very fine hardwood flour, or gypsum. If gypsum is used, it must be free of calcium hydroxide, because it interferes with the acid curing agent.

A very important application for UF, MF, and MUF adhesives is in the manufacture of particleboard. The glue mix is generally composed of a liquid resin to which water has been added to decrease viscosity and to facilitate spraying, plus small amounts of ammonium chloride or sulfate and small amounts ammonia solution. Small quantities of insecticides, wax emulsion, and fire-retarding agents (such as ammonium phosphates) are added before spraying the adhesive onto the wood chips. Pressing temperatures and maximum pressures used in the cycle are in the ranges 150–200°C and 20–35 kg/cm^2, respectively. The moisture content of glued furnish chips is 7–8% for the board core and 10–12% for the surface. The resin contents used (i.e., solids) are 6–8% for board core and 10–11% for board surfaces.

High temperatures and high relative humidity can result in odor problems in a room containing particleboard manufactured with UF resins. The release of formaldehyde from UF particleboard is caused by two factors. It can be due to free formaldehyde present in the board which has not reacted, and it can be due to formaldehyde formed by hydrolysis of the aminoplastic bond as a result of temperature and relative humidity. While the first type of release lasts only a short time after manufacture of the particleboard, the second type of release can continue throughout the entire working life of the board. A considerable number of variables influence the emission of formaldehyde from a UF-glued particleboard. The main ones are the molar ratio of urea and formaldehyde (which influences both types of release), the press temperature, and in service, the ambient temperature and relative humidity.

UF resins for particleboard with urea/formaldehyde molar ratios of 1:1.45, 1:1.32, and 1:1.25 have free formaldehyde contents of 0.8%, 0.3%, and less than 0.2%, respectively [60]. While the current tendency internationally is to use UF resins of urea/formaldehyde molar ratio of 1:1.3 or lower, which release much less formaldehyde, these resins perform less well in the production of UF-glued particleboard [60,63]. In particular, they do not allow as much flexibility in particleboard production as do resins with higher formaldehyde/urea molar ratios [60,63]. This fact stresses the need of greater control and supervision of the production at particleboard plants where UF resins of low molar ratio are used. An example of the variation of properties between particleboard manufactured with different molar ratio resins is given in Table 2.1 and Figures 2.6, 2.8, and 2.9.

It is also necessary to use more hardener when working with UF resin of a lower formaldehyde/urea molar ratio, as the gel time of the resin is slower. Up to 5% urea can sometimes be added to the glue mix to decrease the amount of formaldehyde released during pressing, and to decrease the initial amount of free formaldehyde present in the finished board immediately after manufacture. A more recent development [64]

Table 2.1. Comparison of Particleboard Prepared with UF Resins of Different Molar Ratios

U/F Molar ratio	Approximate density (g/cm)	Internal bond (MPa)	Percent water swelling (2 hr)	Percent HCHO released (perforator method)
1:1.4–1.5	0.680	0.7 –0.8	4	0.08 –0.10
1:1.3–1.35	0.680	0.6 –0.7	4–5	0.04 –0.05
1:1.2–1.25	0.680	0.45–0.55	5	0.025–0.035

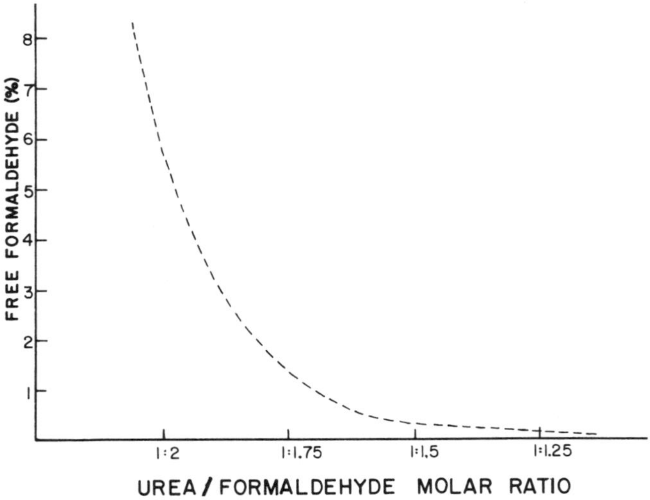

Fig. 2.6. Free formaldehyde content of UF resins as a function of U/F molar ratio.

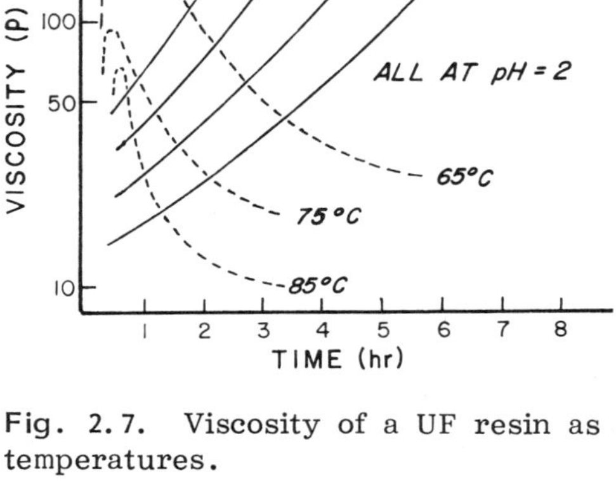

Fig. 2.7. Viscosity of a UF resin as a function of time at different temperatures.

advocates the use of up to 1–2% of tannin extract to decrease the amount of formaldehyde emitted by the board during manufacture, and to decrease the amount of free formaldehyde present in the board after manufacture. Both additives have no influence on the long-term release of formaldehyde due to hydrolysis of the aminoplastic bond. Carbamates appear to be used to control long-term formaldehyde release.

It must be realized that on curing, the viscosity of UF resins changes, not only at a different rate, but also in a different manner according to the temperature. Figure 2.7 clearly illustrates this point. The viscosity gradually increases with temperatures of up to ±50°C. Above 60°C the viscosity quite rapidly reaches a maximum, and then decreases. This indicates that the resin tends to degrade under prolonged heating at high temperatures. To avoid this problem, UF-bonded particleboard must never be pressed for too long, and must never have a "hot-stack" or "postcure" period after pressing. It must preferably be cooled after manufacture to avoid the deterioration of strength and quality. The cured UF resins degrade

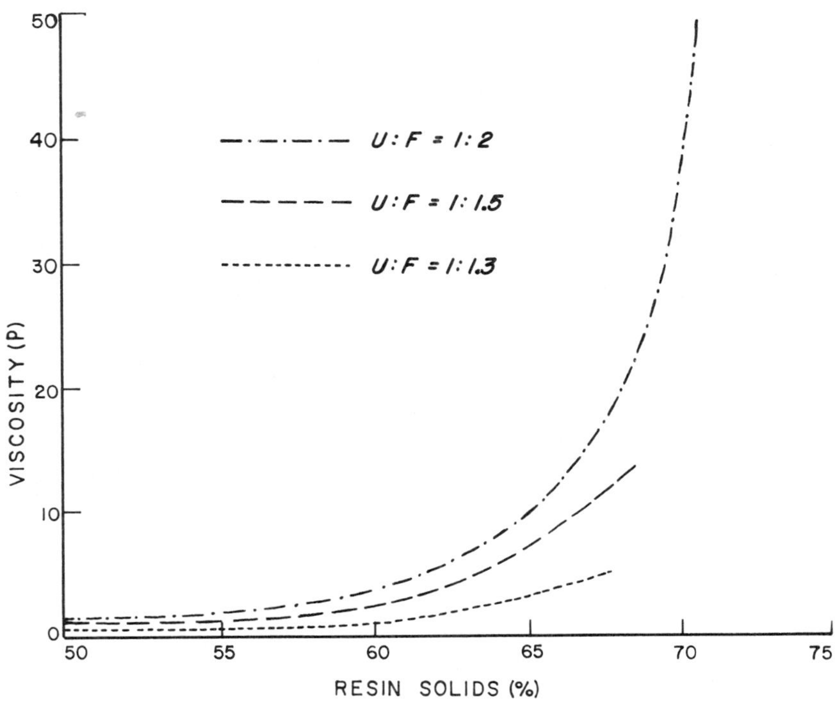

Fig. 2.8. Viscosity of UF resins of different molar ratios as a function of their percentage resin solids.

rapidly at any temperature at a pH lower than 2. The viscosity for a good particleboard resin is of the order of 100–450 cP (at 20°C) (Fig. 2.8) [63].

As important as viscosity is the resin flow, which reflects viscosity under hot-pressing conditions. Resin flow is a determining factor in manufacturing a good particleboard. Excessive flow causes the resin to soak into the wood particles and causes glue line starvation, while insufficient flow causes insufficient contact surface. The gel time generally used at 100°C for glue mixes for UF-bonded particleboard is of 3–12 min, and 30 sec to 5 min for board faces and cores, respectively (Fig. 2.9). The actual gel time in the press depends on the press temperature and is considerably shorter.

Often the ready-to-use resin is modified with a reagent which affects its cured behavior. The most common modification is the addition of some urea to the glue mix immediately before use. Urea is not only a formaldehyde scavenger, but also participates in the reaction by reacting with both the free formaldehyde and methylol groups of the UF resin, and release of heat. The addition of urea also affects the viscosity of the resin, and its rate of viscosity increase is shown in Figure 2.10.

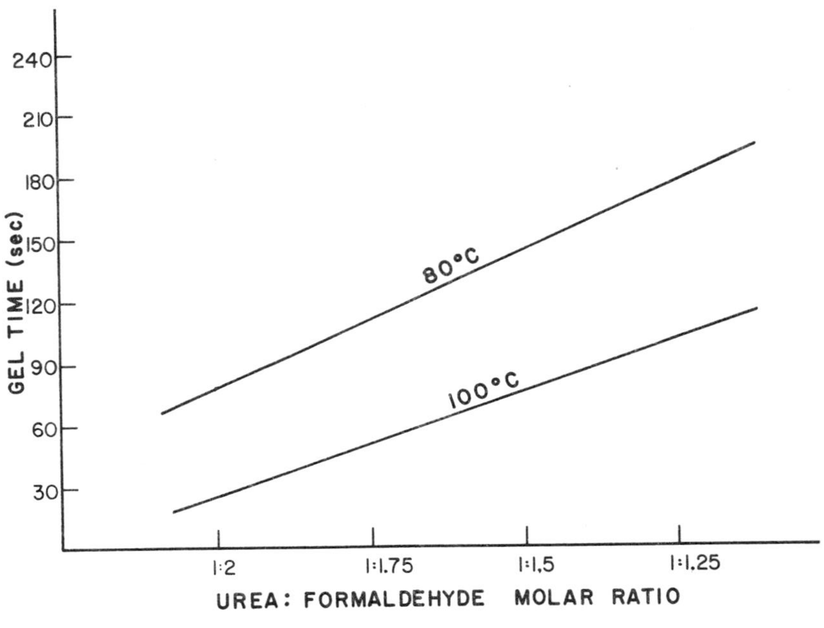

Fig. 2.9. Gel time of UF resins as a function of molar ratio (hardener 2% ammonium chloride).

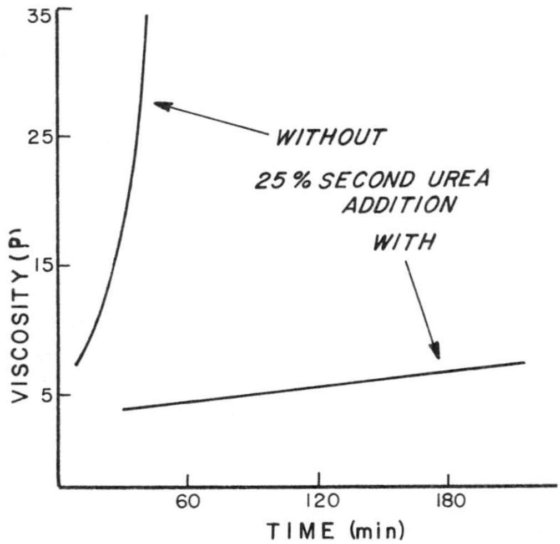

Fig. 2.10 Viscosity change of a plywood UF resin with 5% NH_4Cl as a function of time with second-urea addition of 25%, and without second-urea addition.

D. Melamine Laminates

Melamine-formaldehyde resins are used for the impregnation of paper sheets in the production of laminates for the surfaces of various wood-based panel products. The resin usually used consists of 1 mol of melamine to 2–3 mol of formaldehyde condensed under slightly alkaline conditions to the beginning of hydrophobicity. The impregnation substrate (i.e., α-cellulose paper for very light colors) is thoroughly impregnated by immersing it in the resin solution, squeezing it between rollers, and drying and precuring it to proper flow by passing it through a tunnel oven at 70–120°C. The cured laminated sheets can then be applied to the wood substrate according to two main processes:

1. The sheets of MF-impregnated paper, consisting of one surface layer or a few surface layers, are pressed together with a substrate of paper sheets impregnated with phenolic resins to form laminates of variable thicknesses. These are then glued, in a second operation, to a wood-based panel on which has been spread an aminoplastic or phenolic adhesive.
2. The MF in the impregnated paper is not completely cured, but still has a certain amount of residual activity and is applied directly, in a single sheet, on the wood-based panels. This is done either by self-adhesion, or in the case of an advanced

stage of cure of the MF impregnating resin, by spreading the surface of the wood panel with a liquid aminoplastic resin glue mix.

Correct pressure and temperature are required in both processes. In both processes the cured laminated sheets reproduce on their surface the finish of the molding metal plates (e.g., high-polished gloss, matt, satin, or any other finish).

Press platens are made from stainless steel or chromium-plated brass and copper. The chromium layer preserves surface quality longer than ordinary steel. The MF laminates exhibit a remarkable set of characteristics. Because of their unusual chemical inertness, non-porosity, and nonabsorbance, they resist most substances, such as mild alkalies and acids, alcohols, solvents such as benzene, mineral spirits, natural oils, and greases. No stains are produced on MF surfaces by these substances. This remarkable resistance, in addition to almost unlimited coloring and decorating possibilities, have resulted in the extensive use of MF laminated wood-based panel products for tabletops, sales counters, laboratory benches, heavy-duty work areas in factories and homes, wall paneling, and so on.

For decorative sheets used on a support, a rather interesting buildup has been developed, which produces a laminated, one-sided, decorated sheet in an economical manner. The method consists of combining an inexpensive supporting base, generally an appropriate number of strong superimposed kraft paper sheets, which have been impregnated with phenolic resin, and molding the decorative MF sheet layer on top of it. Its surface is protected with another overlay sheet, which is rich in MF resin. Phenolic sheets produce a dark background on lamination. A covering barrier sheet can be placed on top of them as a second layer, which usually consists of a white α-cellulose sheet containing pigment and impregnated with MF resin. The assembly is then cured between hot-press platens and glued to a support panel such as plywood, particleboard, or fiberboard. The MF layer contains up to 60% MF resin in weight. There are many more variations in the buildup of decorative laminates to meet the particular specifications of the final laminates.

Particular grades of MF-resin-impregnated sheets have been developed for direct gluing on a wood panel support. Two types of MF-impregnated sheets are available. Sheets which are glued onto the panels by the previous application of glue on the panel surface are impregnated at lower resin load by a normal MF resin, which is completely cured or just slightly undercured. The laminated panels obtained are generally used where paint on the surface is required.

The self-adhesive impregnated "one-shot" sheet is produced by adding a plasticizer to the undercured MF impregnating resin, generally a few percent of a toluenesulfonamide-formaldehyde resin. The plasticizer is generally added in the impregnation bath. *p*-Toluenesulfonamide-formaldehyde resins are generally formed by reacting equimolar amounts of *p*-toluenesulfonamide and paraformaldehyde. They are viscous compounds that can be transformed into hard, lightfast, colorless resins by heating them at 110°C. Walter and Blueck [65] have proposed the following linear constitution for these equimolar resins:

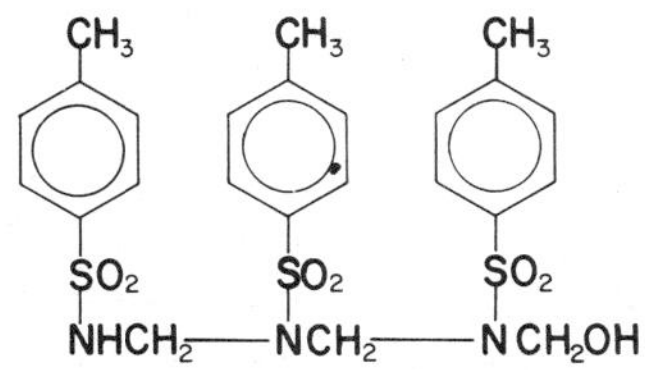

XIV

By reacting a mixture of *o*-toluenesulfonamide with paraformaldehyde at 110–120°C, a soft resin results.

To increase further the binding capacity of the impregnated sheet in the cold before use, and in the hot for postforming, the resins used are usually undercured. This slightly reduces the quality of the surface. The toluenesulfonamide plasticizers also produce a somewhat slower cure, and therefore the impregnated paper is slower to dry, has better storage properties, and displays better flow characteristics on hot pressing. This also produces better transparency of the overlay sheet. The low tendency of such plasticized sheets to craze under the influence of dry heat is also a great advantage.

E. Glulam, Finger Jointing, and Joinery Adhesives

In some countries, an important application for UF cold-setting adhesives is in the manufacture of laminated timber or "glulam" for structural applications and for finger joints used in preparation of glulam. Glulam is generally produced by a simple cold-gluing process, by assembling the glue-coated, solid wood boards in the desired dimensions with strong clamps and fixing their position until cured. Construction members that would be impossible to build of solid wood because of their dimensions can be manufactured in this way. Curved beams and trusses can also be made in this way. Another advantage is the superiority of such wooden constructions to unprotected steel beams in case of fire.

A normal UF resin with an acid-curing agent added can be used for interior-grade applications of this type. MF or MUF resins can also be used, but they generally require curing for about 4–8 hr at 50°C. Another type of UF adhesive which can be used for this application is a UF one-part glue, which is also available in a dry powder. This need only be mixed with water to a suitable gluing consistency to be ready for use. Such one-part gluing mixtures can be prepared by blending a perfectly dry UF resin, usually spray-dried, with a latent catalyst, usually in the form of coarse particles. By adding water, the catalyst dissolves, and the glue mix can be used for normal glue work (preferably in cases when only small quantities of adhesive are needed, such as in joinery at room or elevated temperatures). Such one-part UF glues have good storage stability, especially when the moisture content of the UF resin is kept below 1% or, even better, below 0.5%.

F. Toxicity

UF resins in their cured state are nontoxic. Urea itself is also harmless. However, free formaldehyde and formaldehyde generated by slow hydrolysis of the aminoplastic bond are highly reactive, and easily combine with proteins in the human body. This may cause a painful inflammation of the mucous membranes of the eyes, nose, and mouth. Even a low concentration of formaldehyde vapor in the air can cause disagreeable irritations of the nose and eyes. However, such irritations usually disappear in a short time without permanent damage. Occasionally, allergic or anaphylactic reactions develop and complete removal from exposure is necessary.

Hardened, completely cured, melamine resins are inert and completely nontoxic. Uncured MF resins (i.e., in liquid form) taken internally in high quantities may have damaging effects on the digestive system. However, no cases have been reported, probably as a result of their unpleasant taste.

Aniline is rather toxic, and can be absorbed both through the skin and by inhaling its vapors. It can cause headaches, spasms, and unconsciousness as well as a grayish-blue skin color if absorbed in small doses. It must therefore be handled carefully. Continuous careless exposure may finally result in chronic intoxication, causing nervous and intestinal disorders.

REFERENCES

1. G. Zigeuner, *Fette Seifen Austrichm. 56*:973 (1954); *57*:14, 100 (1955).
2. H. Fahrenhorst, *Kunststoffe 45*:43, 219 (1955).
3. R. H. Glauert, *Ind. Chem.*, Aug. 1957.
4. H. Kadowaki, *Bull. Chem. Soc. Jap. 11*:248 (1936).
5. R. Koehler, *Kunstst.-Tech. 11*:1 (1941); *Kolloid Z. 103*:138 (1943).
6. R. Frey, *Helv. Chim. Acta 18*:491 (1935).
7. H. P. Wohnsiedler, I. H. Updegraff, and R. H. Hunt, *Ind. Eng. Chem. 48*(1):82 (1956).
8. R. Houwink, *Elastomers and Plastomers*, Vol. 2, Elsevier, New York, 1949.
9. G. T. Tiedeman and M. R. Sanclemente, *J. Appl. Polym. Sci. 17*:1813, 1819 (1973).

10. L. E. Smythe, *J. Phys. Colloid Chem.* *51*:369 (1947).
11. G. A. Growe and C. C. Lynch, *J. Am. Chem. Soc.* *70*:3795 (1948); *71*:3731 (1949); 75:574 (1953).
12. L. Bettelheim and J. Cedwall, *Sven. Kem. Tidskr.* *60*:208 (1948).
13. G. Smets and A. Borzee, *J. Polym. Sci.* *8*:371 (1952).
14. J. I. de Jong and J. de Jonge, *Recl. Trav. Chim. Pays-Bas* *71*:661 (1952).
15. J. I. de Jong, J. de Jonge, and H. A. K. Eden, *Recl. Trav. Chim. Pays-Bas* *72*:88 (1953).
16. J. I. de Jong and J. de Jonge, *Recl. Trav. Chim. Pays-Bas* *72*:139 (1953).
17. J. I. de Jong and J. de Jonge, *Recl. Trav. Chim. Pays-Bas* *72*:1027 (1953).
18. J. I. de Jong and J. de Jonge, *Recl. Trav. Chim. Pays-Bas* *72*:207 (1953).
19. J. I. de Jong and J. de Jonge, *Recl. Trav. Chim. Pays-Bas* *72*:213 (1953).
20. H. Sobue and K. Murakami, *Chem. High Polym. (Jap.)* *9*:454 (1952); *Chem. Abstr.* *48*:9097 (1954).
21. A. Takahaski, *Chem. High Polym. (Jap.)* 7:165, 169 (1950); *Chem. Abstr.* *46*:438 (1952); *Chem. High Polym. (Jap.)* *9*:50 (1952); *Chem. Abstr.* *48*:1056 (1954).
22. A. Takahashi, *Chem. High Polym. (Jap.)* *7*:115 (1950); *Chem. Abstr.* *46*:438 (1952); *Chem. High Polym. (Jap.)* *9*:15 (1952); *Chem Abstr.* *48*:1730 (1954).
23. S. Oguri and Y. Sekine, *J. Chem. Soc. (Jap.)* *54*:149 (1951); *Chem. Abstr.* *47*:1427 (1953).
24. Japanese Patent No. 1792 (1952).
25. A. Takahashi, *Chem. High Polym. (Jap.)*, *9*:169 (1952); *Chem. Abstr.* *48*:1056, 1730 (1954).
26. M. Akano and Y. Ogata, *J. Am. Chem. Soc.* *74*:5728 (1952).
27. K. Sato and S. Ouchi, *Polym. J. (Jap.)* *10*(1):1 (1978).
28. K. Sato and T. Naito, *Oikym. J. (Jap.)* *5*(2):144 (1973).
29. K. Koeda, *Chem. High Polym. (Jap.)* *16*:62 (1959).
30. K. Sato, *J. Chem. Soc. Jap., Ind. Chem. Sec.* *72*:1911 (1969).
31. H. Kadowaki, *Osaka Kogyo Gijutsu Shikensho Hokoku* *16*(6) (1935); *17*(3) (1936); M. Hamada, *J. Chem. Soc. Jap., Ind. Chem. Sec.* *58*:286 (1955).
32. R. Steele and L. E. Giddens, *Ind. Eng. Chem.* *48*:110 (1956).
33. A. Takahashi, *Chem. High Polym. (Jap.)* *9*:173 (1952); *Chem. Abstr.* *48*:1730 (1954).
34. P. R. Steiner, *Wood Sci.* *7*(2):99 (1974).
35. J. L. de Jong and J. de Jonge, *Recl. Trav. Chim. Pays-Bas* *71*:890 (1952).
36. J. I. de Jong, *Recl. Trav. Chim. Pays-Bas* *72*:653 (1953).
37. A. Petz and M. Cherubim, *Holz Roh.-Werkst.* *13*:70 (1955).
38. G. Widmer, *Kunststoffe* *46*:359 (1956).

39. C. P. A. Kappelmeiser, Am. Chem. Soc. Meet., Boston, 1951.
40. P. P. Grad and R. J. Dunn, *Anal. Chem. 25*:1211 (1953).
41. B. Tomita and Y. Hirose, *J. Polym. Sci. 14*:387 (1976).
42. R. E. Vogel, *Kunststoffe 31*:309 (1941).
43. M. D. Hurwitz, Chem. Soc. Meet., Boston, 1951.
44. H. Staudinger and K. Wagner, *Makromol. Chem. 11*:79 (1953).
45. R. E. Vogel, *Kunststoffe 44*:335 (1954).
46. H. Staudinger and K. Wagner, *Makromol. Chem. 12*:168 (1954).
47. A. Takahashi, *Chem. High Polym. (Jap.) 7*:56, 62 (1950); *Chem. Abstr. 48*:1056 (1954).
48. G. Widmer, *Paint Oil Chem. Rev. 112*:18, 26, 28, 30, 32–34 (1949); *Kunststoffe 46*(8):359 (1956).
49. R. C. Hirt, F. T. King, and R. G. Schmitt, *Anal. Chem. 26*(8): 1273 (1954).
50. K. Horioka, M. Noguchi, K. Moriya, and A. Oguro, *Bull. Gov. For. Exp. Sta. Tokyo 113*:20 (1959).
51. M. Inoue, M. Kawai, and G. Itow, *Wood Ind. Jap. 11*(2):85 (1956).
52. S. Iwatsuka and H. Tanaka, *Wood Ind. Jap. 11*(11):549 (1956).
53. Y. Nakarai and T. Watanabe, *Wood Ind. Jap. 16*(12):577 (1961).
54. Y. Nakarai and T. Watanabe, *Wood Ind. Jap. 17*(10):464 (1962).
55. J. T. Rice, *For. Prod. J. 15*(3):107 (1965).
56. A. Tanaka and K. Takashima, *J. Jap. Wood Res. Soc. 12*:272 (1966).
57. A. Tanaka and M. Kamitani, *J. Jap. Wood Res. Soc. 13*:285 (1967).
58. G. E. Troughton, *J. Inst. Wood Sci. 4*(5):51 (1969).
59. P. R. Steiner, *For. Prod. J.*, *23*(12):32 (1973).
60. B. Sundin, *Proc. FESYP Int. Particleboard Symp.*, Hamburg, 1978, p. 112.
61. M. A. Howald, British Patent No. 312,343 (1928).
62. Synthetic resins adhesives for wood, British Standard BS 1204, 1965.
63. R. Marutzky and L. Ranta, *Holz Roh.-Werkst. 37*:389 (1979).
64. C. J. Knauff, private communication, 1980.
65. G. Walter and A. Glueck, *Kolloid-Beih. 37*:343 (1933).

3

Phenolic Resin Wood Adhesives

A. PIZZI / *National Timber Research Institute, Council for Scientific and Industrial Research, Pretoria, South Africa*

I. INTRODUCTION

The field of adhesives for wood has been dominated during the last quarter of the century by two main classes of adhesives: aminoplastic and polyphenolic adhesives. The polyphenolic adhesives are the more significant for the production of weather- and boil-proof wood products.

As early as the nineteenth century, it was known that resinous materials could readily be formed by coreacting phenols and aldehydes. From the beginning of this century, much effort has been directed toward their commercial exploitation. Baekeland showed in 1907 that the simplest starting material, phenol and formaldehyde, form products of commercial importance under the correct reaction conditions.

Phenolic resins thus became the first true synthetic polymers ever to be developed. It may have been expected that by now, their chemical and physical structure would have been completely elucidated, but even now their structure is far from being completely clear. The polymers derived from the reaction of phenols and formaldehyde differ in one important aspect from other polycondensation products; polyfunctional phenols can form a variety of isomerides with different chain lengths. Other products derived from polycondensation reactions, such as polyamides and polyesters, are mixtures of molecular chains of various lengths. However, in this case only one structure is possible for a molecule of a certain chain length.

For this reason, kinetic and related studies are feasible if one assumes that the growth of a chain proceeds from one molecule to the next in a smooth and regular manner. Polyfunctional phenols may react with formaldehyde in both the *ortho* and *para* positions to the hydroxyl group. This means that the condensation products exist as numerous positional isomerides for any chain length. This particularly not only makes kinetic studies extremely difficult, but also makes the organic chemistry of the reaction very complex and tedious to unravel.

The result has been that although phenolic resins were developed commercially as early as 1908, and were the first completely synthetic resins ever to be developed, their chemistry is still only partially understood. It may be argued with some justification that such a state of affairs is immaterial, because satisfactory resins for many uses have been developed on purely empirical grounds during the last 70 years. However, it cannot be denied that gradual understanding of the chemical structure and mechanism of reaction of these resins has helped considerably in introducing commercial phenolic resins which have been designed for certain applications, and which are capable of performances undreamed of in formulations developed by the empirical rather than the scientific approach. Knowledge of phenolic resin chemistry, structure, characteristic reactions, and kinetic behavior remains an invaluable asset to the adhesive formulator in designing resins with specific physical properties.

II. CHEMISTRY OF PHENOL-FORMALDEHYDE CONDENSATIONS

A. Reaction Mechanisms

Phenols condense initially with formaldehyde in the presence of either acid or alkali to form a mehtylolphenol or phenolic alcohol, and then dimethylolphenol. The initial attack may be at the 2-, 4-, or 6-position. The second stage of the reaction involves the reaction of the methylol groups with other available phenol or methylolphenol, leading first to the formation of linear polymers [1] and then to the formation of hard-cured, highly branched structures.

Novolak resins are obtained with acid catalysis, with a deficiency of formaldehyde. A novolak resin has no reactive methylol groups in its molecules, and therefore is incapable of condensing with other novolak molecules on heating without hardening agents. To complete resinification, further formaldehyde is added to cross-link the novolak resin. Phenolic rings are considerably less active as nucleophilic centres at an acid pH, due to hydroxyl and ring protonation.

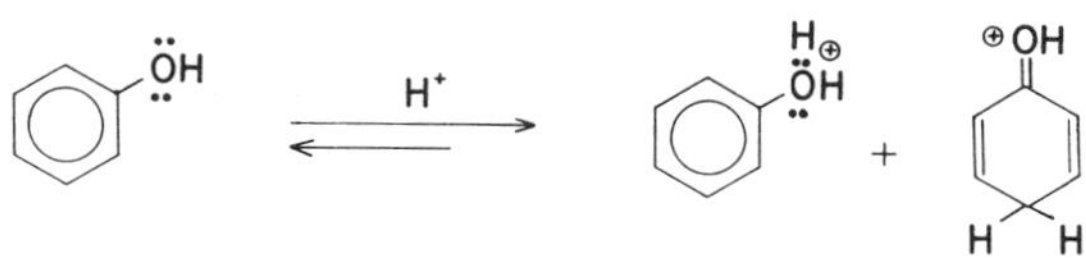

However, the aldehyde is activated by protonation, which compensates for this reduction in potential reactivity. The protonated aldehyde is a more effective electrophile.

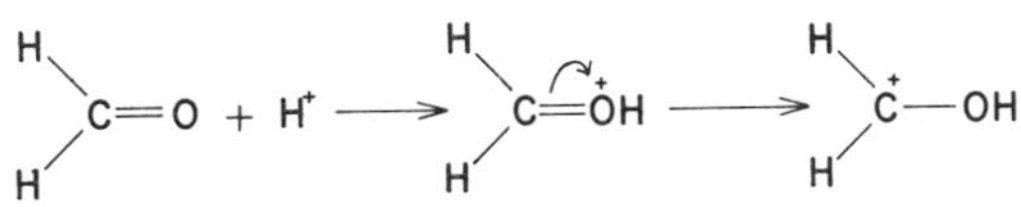

The substitution reaction proceeds slowly and condensation follows as a result of further protonation and the creation of a benzylcarbonium ion which acts as a nucleophile.

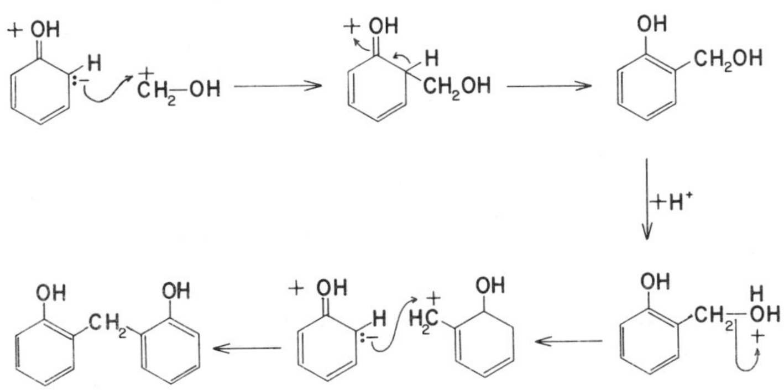

Resols are obtained as a result of alkaline catalysis and an excess of formaldehyde. A resol molecule contains reactive methylol groups. Heating causes the reactive resol molecules to condense to form large molecules, without the addition of a hardener. The function of phenols as nucleophiles is strengthened by the ionization of the phenol, without affecting the activity of the aldehyde.

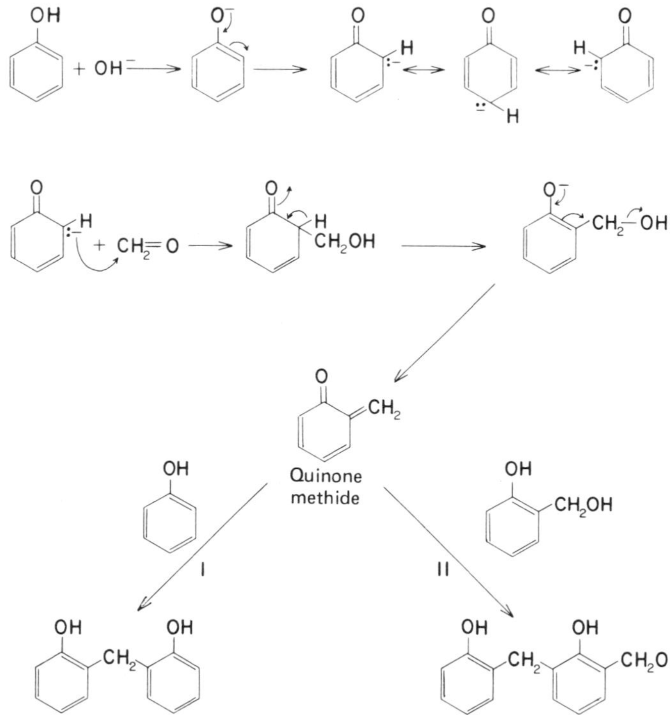

Megson [2] states that reaction 2 (in which resols are formed by the reaction of quinone methides with methylolphenols or other quinone methides) is favored during alkaline catalysis. A carbonium ion mechanism is, however, more likely to occur. Megson [2] also states that phenolic nuclei can be linked not only by simple methylene bridges, but also by methylene ether bridges. The latter generally revert to methylene bridges if heated during curing, with elimination of formaldehyde.

The difference between acid-catalyzed and base-catalyzed processes is (1) in the rate of aldehyde attack on the phenol, (2) in the subsequent condensation of the phenolic alcohols, and to some extent, (3) in the nature of the condensation reaction. With acid catalysis, phenolic alcohol formation is relatively slow. Therefore, this is the step that determines the rate of the total reaction. The condensation of phenolic alcohols and phenols forming compounds of the dihydroxydiphenylmethane type is instead rapid. The latter are therefore predominant intermediates in novolak resins.

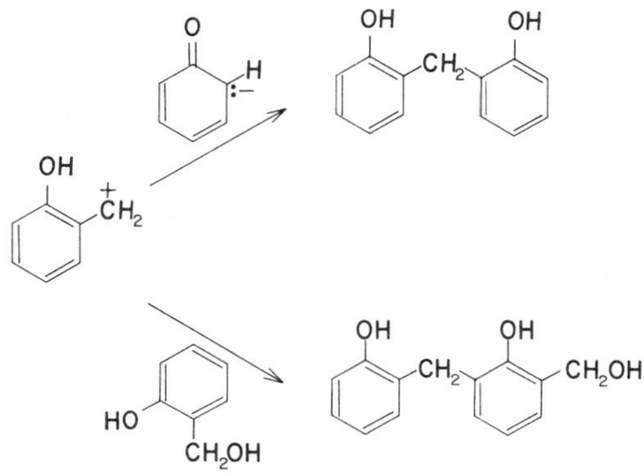

Novolaks are mixtures of isomeric polynuclear phenols of various chain lengths with an average of five to six phenolic nuclei per molecule. They contain no reactive methylol groups and consequently cross-link and harden to form infusible and insoluble resins only when mixed with compounds that can release formaldehyde and form methylene bridges (such as paraformaldehyde or hexamethylenetetramine).

In the condensation of phenols and formaldehyde using basic catalysts, the initial substitution reaction (i.e., the formaldehyde attack on the phenol) is faster than the subsequent condensation reaction. Consequently, phenolic alcohols are initially the predominant intermediate compounds. These phenolic alcohols, which contain reactive methylol groups, condense with either other methylol groups to form ether links, or more commonly, with reactive positions in the phenolic ring (*ortho* or *para* to the hydroxyl group) to form methylene bridges. In both cases water is eliminated.

Mildly condensed liquid resols, which are the more important of the two types of phenolic resins in the formulation of wood adhesives, have an average of less than two phenolic nuclei in the molecule. The solid resols average three to four phenolic nuclei but with a wider distribution of molecular size. Small amounts of simple phenol, phenolic alcohols, formaldehyde, and water are also present in resols. Heating or acidification of these resins causes cross-linking through uncondensed phenolic alcohol groups, and possibly also through reaction of formaldehyde liberated by the breakdown of the ether links.

As with novolaks, the methylolphenols formed condense with more phenols to form methylene-bridged polyphenols. The latter, however, quickly react in an alkaline system with more formaldehyde to produce methylol derivatives of the polyphenols. In addition to this method of growth in molecular size, methylol groups may interreact with one

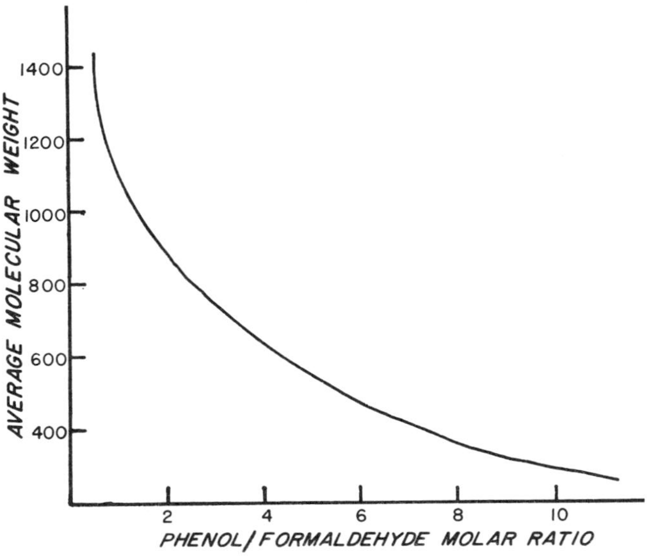

Fig. 3.1. Average molecular weight of PF resins as a function of the phenol/formaldehyde molar ratio.

another, liberating water and forming dimethylene-ether links ($-CH_2-O-CH_2-$). This is particularly evident if the ratio of formaldehyde to phenol is high. The average molecular weight of the resins obtained by acid condensation of phenol and formaldehyde decrease hyperbolically from over 1000 to 200, with increases of the molar ratio of phenol to formaldehyde from 1.25:1 to 10:1 (Fig. 3.1).

The detailed structure of a crystalline chlorinated diphenylmethane (i.e., 3,3'-dichloro-4,4'-dihydroxydiphenylmethane) has been obtained by Whittaker [3] by x-ray diffraction.

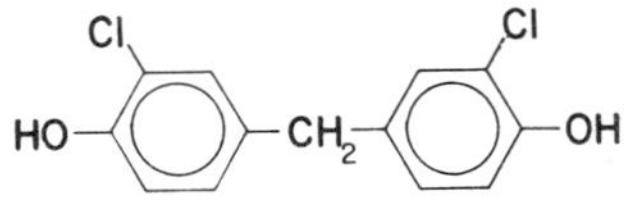

As was expected, the molecule is nonplanar. The benzene rings are inclined equally and in opposite directions at angles of 52° to the plane of the C–CH$_2$–C angle, while the C–CH$_2$–C angle is approximately 10° greater than the tetrahedral angle. The heat evolved during the phenol-formaldehyde reaction is 4.1 kcal/mol of formaldehyde added to phenol, and 16.9 kcal/mol of water is eliminated during the subsequent condensation. This indicates a total heat liberation of some 100 cal/g of anhydrous reactants.

B. Nature of Mechanism: Methylene and Methylene-Ether Bridges

Lilley and Osmond [4] separate the mechanism of condensation of phenol with formaldehyde into two phases. In the early stages, when water is present, condensation is almost definitely ionic. At a later stage, when the reaction proceeds in unreacted phenol as the solvent, after dewatering (or in the final stage when reaction is mainly in molten polymer), it is almost definitely nonionic. Lilley and Osmond [4] formulate the ionic mechanism as follows:

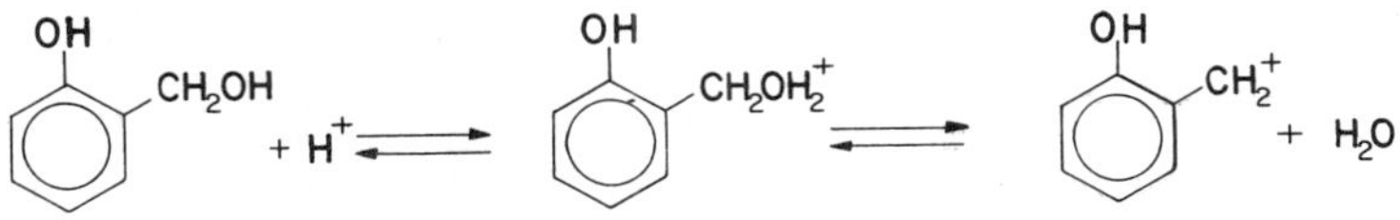

followed by reactions (1) and (2) below:

(1) 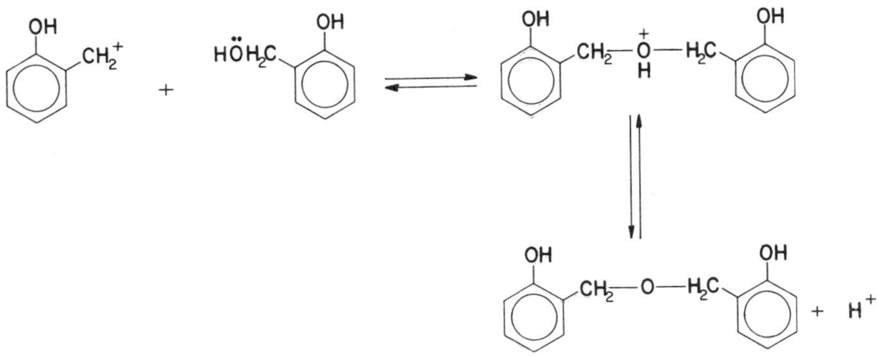

(2) 

The reactions are shown as involving an added proton, but similar arguments apply by solvolysis with a suitable ionizing phenolic medium. The rate of ether formation is

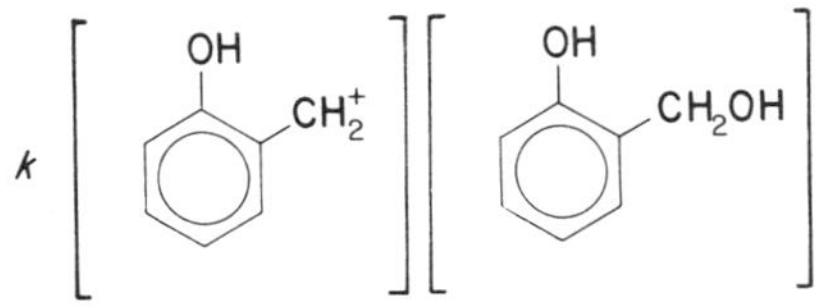

and of the $-CH_2-$ link

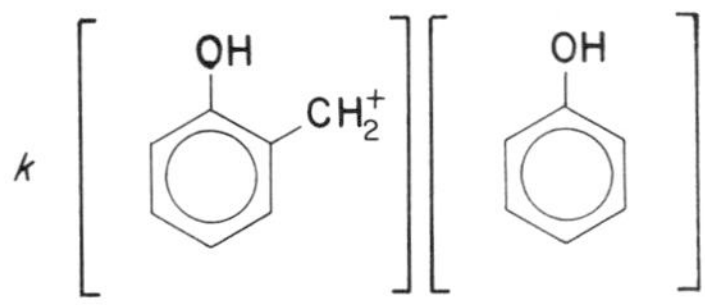

A methylene bridge is more likely to be formed than an ether link under acid conditions and when free positions in uncompletely reacted phenol are available. This can be shown by considering the behavior of the concentration of the carbonium ion when varying the hydrogen ion concentration. In alkaline conditions, the concentration of carbonium ions is low, but by considering the formation of the ions

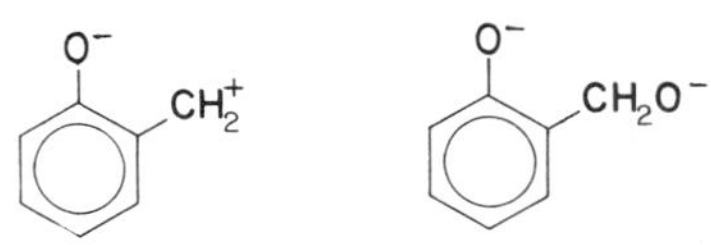

a similar conclusion can be reached.

It appears, therefore, that ethers can be formed only within limited and critical pH ranges, and then only at slow rates. Thus the formation of many ether links is unlikely. These critical ranges do not coincide with technological practice. Lilley points out the need for research on the distribution of methylol groups in phenolic resin syrups, and on the occurrence of ether links in such resins. All available evidence is against their occurrence in large numbers when free nuclear positions are available, as opposed to the monoreactive phenol alcohols used as "models," where ethers form undoubtedly in larger numbers.

Referring to the nonionic mechanism, which is likely to occur in the later stages of phenolic resin formation and in the heat hardening of phenolic alcohols, Lilley shows that since the reaction cannot involve a carbonium ion, it will probably use an intermediate of the quinone methide type. This compound is probably generated by a hydrogen-bonding mechanism. This has been shown to occur in phenol alcohols and dibenzyl ethers [4,5]. In the case of a monoalcohol, the reaction will be

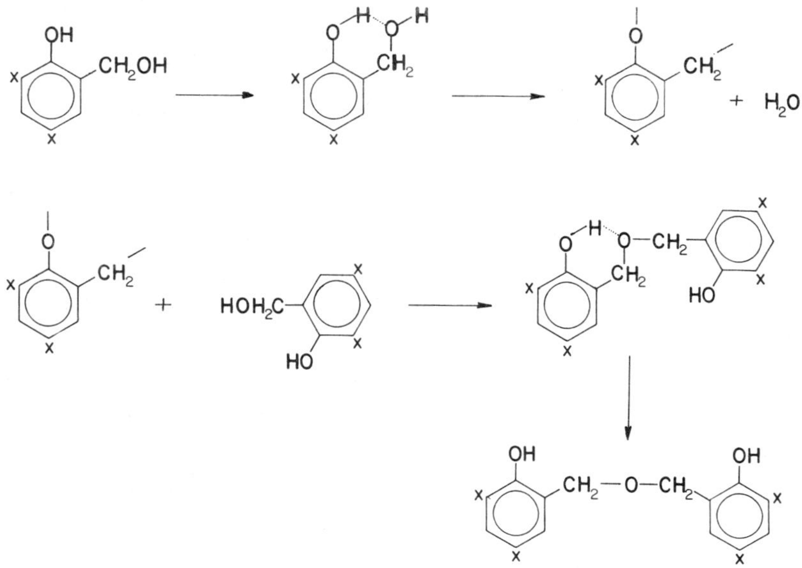

A radical of the quinone methide type is likely to have a considerably long life [6]. It yields a dibenzyl ether easily and a methylene bridge with difficulty. Work done by Van Euler [7] and Hultzsch [8] supports the idea that the intermediate appears through hydrogen bonding. Both investigators found that *ortho* systems such as

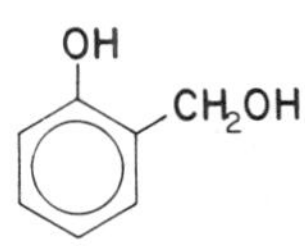

produce ethers more readily that the *para*-alcohols, and that the *ortho* series of ethers is more labile when heated.

This nonionic mechanism not only explains the known reactions of phenol monoalcohols, especially as far as ether formation is concerned, but also covers fully substituted dialcohols. The case is different when partially substituted monoalcohols such as

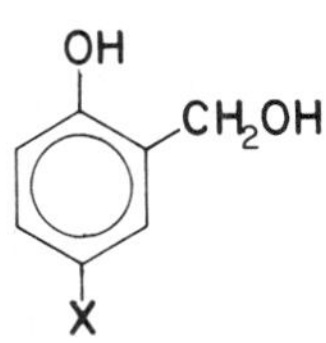

are considered. Here the intermediate has the form

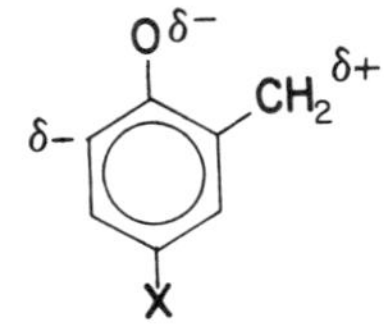

and there is a definite choice of reactions:

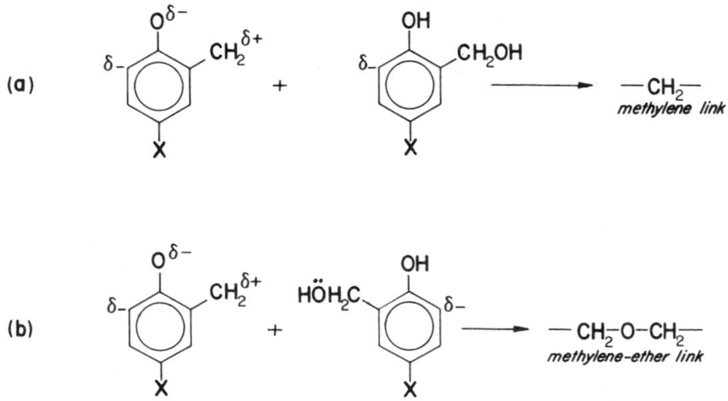

The former reaction is much more likely to occur and the formation of ethers is improbable. Experimental evidence supports this view. Although fully blocked monoalcohols undoubtedly lead to dibenzyl ethers in good yield, this is not the case of partially or nonsubstituted phenol alcohols. Thus ether forms with great difficulty from saligenin and homosaligenin [9] and in very poor yield [10].

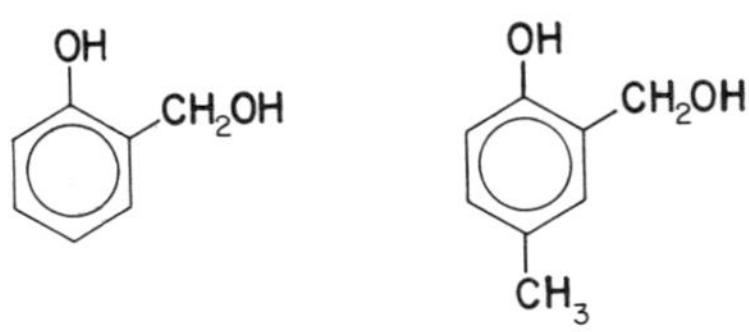

It is true that the dinuclear monoalcohol

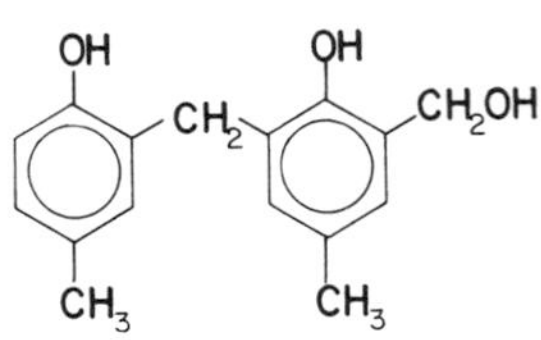

produced a high yield of ether [11]. In this case, however, the free *ortho* position is almost definitely deactivated because of interaction between hydroxyl groups. This belief is supported by similar results on the formation of ethers from etherified or acetylated phenolic alcohols.

Lilley believes that the formation of dibenzyl ethers depends on the distribution of activation in the phenol alcohol. He suggests that for an alcohol containing a free position, subject to activation by the phenolic hydroxyl group, such activation will oppose ether link formation. Substituents in the benzene ring also affect ether stability. The basis for these ideas was obtained by comparing the behavior of two alcohols, 4-chlorosaligenin and 4-nitrosaligenin.

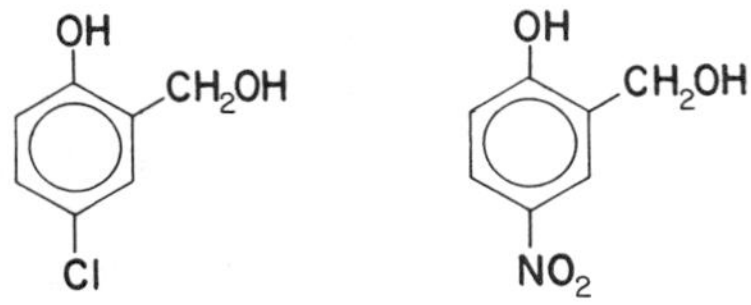

Each of these has a free *ortho* position, and they are comparable as regards the number of orientation of available reactive positions. However, the activations of their *ortho* positions are opposite in polarity. It has been shown that the formation of methylene groups is bound up with the δ-charge effect of the free *ortho* position. In the nitrocompound, the *ortho* position is more positively charged. The lone pair of the methylol group provides the most nucleophilic point for the attack of the carbonium ion, and therefore ether formation should predominate.

In the chloro derivative, the negative charge effect is increased and methylene linkages should predominate. Other factors must be taken into consideration. The nitro group directly deactivates the 6-position, so that reaction there is unlikely. However, the chloro group does not affect that position. The nitro group opens positions 3 and 5 for attack, so that the chances of methylene bridge formation are increased. Similar effects were observed on *p*-cresol [12]. Therefore, the ether is more likely to form in the case of nitrosaligenin and less likely to form with chlorosaligenin than at first expected. However, results obtained clearly demonstrate that the proportion of ether links formed depend a great deal on the activity of the free *ortho* position.

One can conclude that in the hardening of phenol alcohols which have free positions for substitution by methylol or its equivalents, a methylene bridge rather than an ether link will preferably be formed. This happens even more for fast-reacting phenols such as *meta*-cresol, resorcinol, and phloroglucinol, where the charge in *ortho* position is considerably increased. It gives rise to a nonionic reaction condition. Under ionic conditions, there is a limited range of pH values in which ether can be formed. However, this does not correspond with the working conditions of generally used pH. The rate of such a formation is slow whatever the position of equilibrium. The case of a fully substituted phenol monoalcohol, where ether link formation is more probable, is therefore not parallel to that of one with free positions. This can be seen by comparing the low yield of ether from 4-chlorosaligenin with 4,6-dichlorosaligenin (Zinke [20] and Ziegler [15,16]).

This yields 70% of the corresponding ether if heated to 130°C. The use of phenol alcohols as models is therefore of doubtful validity. Phenolic resins occurring in typical practice are likely to contain far fewer ether links than suggested by conclusions drawn on the basis that monoreactive monoalcohols are reliable models for study.

Unfortunately, there is little experimental proof of the effect of pH on yields of ethers and methylene compounds. Lilley [4] states that the reactions of phenolic resins were deduced from those of urea-formaldehyde resins. The two systems contain identical electronic configurations, and it therefore appears justifiable to suggest that a carbonium ion mechanism covers both cases. During attempts to isolate the alkyl ethers of methylolureas, Lilley found the following:

1. At pH 3, methylene compounds predominated.
2. At pH 6, ethers can be obtained.
3. At pH 7, reaction is slow but favors methylene links.
4. At pH 8, ethers again form, this time readily.
5. At pH 10, methylene links predominate.

In a later study, Rossouw, Pizzi, and McGillivray [13] found that ether links form at pH 4.3–5.0, and that no ether links form at pH 9 when phenol, resorcinol, and phloroglucinol are used as the model compounds. The rate of the formation and decomposition of the ethers and the amount formed was observed during kinetic experiments and was found to decrease for phloroglucinol, resorcinol, and phenol.

It is important to note that extremely fast-acting phenols, such as phloroglucinol, do form ether links. (It was thought that they do not form ether links because of the instability of their hydroxybenzyl alcohols.) The reason why it is impossible to isolate, for instance, the phloroglucinol ether links is because such ethers form and decompose in the first half hour of reactions at ambient temperature. However, the ether links formed by phenol are much slower to form and to decompose. They are therefore stable enough to be isolated and detected.

The range of ether formations given by Lilley on deductions derived from aminoplastic resin chemistry are probably valid. A pH of 5 is still in the range of ether formations advocated by Lilley (explain pH 6), and pH 9–9.5 is too far from the pH 8 advocated for their formation. Whatever the case, it is certain that ether links form in very narrow pH ranges.

C. Acid Catalysis

Consideration must be given to the possibility of direct intervention by the catalyst in the reaction. Hydrochloric acid is the most interesting case of acid catalyst, and ammonia of an alkaline catalyst. When the phenol-formaldehyde reaction is catalyzed by hydrochloric acid, two mechanisms possibly come into operation. Vorozhtov has proposed a reaction route which passes through the formation of bischloromethyl ether ($Cl—CH_2—O—CH_2—Cl$) [14]. Ziegler [15,16] has suggested a route through the formation of a chloromethyl alcohol ($Cl—CH_2—OH$) as intermediate. The second route appears to be the more probable one. Both hypotheses agree that chloromethylphenols are the principal intermediates. The chloromethylphenols have been prepared and isolated by various means. They are highly reactive compounds which, with phenols, form dihydroxydiphenylmethanes and complex methylene-linked multiring polyphenols. Reaction is highly selective and takes place in the *para* position.

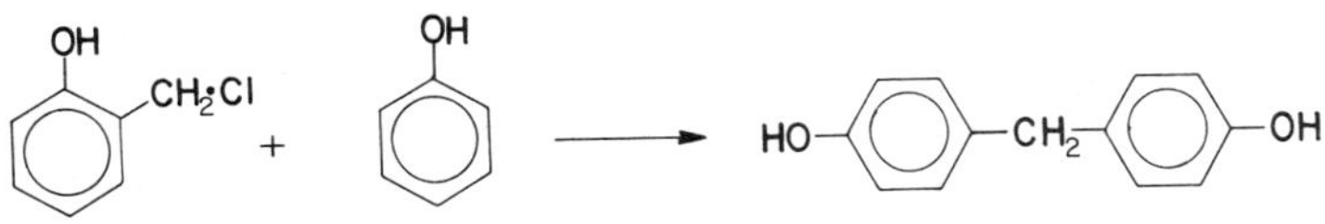

D. Alkaline Catalysis

Different mechanisms of alkaline catalysis have been suggested according to the alkali used. In the case where caustic soda is used as the catalyst, the type of mechanism which seems the most likely is that which involves the formation of a chelate ring similar to that suggested by Price and Sachanan [17]. The chelating mechanism may initially cause the formation of a sodium-formaldehyde complex or of a formaldehyde-sodium phenate complex.

When ammonia is used as a catalyst, the resins formed are very different in some of their characteristics from other alkali-catalyzed phenol-formaldehyde resins. The reaction mechanism appears to be quite different to that of sodium hydroxide-catalyzed resins. An obvious deduction is that intermediates containing nitrogen are formed. Several such intermediates have been isolated from ammonia-catalyzed phenol-formaldehyde reactions by various researchers [18–20]. Similar types of intermediates are formed when amines or hexamethylenetetramine are used instead of ammonia. In the case of ammonia, the main intermediates are dihydroxybenzylamines and trihydroxylbenzylamines.

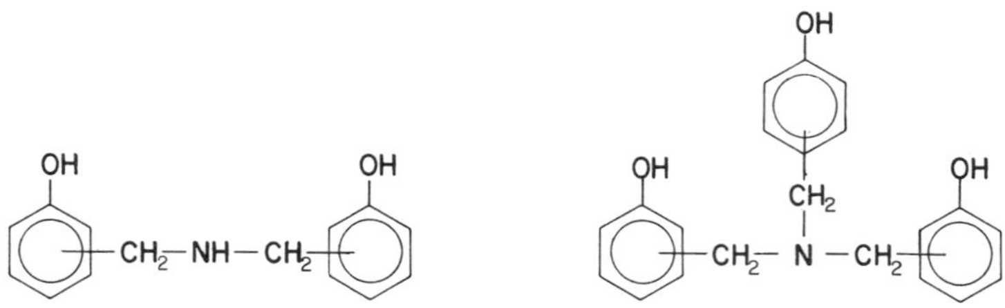

These intermediates contain nitrogen and have polybenzylamine chains. They react further with more phenol, causing the splitting and elimination of the nitrogen as ammonia or amines, and producing nitrogen-free resins. This, however, requires a considerable excess of phenol and a high temperature. With phenol-hexamethylenetetramine resins of molar ratio 3:1, the nitrogen content of the resin cannot be reduced to less than 7% when heated at 210°C. When the rate is increased to 7:1, the nitrogen content on heating at 210° can be reduced to less than 1%.

Ammonia-, amine-, and amide-catalyzed phenolic resins are characterized by greater insolubility in water than that of sodium hydroxide-catalyzed phenolic resins. The more ammonia used, the higher the molecular weight and melting point which are obtained without cross-linking. This is probably due to the inhibiting effect of the nitrogen-carrying groups (i.e., $-CH_2-NH-CH_3$ or $-CH_2-NH_2$), which is caused by their slow rate of subsequent condensation and loss of ammonia. Ammonia, amines, and amides (particularly dimethylformamide) are sometimes used as accelerators during the curing of phenolic adhesives for wood products.

E. Metallic Ions Catalysis and Orientation of the Reaction

Bender [21] shows that resins made under special conditions, where a high proportion of *ortho-ortho* links are formed in the phenol-methylene chains, can be cured by hexamethylenetetramine at a much higher rate than resins made with conventional acid or alkali catalysis. This involves condensation within the pH range 4–7 by oxides or hydroxides of alkaline or alkaline earth metals such as zinc, magnesium, and aluminium.

Fraser, Hall, and Raum [22] also described the effect of a wide variety of organic salts, oxides, and hydroxides of various metals. They came to the following conclusions: (1) that the orientation effect also occurs with 2,2'-dihydroxydiphenylmethane-formaldehyde condensation; (2) that the directive effect occurs in both the initial condensation and in the subsequent condensation of phenol alcohols; (3) that molecular weight distribution of "high-*ortho*" novolaks differs from those made during normal catalysis; (4) that a molar excess of phenol in the original condensation is necessary; (5) that an apparent pH of the reaction mixture between 4 and 7 is needed; (6) that the presence of electropositive bivalent metallic ions such as Mn^{+2}, Zn^{+2}, Cd^{+2}, Mg^{+2}, Co^{+2}, Ca^{+2}, and Ba^{+2} (in decreasing order of usefulness) is necessary; and that (7) the order of the efficiency of the metal ions used appears to be indifferent to the stability of the unidentified probable complex formed as reaction intermediate.

Fraser and other researchers conclude that the high rate of curing of phenolic resins prepared by metal ions catalysis is due to the preferential *ortho* methylolation and therefore also to the high proportion of *ortho-ortho* links in the uncured phenolic resins prepared by metal ions catalysis. Normal condensation yields roughly equal quantities of *para-para* and *ortho-para* links, but uncured resins produced by metal ions catalysis yield approximately one *ortho-para* to two *ortho-ortho* links. They ascribe faster curing rates of phenolic resins prepared by metallic ions catalysis to the higher proportion of the free higher-reactive *para* positions available for further reaction during the curing of the resin.

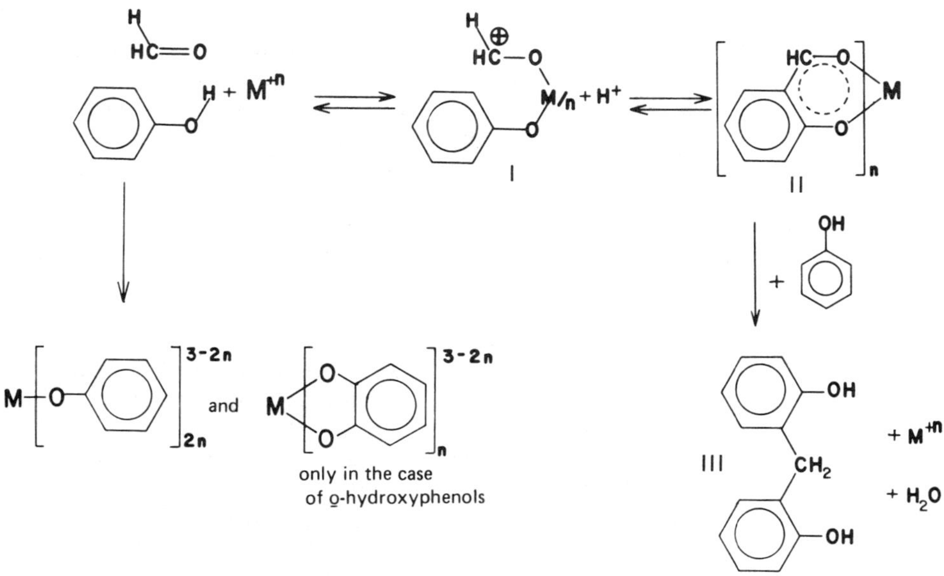

More recently, Pizzi [23] explained the mechanism of the reaction and corrected some of the conclusions which other researchers deduced from the work of Fraser. The mechanism presented is similar to that of the formation of metallic acetylacetonate complexes [24]. It involves the formation of chelate rings between metal, formaldehyde, and phenols similar to that suggested by price and Sachanan for alkaline phenol-formaldehyde reactions.

Pizzi substantiated the possibility of such a mechanism by isolating and characterizing the chromium-resorcinol-formaldehyde and chromium-phenol-formaldehyde complexes.

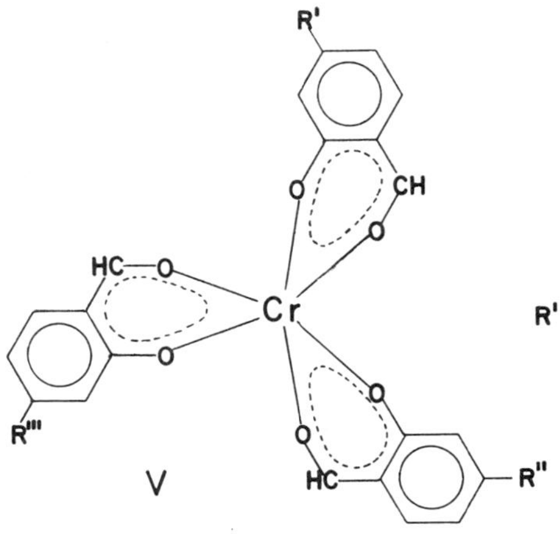

V

R' = R" = R"' = —H , —OH

The rate of metal exchange in solution [23,25] and the instability of the complex formed determine the accelerating or inhibiting effect of the metal in the reaction of phenol with formaldehyde. The more stable complex II is, the slower the reaction proceeds to the formation of resin III. A completely stable complex II should stop the reaction proceeding to resin III. If complex II is not stable, the reaction will proceed to form phenol-formaldehyde resins of type III. The rate of reaction is directly proportional to the instability, or the rate of metal exchange in solution of complex II.

As in the case of equivalent acetylacetonates [24], the bonding between metal and ligands in these chelates is partly covalent and their stability can be ascribed to their partly aromatic character. The metal catalyst does not change its valence, but interacts with molecules or ions containing electron donor groups, and accelerates reactions like hydrogen ions do [24]. The acid catalysis due to the metal ion differs only in degree from that of the hydrogen ion [24].

The effect of the metal is stronger than that of hydrogen ions, because of higher charge and greater covalence, since its interaction with donor groups is often much greater [24]. This is important, because in the preparation of phenol-formaldehyde adhesives for wood, very acid pH values cannot be used because they cause wood deterioration. Metal ion catalysts have the same catalytic effect on the preparation and setting of the resins at a higher pH, which is shown by a higher concentration of hydrogen ions and therefore lower pH values.

This allows phenolic resin adhesives to set in milder acid conditions, with neither extention of the setting nor any wood deterioration. The gel times for formaldehyde obtained for resorcinol and catechol indicate that the metal catalysis effect is probably valid for most phenols with a free *ortho* position to the hydroxy group [23]. Most bivalent metallic ions accelerate phenol-formaldehyde reactions.

The extent of this effect is directly proportional to the quantity of metallic ion present [23]. First, bivalent metals do not inhibit the reaction, as the complexes they form are unstable and their rate of exchange in solution is high. Second, they accelerate the reaction in a manner similar to that of hydrogen ions, although it is even faster, due to their stronger charge and greater covalence. The formation of the identified stable complexes of type II and V slows down or inhibits the reaction from proceeding to the formation of phenol-formaldehyde resins, when trivalent metals are used.

An exception is Fe^{III}, as shown by the effect of ferric acetate on the resorcinol-formaldehyde gel time [23]. This indicates that Fe^{III} is not capable of producing stable complexes of type II, and that the rate of metal exchange between complex and solution is high. This is consistent with the scale of rate of metal exchange given by Kluiber [25] for metal acetylacetonates in which a few metals are grouped in the following order of decreasing rate of exchange:

$$Cu^{II} > Fe^{III} \gg Al^{III} > Cr^{III}, Co^{III}$$

The rate of exchange, expressed as a percentage in a given unit of time for the different metals, is the following [24, 26]:

$$Cu^{III}, 100\%;\ Fe^{III}, 90\%;\ Cr^{III}, 1\%;\ Co^{III}, 0.0\%$$

The order of the metals in the scale corresponds to the gel times obtained for resorcinol-formaldehyde condensation [23]. It is consequently better to express the accelerating or retarding effect of the metal ions in the order of the rate at which they exchange in solution, rather than in the order of their stability [23]. The metals complexes stability does not bear any relationship to the gel times obtained for phenol-formaldehyde condensation [22]. The extent of the metal ions accelerating effect is

$$Pb^{II}, Zn^{II}, Cd^{II}, Ni^{II} > Mu^{II}, Mg^{II}, Cu^{II}, Co^{II}, Co^{II} > Mn^{III}, Fe^{III}$$

$$\gg Be^{II}, Al^{III} > Cr^{III}, Co^{II}$$

To conclude, Pizzi [23] found that

1. Contrary to Fraser's conclusions, it is not necessary to have a molar excess of phenol for the effect to take place.
2. Electropositive bivalent metallic ions accelerate the phenol-formaldehyde reaction, and electropositive trivalent metallic ions generally inhibit or retard it.
3. The efficiency of the metal ions on a molar basis is fairly independent of the stability of the complexes formed, but is directly dependent on the rate of metal ion exchange in solution.

4. The accelerating effect due to the metallic ions catalysts is similar to that of hydrogen ions in acid-catalyzed phenolic resins. It allows the use of acid-curing phenolic resins in the gluing of exterior-grade wood products with less fear of massive wood deterioration due to too high an acid content.

However, the most important conclusion to be drawn is that the accelerating effect is indeed present in both the manufacture of the phenol-formaldehyde resin and its curing. Therefore, the fast rate of curing on high-*ortho* phenolic resins can be only partially ascribed to the high proportion of *para* positions available. The other reason for the fast rate of curing is that the metallic ion catalyst is still present, and free to act, in the resin at the time of curing. In such a resin, a considerable number of *ortho* positions (especially of methylol groups in *ortho* positions to the phenolic hydroxyls) are still available for reaction, and capable of complexing. This concept is also suggested in Fraser's articles on his discovery of the accelerating effect in the condensation of dihydroxydiphenyl methanes, or phenolic "dimers", with formaldehyde.

The accelerating effect is therefore a result of the stronger protonation effect of the formaldehyde, which causes a much faster attack of the aldehyde onto the phenol. The initial HCHO attack determines the rate of acid-catalyzed phenolic reactions, as it is the slower step. The retarding effect is due to the effective blocking of the methylol group by the metal. Subsequent kinetic studies clearly confirm this. They also show that the trivalent metallic ions retarders accelerate the initial formaldehyde attack [26]. The accelerating effect of the metal ions is also present throughout the whole pH range, but is particularly evident in the range 2.5–7.

The decrease in the effect at lower and higher pH values can be ascribed to the "drowning" of the metal ion effect by the increased reactivity of the phenol-formaldehyde system. At lower pH values, the hydrogen ions are very abundant and consequently the metal ions contribute only minimally to the protonation of the formaldehyde and the acceleration of the reaction. At higher pH values the reaction of the phenol-formaldehyde system depends on formaldehyde protonation, and consequently the metallic ions have only a minimal bearing on the rate of the reaction.

The anion also influences the effectiveness of the metal as an accelerator of the reaction. For example, the following scale represents the decreasing accelerating effect for Zn^{2+} anions:

$$Zn(CH_3COO)_2 > ZnCl_2 + CH_3COONa > ZnCl_2$$

The anion effect can be explained in terms of the relative affinities of the metal for the anion used and the complex formed. It should therefore be more difficult for a strong acid to release Zn^{2+} than a weak acid.

A strong acid salt will slow down considerably the rate of exchange in solution of the metallic ion.

Another interesting way of producing high-*ortho* phenolic resins [27] without using a metallic ion catalyst is by reacting formaldehyde gas onto crystalline solid phenol with temperatures between 20 and 40°C in the presence of naphthalene. The naphthalene does not react with the phenol, but forms relatively solid coordination complexes with it. The presence of these complexes renders the phenol more reactive toward formaldehyde, allowing the reaction to happen at ambient temperature. When crystalline phenol is used at ambient temperature rather than molten phenol, the *ortho-ortho* configuration seems to be the favorite one, as it decreases the molecular disorganization of the phenol. The advantages of this type of preparation are the reduced energy used in preparation, and the elimination of dewatering of the final resin. No polycondensation water is produced on the surface of the resins.

F. Reaction Kinetics

Various researchers have investigated the reaction of phenols with formaldehyde and the results of earlier investigations have been controversial. Table 3.1 shows the results obtained by various investigators, and confirms that the phenol-formaldehyde reaction is bimolecular under acid conditions. The only exception is Jones's finding that such reactions can be monomolecular at low temperatures. However, under alkaline conditions, both mono- and bimolecular reactions are reported. Complex orders that do not conform to either of these are also present.

Until 1954, all investigations, whether qualitative or quantitative, had included analyses of the overall reaction of phenol with formaldehyde. Work of this kind provided only an overall kinetic rate constant representing the numerous reactions, which occurred either consecutively or concurrently, such as the formation of five methylol phenols. This could constitute one reason for the conflicting results obtained by various workers. Once all the methylol phenols had been synthesized, it became possible to examine the individual steps in detail. This was done by Freeman and Lewis [35] during an investigation which is considered to be one of the most comprehensive and detailed ever carried out.

The method adopted by Freeman and Lewis [35] consisted by reacting phenol, saligenin, *para*-hydroxybenzyl alcohol, or 2,4- or 2,6-dihydroxymethylphenol at 30°C with sufficient formaldehyde (as formalin) to bring about complete substitution, in the presence of 1 mol of caustic soda per mole of the phenol. The compounds obtained included all five methylol phenols, as well as tetramethyloldihydroxydiphenylmethane toward the end of the reaction.

Table 3.1 Orders of Reaction Reported for Phenol-Formaldehyde Reactions

Investigator	Phenol	P/F ratio	Catalyst	Order of reaction
Nordlander [28]	Phenol	—	NH_3	1
			Very little NH_3	2
Sprung [29]	Phenol	1:0.88	Triethanolamine	1
von Euler, Adler, and Bergström [30]	2,4-Xylenol	1:0.5	HCl	2
	2,6-Xylenol	1:0.5	HCl	2
	Phenol	1:0.5	HCl	2
	2,4-Xylenol	1:0.5	NaOH	2
Jones [31]	Phenol	1:1	HCl (80 and 100°C)	2
	Phenol		HCl (30°C)	1
	Phenol		NaOH	1
Goldblum [32]	Phenol	—	NaOH	2
Debing, Murray, and Schatz [33]	Phenol	1:0.38	Hexa	1
		1:1	NaOH	1–2
			Amines	1–2

Yanagita [34]	Phenol	—	Hexa	1
Freeman and Lewis [35]	Phenol	1:3	NaOH (equimol.)	2
	Phenol monoalcohols	1:2	NaOH (equimol.)	2
	Phenol dialcohols	1:1	NaOH (equimol.)	2
de Jong and de Jonge [36]	Phenol	Various	NaOH	2
	para-Cresol		HCl	2
			Triethanolamine	2
Fitzgerald and Martin [37]	2,6-Xylenol	1:1	NaOH (excess)	Complex
	Prehnitenol	1:1	NaOH (excess)	Complex
Kämmerer [38,39]	*para*-Cresol dialcohol	None	Heating only	1
	Various dialcohols	None	Heating only	2
	Phenol trialcohol	None	Heating only	2
	One dialcohol at 180°C	None	Heating only	3
Malhotra and and Kumar [40]	*para*-Cresol			2
	Phenol			2
Rossouw, Pizzi, and McGillivray [13]	Resorcinol	1:1 and 1:3	NaOH and heating only	2
	Phloroglucinol	1:1 and 1:3	NaOH and heating only	2

The reactions that occurred are summarized by Freeman and Lewis in the following diagram, which shows the symbols used for the individual rate constants.

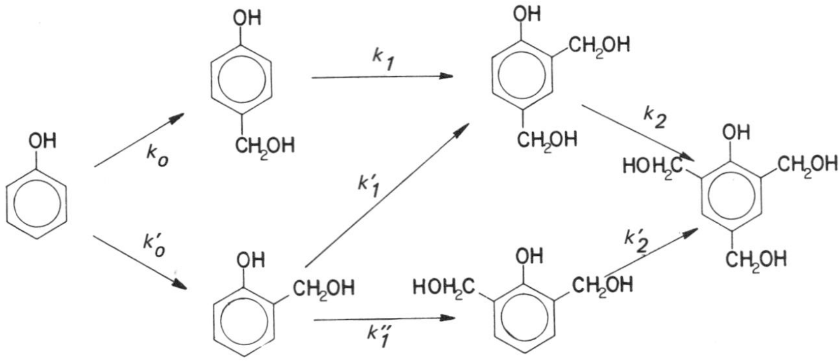

Because of the complexity of the system, Freeman and Lewis simplified the calculations by separately examining the mono- and dimethylol phenols. They assumed that each reaction was of second order. Evidence for the correctness of this assumption was obtained in the case of 2,4- and 2,6-dihydroxymethylphenols, both of which produced only the trimethylol derivative. Plots of log b(a - x)/a(b - x) against time led to straight lines. From the slopes of these lines, k_2 and k'_2 were derived. It was also possible to calculate the other constants by a method of successive approximations, from an examination of the appropriate rate equations. The values given in Table 3.2 are quoted from Freeman and Lewis [35]. From this study, Freeman and Lewis drew the following conclusions:

1. In phenol, the *para* position has a slightly greater affinity for formaldehyde than the *ortho* position (i.e., 6.2 as compared with 5.25). Nevertheless, because two *ortho* positions are available, saligenin is produced 1.7 times more quickly than *para*-isomeride.
2. The introduction of a methylol group decreases the reactivity of the remaining nuclear positions toward formaldehyde. The decrease is bigger than one expects. Thus *para*-hydroxybenzyl alcohol is 50% less reactive than phenol (i.e., 7.5 as compared with 16.7), even though it has two-thirds the number of positions still available for reaction.
3. The introduction of a methylol group into the *ortho* position causes little deactivation. Therefore, saligenin is twice as reactive as *para*-hydroxybenzyl alcohol, and almost as reactive as phenol, even though it has one less position available. Similarly, 2.4-dihydroxymethylphenol is more reactive than *para*-hydroxybenzyl alcohol, and slightly more than half as reactive as saligenin.

Table 3.2. Reaction Rate Constants for Methylol Phenols + Formaldehyde

Reaction	Second-order rate constant × 10^6
Phenol — $\xrightarrow{k_0}$ *p*-monomethylol phenol	k_0 = 6.2
Phenol — $\xrightarrow{k'_0}$ *o*-monomethylol phenol	k'_0 = 10.5
p-Monomethylol phenol $\xrightarrow{k_1}$ 2,4-dimethylol phenol	k_1 = 7.5
k'_1	k'_1 = 7.3
o-Monomethylol phenol $\xrightarrow{k''_1}$ 2,6-dimethylol phenol	k''_1 = 8.7
2,4-Dimethylol phenol $\xrightarrow{k_2}$ 2,4,6-trimethylol phenol	k_2 = 9.1
k'_2	
2,6-Dimethylol phenol	k'_2 = 41.7

The relative reactivities of the *ortho* and *para* positions of phenol are 1.4 and 1.7, respectively, and are relative to a value of the *ortho* position of *p*-hydroxybenzyl alcohol of 1.0. Saligenin *ortho* and *para* positions have the relative reactivities of 2.3 and 2.0. The 2,4-dimethylolphenol *ortho* position has a value of 2.4. The 2,6-dimethylolphenol *para* position has a value of 11.1.

Predictions based on these figures have been made by Freeman and Lewis on the likely course of the reaction. They predict that when less than three equivalents of formaldehyde are used, saligenin is the first compound formed in large amounts. Its forward reaction will, however, be nearly as rapid as its rate of formation, and therefore its concentration will remain approximately the same. Ultimately, it will be the first phenolic alcohol to disappear. The *para* isomeride will appear later than saligenin, but because of its low reactivity, will accumulate in the mixture. 2,4-Dihydroxymethylphenol will appear even later, but because of its low reactivity, will also accumulate. 2,6-Dihydroxymethylphenol reacts so quickly that it will rarely be detected in resin mixtures. This deduction confirms the findings of Sprengling and Freeman [41]. The amount of 2,4,6-trihydroxymethylphenol will increase rapidly to a quantity determined by the amount of formaldehyde available.

A contribution by de Jong and de Jonge [36] is important because it deals with kinetic effects in dilute aqueous solutions during the early stages of reaction. The authors studied the hydroxymethylation of phenol, and to a less extent, *p*-cresol, over the pH range 1–11 and over the temperature range 70–130°C. They have found that the reaction is invariably of the second-order type. The rates of reaction at 70°C are proportional to the hydrogen ion concentration below pH 3.6. A plot of log k against pH produces a straight line with a negative slope. The rates are also proportional to hydroxyl ion concentrations above pH 5. (A straight line with a positive slope was obtained.) Similar results were recorded at 90°C (pH 1–2), 110°C (pH1.3), and 90°C (pH 8). The direct relationship between reaction rate and hydroxyl concentration does not hold when the concentration of caustic soda is of the same order of magnitude as that of the phenol. The addition of buffer solutions at pH values of 8.0 (phosphate), 4.0 (acetic acid-acetate), and 2.0 (phosphoric acid) do not affect the results. When triethanolamine is used as a catalyst under diluted conditions a second-order rate law is still observed.

De Jong and de Jonge made the observation that the rates of hydroxymethylation of *ortho*-methylol-*para*-cresol and saligenin were of the same order of magnitude, under identical conditions, as the initial hydroxymethylation of *para*-cresol and phenol, respectively. In other words, the introduction of a methylol group has no effect on the reactivity of the remaining reactive positions. This finding is in conflict with other observations.

Calculations of the effect of temperature on reaction rates of *para*-cresol reactions indicate that the activation energy is about 22 kcal/mol. The authors also propose mechanisms for the reactions occurring during the early stages of reaction which are consistent with their findings. They suggest that in strongly acid solutions, the course of reaction will involve the formation of a conjugate acid of formaldehyde:

$$HCHO + HA \underset{k_{-1}}{\overset{k_1}{\rightleftharpoons}} H_2\cdot\overset{+}{C}OH + A^-$$

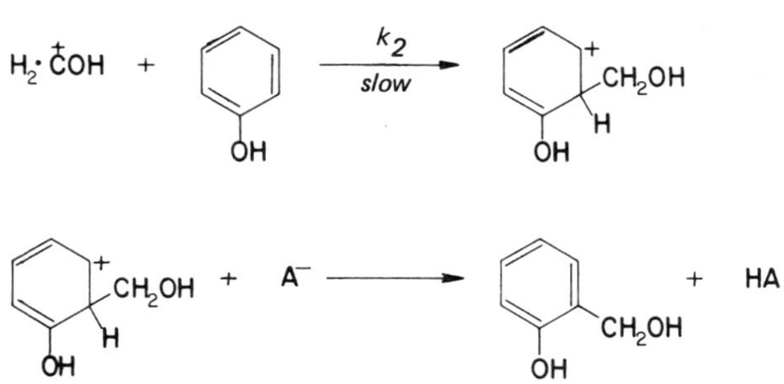

In this scheme, HA represents any acid. The second of these equations is the rate-determining step. It leads to the following expression for the reaction rate:

$$v = k_2[P][FH^+] = \frac{k_2 k_1}{k_{-1}} \frac{[HA]}{[A^-]}[P][F]$$

$$= \text{constant}[H^+][P][F]$$

This expression agrees with the observed bimolecularity of the reaction and the specific catalysis by hydrogen ions.

For pH values above 4, it is suggested that formaldehyde reacts with phenolate ions according to the following scheme:

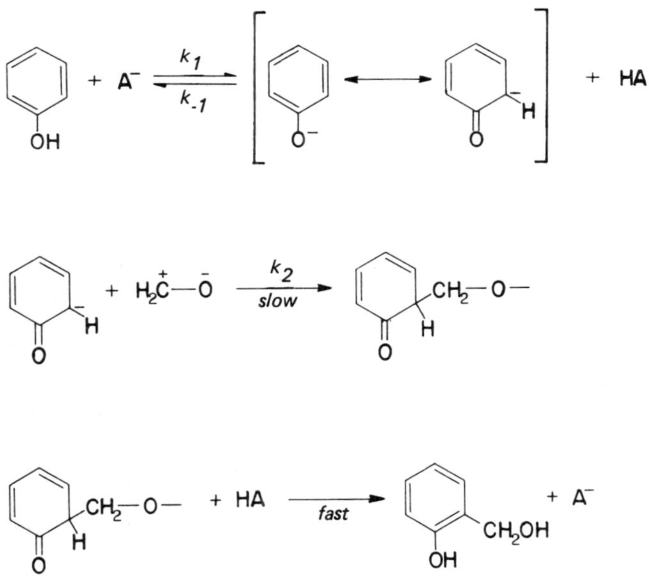

A^- denotes any base present in the solution. The reaction rate can be represented by the following expression:

$$v = \frac{k_1 k_2}{k_{-1}} - \frac{[A^-]}{[HA]} [P][F] = \text{constant}[OH^-][P][F]$$

This expression agrees with the bimolecularity of the reaction observed and the specific catalysis by hydroxyl ions. If the concentration of hydroxyl ions is high, the concentration of phenolate ions is no longer proportional to, but almost independent of, base concentration.

A new field was opened by Kämmerer [38,39], who investigated the kinetics of the reaction that occurs when phenolic alcohols are heated and produce dibenzyl ethers. Kämmerer [38] demonstrated that the equations developed by Flory for polyesterification reactions are equally applicable to polyetherifications. Flory's equation can be stated as follows:

$$\frac{\text{Number of reacted methylol groups}}{\text{Number of original methylol groups}} = \frac{n_0 - n}{n_0} = \frac{c_0 - c}{c_0} = p$$

Where n_0 and n are the number of molecules which exist at time = 0 and at time t; c_0 and c are the concentrations of methylol groups originally and at time t, and p is extent of reaction. The relative number of nonreacted groups is

$$1 - p = \frac{n}{n_0} = \frac{c}{c_0}$$

Since two molecules, or two methylol groups, yield one molecule of water, it follows that

$$n_o - n = 2n_{H_2O} \quad \frac{n_0 - n}{n_0} = \frac{2n_{H_2O}}{n_0} = p$$

p can be estimated experimentally from this relationship, since n_0 is known and $2n_{H_2O}$ can be determined. Using the relationships above, Kämmerer showed that by plotting the functions of p against time, the order of reaction could be ascertained. He checked the validity of his deductions on the evolution of water on heating the phenolic alcohols, and showed that for experiments done at 165°C, the reactions were all second order. At 180°C, some compounds underwent a third-order reaction over the range 70–93% conversion.

Kämmerer next examined the hardening of *para*-cresol dialcohol, in detail, as estimated by loss of water and formaldehyde, and showed that it depended very much on experimental conditions. He also found the reaction was of first order for the purest dialcohol, for heating between 40 and 240 min, corresponding to between 50 and 90% conversion to the dibenzyl ether. For times below 40 min, the order was different, and appears to be associated with a solid-state reaction.

Kämmerer [38] investigated the hardening of phenol trialcohol by heating it at 130°C under nitrogen. When it was heated for up to 7 hr, corresponding to 90% conversion, the kinetic values indicated that a second-order reaction had occurred. By using the bromination method, he showed that more than 95% of the molecular chains were bonded by ether groups.

The use of ammonia leads to monomolecular reactions. However, Nordlander's findings show that the order of reaction can be raised to 2 by a reduction in the ammoniacal catalyst. In addition, the order of

Table 3.3. Heats of Activation Reported for the Phenol-Formaldehyde Reaction

Investigator	Heat of reaction (cal/mol)
Sprung [29]	13,800
Debing, Murray, and Schatz [33]	15,400
Yanagita [34]	17,000
De Jong and de Jonge [36]	22,000 (*para*-cresol)
Lomakin et al. (Ref. given in [2])	22,000 (acid catalyst) 13,600 (alkaline catalyst)
Jones [31]	16,900 (condensation reaction) 4,100 (substitution reaction)

reaction is 1 when triethanolamine is used as the catalyst under anhydrous conditions, but is 2 under aqueous conditions.

Activation temperatures show a fairly wide range, as reported by different researchers. They are summarized in Table 3.3. The variations observed must again be attributed to differences in reaction conditions.

A later investigation was carried out by Rossouw, Pizzi, and McGillivray on the reaction kinetics of phenol, catechol, resorcinol, and phloroglucinol with formaldehyde and other aldehydes in a water medium at pH values of 4.9 and 9.0–9.5. The reactions at phenol-to-formaldehyde molar ratios of 1:1 and 1:3 in conditions as close as possible to those used for the manufacture of phenol-formaldehyde resins were also observed. They found that the reaction of phenols with aldehydes under alkaline conditions follows a second-order kinetic and that the rate of reaction is

$$v = k_1[\text{aldehyde}][\text{phenol}] + k_2[\text{aldehyde}][\text{phenate ion}]$$

This is due to the phenol $\rightleftarrows$ phenate ion equilibrium. The k_1 term of the equation is negligible with pH values of 8.0–8.2, where the reaction rate is determined by the reactivity of the phenate formed. The k_2 term is negligible at lower pH values. The reaction of the aldehydes with the "polymeric" condensates formed under alkaline conditions during the initial part of the reaction follows a second-order kinetic: namely, for formaldehyde,

$$v = [\text{aldehyde}] \sum_n k_n [\leftarrow CH_2\text{—phenol}\rightarrow_n]$$

This is valid with different k values for heat condensates in the phenol or phenate ion form, according to the pH used. The total reaction rate equation, in alkaline conditions, was summarized in a single equation:

$$v = k_{1+2}[\text{aldehyde}][\text{phenol}]^{a}[\text{phenate ion}]^{1-a} + [\text{aldehyde}] \sum_n k_n [(\text{CH}_2\text{—phenol})_n]$$

In acid conditions, they also found the reaction to be of second order. The reaction is very complex in acid conditions as a great variety of species contribute to the total kinetic curve. Depending on the acid pH range used, some of the terms in the equations are negligible. For all practical purposes the equation in acid conditions can be expressed as

$$v = k_3[\text{HCHO}]^{a}[\text{CH}_2\text{OH}^+]^{1-a}[\text{phenol}] + k_4[\text{phenolic ether}] + [\text{HCHO}]^{a}[\text{CH}_2\text{OH}^+]^{1-a} \sum_n k_n [(\text{CH}_2\text{—phenol})_n]$$

where the contribution of the k_4 term is negative, due to ether decomposition. This indicates that the formaldehyde has been released. The kinetics of the phenol-formaldehyde reactions with the method used by these authors indicate that two types of reactions are present during the formation of a phenolic resin:

1. The reaction of formaldehyde with the phenol, and with the hydroxydiphenylmethanes or higher condensates formed during the reaction
2. The liberation of formaldehyde, due to the passage of unstable, methylene-ether links to methylene bridges

Methylene ethers have been isolated for the phenol-formaldehyde reaction quite a few times previously. However, this researchers have found by kinetic means that these also form for fast-reacting phenols, such as resorcinol and phloroglucinol [13]. The hydroxybenzyl alcohols formed by the reaction of formaldehyde with these two phenols was previously considered too unstable for the formation of the ether bridges. They were converted to hydroxybenzylcarbonium ions or quinone methides [2] instead. The ethers decompose very fast, at ambient temperature, generally in the first half hour of reaction, and this explains why they have never been detected before. It also explains the equality of the final, stable, reaction products of the phloroglucinol-formaldehyde and resorcinol-formaldehyde condensations in acid and alkaline conditions.

Pizzi and van der Spuy [26] also examined the effect of the presence of an accelerating or retarding metallic ion on the kinetics of phenol-formaldehyde reactions. Second-order kinetics as a function of the concentration and pH were also found to fit the metallic ion catalyzed reactions. In the presence of any accelerating or retarding metallic ions, the reaction was considerably less sensitive to variations of $[H^+]$ or $[OH^-]$ concentration, and therefore the pH. The dependence of the reaction rate constant on the metallic ions concentration was found to obey a law of the following type:

$$k = k_0 + k_{M^{+n}}[M^{+n}]$$

at constant pH values (k_0 = rate constant in the absence of M^{+n}, $k_{M^{+n}}$ = catalytic constant, and k = rate constant in the presence of M^{+n}). An anomaly was found in the use of Cr^{+3} as retarder of the reaction. Cr^{+3} can accelerate the first part of the phenol-formaldehyde reaction, and is therefore able to accelerate the attack of formaldehyde on the phenol. The formation of the methylol alcohol is due to its strong charge and strong protonation of the formaldehyde. The rest of the reaction was retarded due to the formation of the complexes as discussed. This is further proof of the accelerating and retarding mechanism proposed.

G. Hardening

When resols, or A-stage resins made from phenols with three reactive positions, are heated to temperatures above 100°C, they become more viscous and eventually rubbery. They are then in the resitol or B stage and they are insoluble in the usual solvents, such as alcohols and water. However, they will swell in acetone and will flow under heat and pressure. Further heating causes further hardening of the resins until they reach the resite or C stage, when they are insoluble and infusible.

This change is accompanied by a change in structure from the low molecular weight mono- and polyalcohols of the A stage, to the partly linked polymers of the B stage. They finally change to the branched and cross-linked three-dimensional networks characteristic of the C stage. Novolaks also pass through the same sequence of changes if heated with methylene-supplying agents under alkaline conditions.

Branched-chain novolaks are more reactive than linear novolaks, because of the higher proportion of monosubstituted phenol in branched structures. Branched novolaks cure faster than linear novolaks. The same changes can be obtained by adding strong acids to low molecular weight liquid resols. The acid-initiated condensation of the methylol groups with the phenolic nuclei may be exothermic enough to reach temperatures high enough to complete the hardening. With sufficient acid catalysis, resols will harden, if they are in thin films, at room temperature.

Various groups of researchers have carried out research on the hardening mechanism of phenol-formaldehyde resins [20, 7, 42]. They have found that the hardening of the resin takes place in two stages. All the research groups involved agree that in the first stage of hardening, ether-linked polymers are formed, with loss of water, by the condensation of two adjacent methylol groups. Although it is certain that methylene-ether links form in the first stage of hardening, the use of "blocked" methylolphenols used by these researchers is not conclusive. Ether formation appears to occur only to a limited extent in phenolic resins used in industry.

The secondary stage of hardening at higher temperatures is accompanied by a change of color from pale yellow-red to dark brown. This takes place either in air or in inert atmosphere. There is a reduction of solubility in aqueous alkali, chloroform, and benzene. The products at first become partially soluble and partially insoluble. Only the soluble portion contains ether links.

There is some disagreement among various researchers as to what happens during the secondary breakdown of the primary ether-linked network, but it seems that several different reactions occur at the same time. If the ether links break down to methylene bridges, one molecule of formaldehyde should be set free from each link. Since only about half of this is obtained, the Zinke group concluded that the formaldehyde reacts with the remaining reactive positions on the phenolic nuclei, if any are still present; or if there are none, with the methylene bridges or the phenolic hydroxyl groups. This gives rise to the following structures:

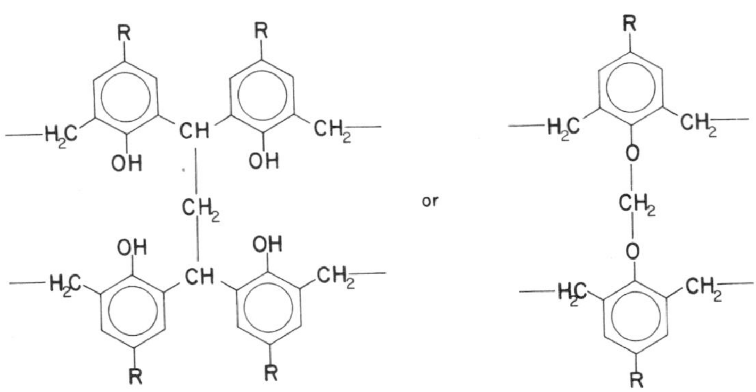

These reactions involve further water formation. This is in fact observed. This hypothesis is supported by the fact that polynuclear dimethylolphenols containing methylene bridges give off less formaldehyde and more water per mole when hardened at high temperatures than do simple mononuclear phenolic dialcohols. Zinke also found small amounts of methoxy groups in the hardened resins, indicating a slight methylation of phenolic hydroxy groups. Crystalline dimethylphenols (xylenols) and phenolic mono- and dialdehydes, presumably formed by the cracking of the ether chains, were obtained in small amounts.

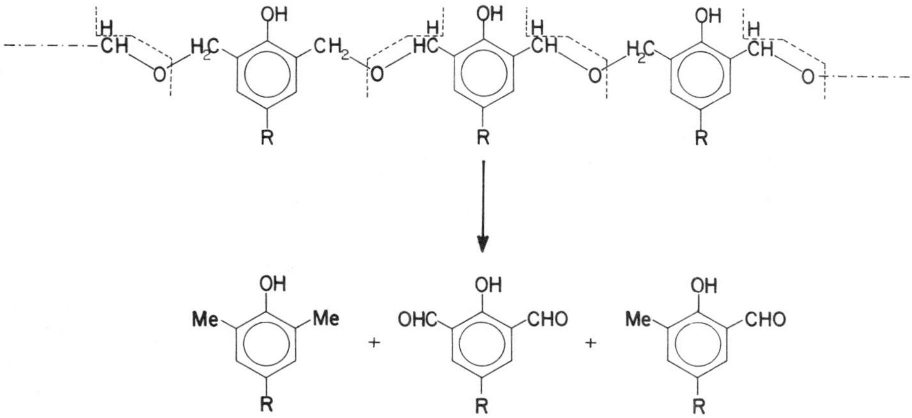

Evidence that the hydroxy groups are essential for the secondary reactions of high-temperature hardening was provided by the hardening of mononuclear dimethylolphenols which had no free reactive positions, and which had the phenolic hydroxy group esterified. Hardening of these esters produces only ether-linked chains, even at 200°C. One mole of water is evolved for every mole of dialcohol, but no formaldehyde is liberated. Acylation of the phenolic hydroxyl groups reduces the hardening rate only slightly, but esterification (e.g., methylation) greatly reduces the hardening rate.

Hultzsch [42,43] disagrees with Zinke's theory that liberated formaldehyde causes cross-linking by condensing with methylene bridges. He found that the dihydroxydiphenylmethanes did not react with formaldehyde at 200–240°C. He also found that monomethylolxylenols, which have no free reactive positions, resinify on heating with a loss of water and formaldehyde. The initial product obtained on heating at 130–170°C is an ether, namely

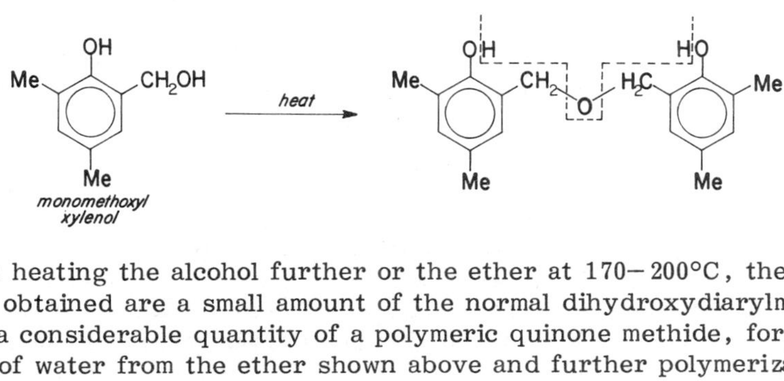

On heating the alcohol further or the ether at 170–200°C, the products obtained are a small amount of the normal dihydroxydiarylmethane and a considerable quantity of a polymeric quinone methide, formed by loss of water from the ether shown above and further polymerization.

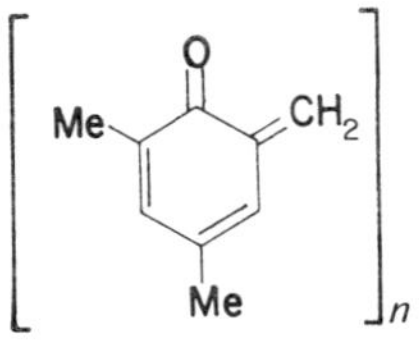

polymeric quinone methide

The polymeric quinone methide, which has a melting point of 200°C, has an average degree of polymerization of between 2 and 3. The polymeric quinone methide resinifies at 220°C and turns black without liberating water or formaldehyde. Another product of the hardening of the ether is the dihydroxydiarylethane. It is likely that hydrogenation and dehydrogenation at these reactions occur at fairly high temperatures (of the order of 200°C), simultaneously converting the polymeric and monomeric quinone metides to diphenylethanes, aldehydes, alkyl phenols, and undistillable resins [43].

Von Euler proposes that the polymerization of the quinone methides occurs through the oxygenation of the substituted methylene group and not only through its carbon atom, and that the trimer is probably an annular compound of the following type:

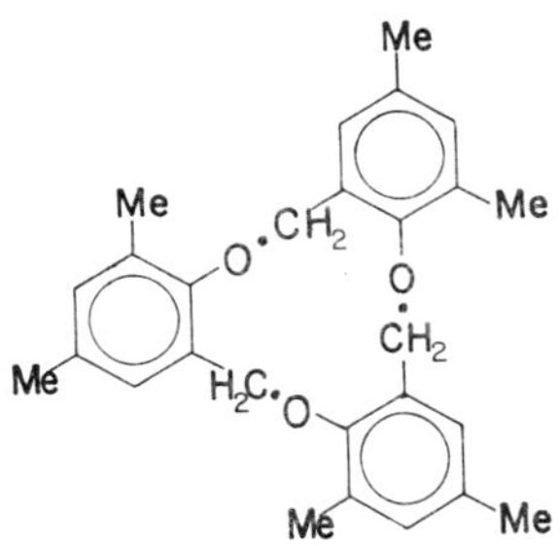

The trimer with the proposed annular structure is extremely resistant to acid and alkali attack, cannot be esterified, has no color reaction to ferric chloride, and cannot be reacted with ketonic reagents. All this indicates that the phenolic hydroxyl has etherified. Von Euler also states that the quinone methides can polymerize directly, or they may first produce dihydroxystilbenes (HO—ArCH═CH—Ar—OH). These produce diarylethane derivatives (HO—Ar—CH_2—CH_2—Ar—OH) and polymerized stilbene quinones (O═Ar═CH—CH═Ar═O), by simultaneous oxidation and reduction. He also found that *ortho*-quinone methides (from *ortho*-phenolic alcohols) polymerize directly to their stable polymers, and *para*-quinone methides (from *para*-phenolic alcohols) form dihydroxystilbenes and other reaction products.

Furthermore, von Euler believes that the liberation of formaldehyde during high-temperature hardening accompanies the formation of diphenylmethanes. He thinks that the liberation of formaldehyde takes place directly from the alcohols, with subsequent condensation, rather than from ethers. This is because monoalcohols without free reactive positions lose formaldehyde faster and at a lower temperature than do the corresponding ethers.

Some indication of the proportion of the different products given off is shown by the following results with *para*-hydroxymesityl alcohol. After 0.5–1.0 hr at 140°C the yields were:

$R—CH_2OH$ (unchanged), 55–57%
$R—CH_2—R$, 10–13%
$R—CH_2—O—CH_2—R$, 1–5%
$R—CH_2—CH_2—R$, 0–5%
HCHO, traces

At 170–180%, a good yield of $R—CH_2—CH_2—R$ was obtained and the stilbene R—CH=CH—R was also obtained. By heating for a relatively short time at 240°C in a carbon dioxide atmosphere, $R—CH_2—CH_2—R$, R—CH=CH—R, a considerable amount of R—CHO and traces of $R—CH_3$ were obtained. The quantities of different products obtained varies greatly according to the phenols used.

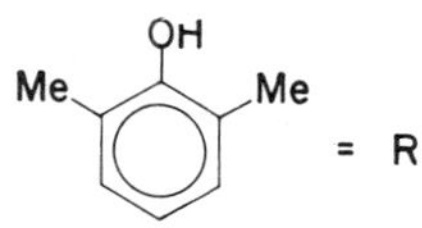

A schematic representation of Hultzsch's ideas on hardening of phenolic resins is shown in Figure 3.2.

To summarize, during the process of hardening at 140°C, methylol- and dimethylolphenols give products containing methylene bridges ($—CH_2—$) and methylene-ether links ($—CH_2—O—CH_2—$), with molecular weights of approximately 5000. They do not produce quinone methides. The methylene links are stable irrespective of further increases in temperature. The ether links are stable only up to 160°C. At higher temperatures (i.e., 170–200°C), quinone methides are formed nearly quantitatively. These polymerize into inert resins at 200–230°C, forming traces of oxidation and reduction products of the quinone methides. Hydroxyaldehydes are also formed through the quinone methides or by direct cracking of the ether chains, especially above 200°C.

Phenol-formaldehyde resols harden in an alkaline environment as a result of the formation of mainly methylene links ($—CH_2—$). Very few quinone methides and no ether links are formed [44]. The fully hardened resins are rigid three-dimensional structures in which cross-linking is random. There is no evidence of crystallites in the network, but it is presumed that there are relatively large holes at certain points throughout the mass [45]. These form points of weakness because all the cross-links theoretically possible have formed. This is a result of the special arrangement and cumbersome structure of the original molecules, and perhaps as a result of trapped water or catalyst molecules.

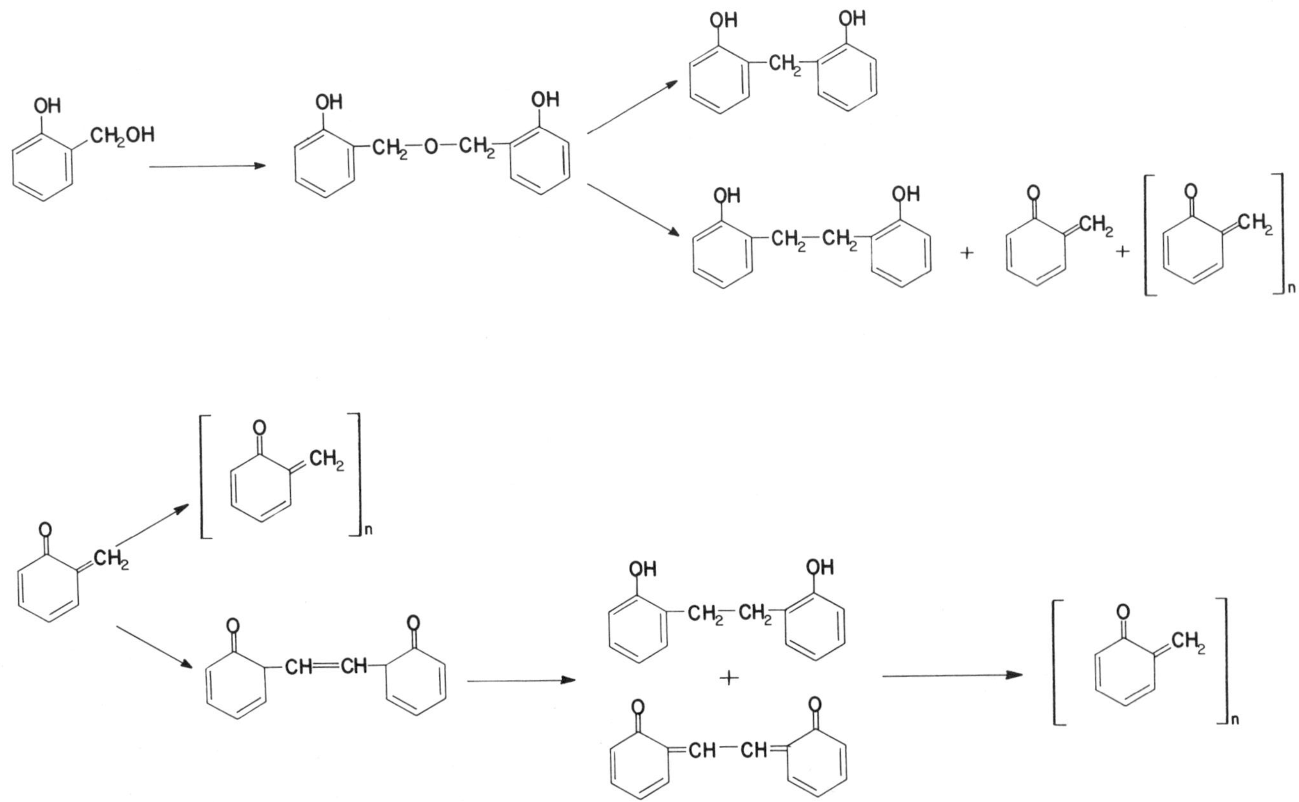

Fig. 3.2. Schematic representation of Hultzsch's ideas on hardening of phenolic resins.

These points of weakness account for the fact that the tensile strength of the hardened resin is very much less than in theory.

Several researchers have concluded from these data that phenol-methylene chains are very irregular, and that they have only a few cross-links. During the hardening process, the resin molecule is first a simple chain which grows slowly. Other chains subsequently condense onto it to form branches [46]. This process continues until separate molecules begin to entangle and form a gel. As the hardening continues, there is further growth and entanglement. The resin molecules become immobilized and therefore insoluble and infusible [46]. The fact that saligenin readily hardens when heated in the presence of traces of acid or alkali supports the above theory. However, it is not possible to obtain tridimensional cross-linking through methylene bridges in a molecule build up from saligenin.

When the resin has hardened, the impact strength and flexural strength of phenolic resins increase with the number of methylene groups between the phenol nuclei. When the number of methylene groups is increased from 1 to 10, a more flexible cross-linked network is obtained [47]. The hardness of phenolic resins also decreases as the length of the polymethylene chain in the phenol increases. Several authors believe that the hardened resin first passes through two phases: first, the formation of a highly condensed macromolecular gel structure, and second, a viscous liquid phase of lower molecular weight, or better a continuous isogel structure, In the latter, the cavities of a continuous net- or spongelike skeleton have filled with a viscous liquid condensate [48].

H. Resorcinol and *Meta*-Aminophenol Condensates

Resorcinol reacts readily with formaldehyde to produce novolak resins, which harden at ambient temperatures if formaldehyde is added. The initial condensation reaction, in which A-stage liquid resins are formed, leads to the formation of only linear condensates when the resorcinol/formaldehyde molar ratio is approximately 1:1 [1]. This reflects the reactivity of the two main reactive sites (4- and 6-positions) of resorcinol.

However, reaction with the remaining reactive, but sterically hindered site (2-position) between the hydroxyl functions also takes place [1]. In relation to the weights of resorcinol-formaldehyde condensates which are isolated and on a molar basis, the proportion of 4- plus 6-linkages relative to 2-linkages is 10.5:1. However, cognizance must be taken of the fact that the first-mentioned pair represents two condensation sites relative to one. The difference in reactivity of the two types of sites (i.e., 4- or 6-position relative to the 2-position) is then 5:1 [1]. Linear components always appear to form in preference to branched components in A-stage resins [1]; that is, terminal attack leads to the preferential formation of linear rather than branched condensates. This fact can be attributed to:

1. The presence of two reactive nucleophilic centers on the terminal units, as opposed to single centers of doubly bound units already in the chain.
2. The greater steric hindrance of the available nucleophilic center (nearly always at 2-position) of the doubly bound units as opposed to the lower steric hindrance of at least one of the nucleophilic centers of the terminal units (a 4- or 6-position always available). The former is less reactive as a result of the increased steric hindrance. The latter are more reactive.
3. The lower mobility of doubly bound units, which further limit their availability for reaction.

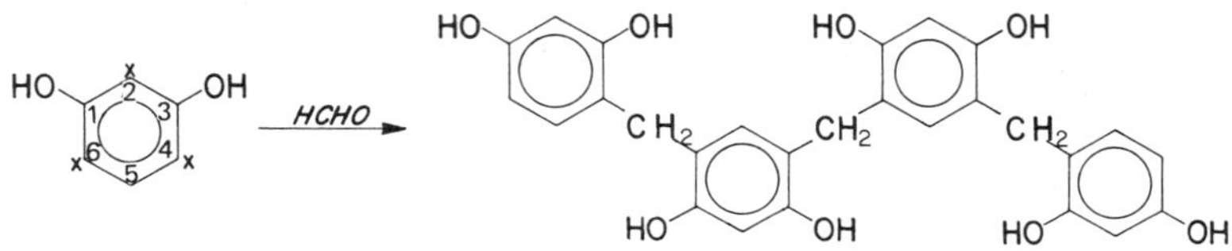

most common type of resorcinol/formald. "tetramer"

The absence of methylol (—CH_2OH) groups in all six lower molecular weight compounds isolated [1] reflects the high reactivity of resorcinol under acid or alkaline conditions. It also shows the instability of its *para*-hydroxybenzyl alcohol groups and their rapid convension to *para*-hydroxybenzyl carbonium ions or quinone methides. This explains how identical condensation products are obtained under acid or alkaline reaction conditions [1]. In acid reaction conditions methylene-ether-linked condensates are also formed, but they are highly unstable and decompose to form methylene links in periods of between 0.25 and 1 hr at ambient temperature [13].

From a kinetic point of view, the initial reaction of condensation to form "dimers" is much faster than the later condensation of these "dimers" and higher polymers. The reaction of resorcinol with formaldehyde, on an equal molar basis and under identical conditions, also proceeds at a rate which is approximately 10–15 times faster than that of the equivalent phenol-formaldehyde system [13]. The high reactivity of the resorcinol-formaldehyde system renders it impossible to have these adhesives in resol form. Therefore, only resorcinol-formaldehyde novolaks can be produced. Thus all the resorcinol nuclei are linked together through methylene bridges with no methylol groups and methylene-ether bridges.

The reaction rate of resorcinol with formaldehyde is dependent on the molar ratio of the two constituents, the concentration of the solution, the temperature, pH, the presence of various catalysts, and the amount

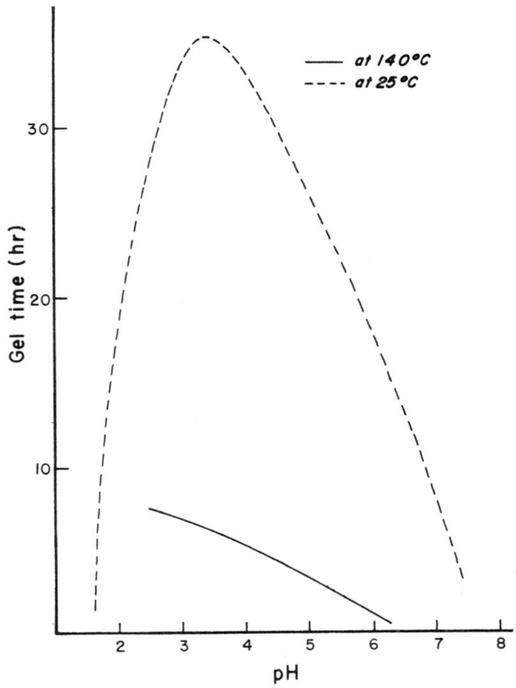

Fig. 3.3. Gel times of aqueous resorcinol-formaldehyde adhesives as a function of pH.

of certain kinds of alcohols present [49–52]. The effect of pH and temperature on the reactivity of the resorcinol-formaldehyde system is shown in Figure 3.3.

Methanol and ethanol slow down the rate of reaction. Other alcohols behave similarly, the extent of their effect being dependent on their structure. Methanol lenghtens gel time more than other alcohols; higher alcohols are less effective. The retarding effect on the reaction is due to temporary formation of hemiformals between the alcohols and the formaldehyde. This reduces the reaction rate because of the lower concentration of free formaldehyde [50-53]. Other solvents, such as acetone, dioxane, dimethylformamide, and tetrahydrofurane, also affect the rate of reaction by forming complexes, or by hydrogen bonding with the resorcinol.

In the manufacture of pure resorcinolic resins, the reaction would be violently exothermic unless it is controlled by the addition of alcohols. Because the alcohols perform other useful functions in the glue mix, they are left in the liquid glue. The tendency for resorcinol-formaldehyde resin solutions to precipitate insoluble resinous polymers from acid aqueous solutions can be eliminated by carrying out the initial reaction at pH values lower than 1 [54].

Meta-aminophenol (or *meta*-hydroxyaniline) is sometimes used to modify phenolic resins for wood to enable them to condensate more quickly at ambient temperature. The reactivity of *meta*-hydroxyaniline with formaldehyde is approximately 5–10 times higher than that of resorcinol, which is already high. This is caused by the presence of an amino group substituent on the aromatic ring instead of a second hydroxy group as in the case of resorcinol. This substituent activates the aromatic ring considerably toward electrophilic reagents such as formaldehyde. An aliphatic amine reacts by necessity with a phenolic methylol group via a benzylic carbon-nitrogen bond. *Meta*-hydroxyaniline has the advantage over other phenols of having three reactive sites on the aromatic ring and theoretically two more on the amino group. Therefore, *meta*-hydroxyaniline can be bound to phenolic resins by ring-to-ring methylene bridges, or by phenolic ring to amino-group methylene bridges.

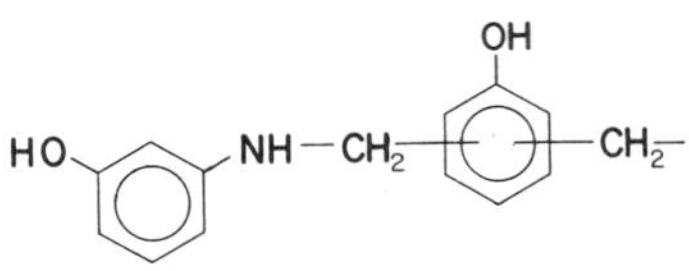

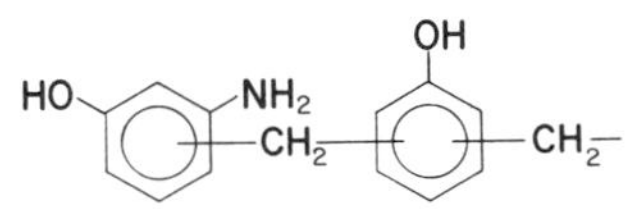

The presence of large amounts of primary amino groups in phenolic resins modified by *meta*-hydroxyaniline implies the carbon-carbon bonding of the phenolic ring to the ring of the aromatic amine. The presence, in the same resins, of secondary amino groups implies carbon-nitrogen bonding.

Tertiary amino groups $\left(-\mathrm{N}\begin{matrix}\diagup CH_2- \\ \diagdown CH_2-\end{matrix}\right)$ are not present. The proportion of phenolic ring-to-ring bonds on a molar basis is from 5 to 14 to one phenolic ring-to-amino group bond [55, 56].

I. Physical Properties of Phenol-Formaldehyde Resins

Hardened phenol-formaldehyde resins have a specific gravity of approximately 1.2–1.3, a refractive index of 1.6, and a specific heat of 0.5. They are typically brown in color, and novolaks are lighter than resols. Resols are dark yellow, orange reddish, or brownish, even when made with pure raw materials. However, if the alkali is neutralized, resols become almost colorless. The best results were obtained with citric, lactic, and phosphoric acids. Pale-colored, hardened resins can be prepared with them [57].

Phenolic resins are relatively stable up to about 200°C, although oxidative degradation takes the form of attack at the methylene bridges to produce substituted dihydroxybenzophenones [58]. Above this temperature, they begin to char slowly, and at higher temperatures charring is more rapid. At about 400°C decomposition is rapid, yielding phenols and aldehydes, leaving a cokelike residue.

In the A stage, simple phenol-formaldehyde resins are readily soluble in alcohol, esters, ketones, phenols, and some ethers, and insoluble in hydrocarbons and oils. As a class, resols tend to be more soluble in alcohols and water, and novolaks tend to be more soluble in hydrocarbons. In the early stages of condensation, resols are often soluble in water, owing to the presence of methylolphenols, especially polyalcohols. This is more pronounced with resols that are derived from phenol. Cresilic resols are less soluble, and xylenolic resols almost insoluble in water. The solubility of A-stage resins in dilute aqueous sodium hydroxide or in mixtures of water and alcohols follows the same trend.

Solubility in alcohols and insolubility in hydrocarbons appear to go together. The alcohol and water solubility can be reduced only by using acetaldehyde, or other aldehydes in the place of formaldehyde, and by introducing hydrocarbon chains, particularly in the *ortho* or *para* position, in the aromatic ring. B-stage resins are soluble in only a few solvents, such as boiling phenols, and acetone, aqueous sodium hydroxide, and deca- and tetrahydronaphtalenes.

Resins in the hardened or C stage are very resistant to most chemical reagents. They are unaffected by all ordinary organic solvents and water, although a few percent of water may be absorbed in filled material, mainly by the filler, thus causing slight swelling. The C-state resins dissolve slowly in boiling phenols such as naphtols. Resins from the simplest phenols can also be broken down and dissolved by hot, strong, alkali solutions.

Simple phenol-formaldehyde resins are readily attacked by sodium hydroxide. However, cresol-formaldehyde and especially xylenol-formaldehyde are much less susceptible to attack. Resins are often more resistant to strong alkaline solutions (i.e., 15–20%) than to dilute solutions (i.e., 5%). The filler has a considerable influence on the chemical resistance of the resins. Inert mineral fillers have a better resistance than cellulosic fillers. C-stage resins are resistant to most acids, except sulfuric acid stronger than 50%, formic acid, and oxidizing acids such as nitric and chromic acids.

The insolubility of hardened resins in acetone is used to test the degree of cure of the resin. The curing temperature influences the amount of matter which is insoluble in acetone, after prolonged heating [59]. The higher the hardening temperature, the lower the amount of acetone extractives. The mechanical properties of hardened phenol-formaldehyde resins are greatly influenced by the moisture content. This applies even more to resins containing fillers, plasticizers, and other ingredients. The rate of water absorption decreases with time, but thick samples may not reach an equilibrium even after several months in water.

Therefore, in measuring the mechanical properties of resins, it is necessary to condition the test pieces under carefully controlled temperature and humidity prior to making the tests. In many cases, the mechanical properties of hardened resins are largely dependent on the type and orientation of the filler. This applies particularly to water absorption, tensile strength, and impact strength. It also applies to shear strength, with the condition that in the plane of the laminations the shear strength depends on the adhesion between the laminae of sheet material. The properties of the resin are more important in determining the compression strength than those of the filler.

III. CHEMISTRY AND TECHNOLOGY OF APPLICATION OF PHENOLIC RESIN ADHESIVES FOR WOOD

A. General Principles of Manufacture

A typical phenolic resin is made in batches in a jacketed, stainless steel reactor, which is equipped with an anchor-type or turbine-blade agitator, a reflux condenser, vacuum equipment, and heating and cooling facilities. Molten phenol and formalin (containing 37–42% formaldehyde, or paraformaldehyde) are charged into the reactor in molar proportions between 1:1.1 and 1:2, and the mechanical stirring is started.

To make a resol-type resin (such as that used in plywood or particleboard manufacture), an alkaline catalyst, such as sodium hydroxide, is added to the batch. It is then heated to 80–100°C. Reaction temperatures are kept under 100°C by applying vacuum, or by cooling water in the reactor jacket. Reaction times vary between 1 and 8 hr, according to the pH, the phenol/formaldehyde ratio, the presence or absence of reaction retarders (such as alcohols), and the temperature of the reaction.

Since a resol can gel in the reactor, dehydration temperatures are kept below 105°C, by applying vacuum. Tests have to be done to determine first, the degree of advancement of the resin, and second, when the batch should be discharged. Examples of methods of such tests are the measurement of the gel time of a resin on a 150°C hot plate, or at 100°C in a water bath. Another method is measuring the turbidity point, that is, precipitating the resin in water or solutions of a certain concentration.

Resins which are water soluble and which have a low molecular weight are finished at as low a temperature as possible, usually around 40–60°C. It is important that the liquid, water-soluble resoles retain their ability to mix with water easily since when they are used as particleboard adhesives, they are diluted further with water before they are sprayed onto the wood chips. The resoles based on phenol are considered to be stable for 3–9 months. Properties of a typical resin are a viscosity of 100–200 cp at 20°C, a solids content of 60%, a water mixibility of a minimum of 2500%, and a pH of 7-13.

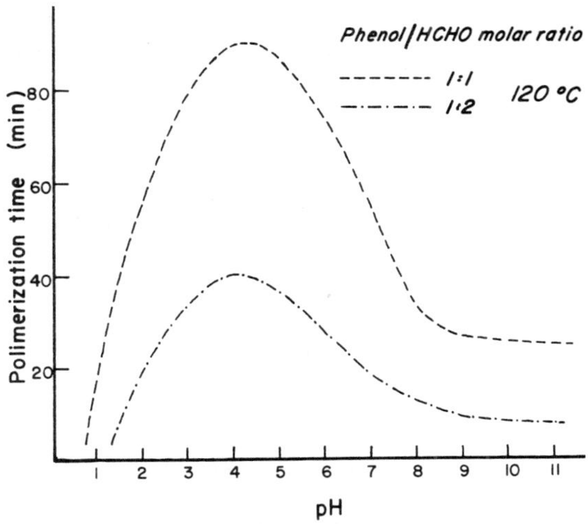

Fig. 3.4. Rate of polymerization as a function of pH for phenolic resols of different molar ratios at 120°C.

The manufacture of novolaks is of less interest in the manufacture of resins for wood products. The novolaks used in the wood industry are phenol-formaldehyde resols that have been transformed into novolaks by blocking the methylol groups present with resorcinol, or resorcinol-formaldehyde novolaks. Both need the addition of paraformaldehyde as a hardener for curing. They are prepared in the same way as for resols at acid or alkaline (or, alternatively, both), pH values. A pH of about 4 is the point of lowest reactivity for all phenolic resins. The effect of the pH and the phenol/formaldehyde molar ratio on the rate of polymerization of phenolic resols is shown in Figure 3.4. Tolerance to dry-out or to longer assembly times in plywood resins to be used for wood products is achieved by adding either the phenol or the formaldehyde in different amounts at different stages in the reaction, rather than charging their total amount in the reactor at the start of the reaction.

B. Plywood and Particleboard Adhesives and the Factors Regulating Their Application

Phenolic resins are used as binders for exterior-grade plywood and particleboard, which need the superior water resistance provided by these resins. In the manufacture of plywood, the phenolic resins adhesive is usually applied to the wood veneers by roller or extrusion coating. The coated veneer is then cross-grained, stacked, and cured in a multidaylight press for 5–10 min at 120–130°C, and at 12–16 kg/cm^2.

In the manufacture of particleboard, the phenolic resin adhesives are sprayed onto the wood chips, or sprayed plus spread by continuous blenders. The glued wood chips are formed in a mat and then pressed for 4–7 min, according to thickness and moisture content, at 170–195°C and 25–35 kg/cm^2.

The only type of phenolic resins commercially used for this application are resol-type resins, which have the following structure:

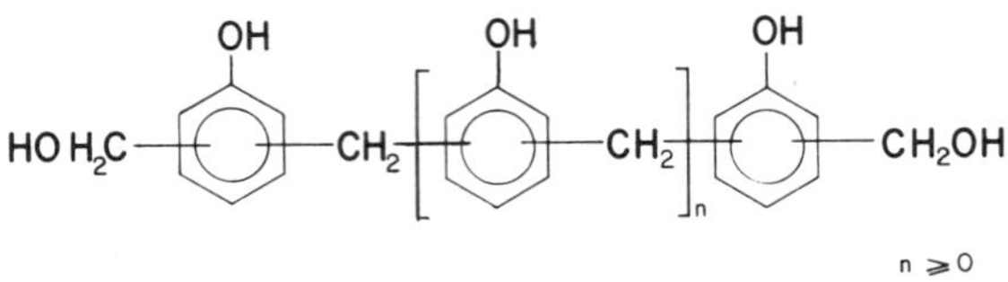

These are hardened by heating after the addition of small amounts of wax emulsion and insecticide solution in the case of particleboard, and of vegetable or mineral fillers and tackifiers in the case of plywood. Accelerators are often added in both types of glue mixes. These are aminic or amidic (e.g., dimethylformamide) in the case of particleboard, and metallic salt catalysts (e.g., zinc acetate), in the case of plywood. Other accelerators are also sometimes used. The pH of these resins varies between 10 and 13.5, and is generally between 12 to 12.5.

In dealing with wood-related factors that affect glue bonds, it is important to remember that adhesion is at least 95% chemical in nature. The mechanical aspects of bond formation (such as keying cured adhesive solids into the wood surface) contribute negligibly to the bond strength or wood failure. The main chemical forces in thermosetting resin adhesion are primary valence bonds, covalent bonds, and hydrogen bonds, plus secondary forces such as van der Waals' and London's forces, and any other types of electrostatic, dipolar, and associative forces. It is therefore essential that the resin contain significant numbers of functional groups and that the wood surface present significant number of reactive sites to enable the resin to bond. Any factors that limit resin functionality or block reactive sites on the wood structure necessarily impede adhesion.

C. Properties of Phenolic Adhesives for Plywood [60]

Certain attributes of phenolic resins have been designed to give the strongest and most durable plywood bonds. Laboratory and field experience have demonstrated that certain types of phenol-formaldehyde plywood resins perform significantly better on veneers than do others. These superior resins have several properties in common:

1. They are relatively low in alkali content, generally about one-third to not more than one-half molar.

2. They have a lower molecular weight for hardwood veneers than do phenolic resins designed for softwood gluing.
3. They are high in methylol group content. Alternatively, they may contain free formaldehyde, or require a matching catalyst which contains paraformaldehyde.
4. Even in dried adhesive carrier films or powdered resins, phenolic resins for hardwoods share the B-stage characteristic of reliquefying briefly under heat and pressure to allow transfer and flow on the glue lines of a plywood panel. This liquefaction can occur without water. This is unlike all but the most recently developed softwood phenolic resins.
5. Phenolic resins for hardwoods have higher thermal softening points than many other conventionally prepared resins. This indicates a network which has more cross-links after final cure and also greater durability.
6. They have a 40–45% solids content and 150–600 cp of viscocity at 25°C.

In general, lower resin alkali content and lower molecular weight are associated with slower cure, which explains why softwood plywood adhesive resins are both more alkaline and more condensed. Adequate exterior-grade adhesion can be attained on softwood veneer if these considerably condensed phenolic resins, which also cure more quickly, are used. Conversely, the increased functionality of hardwood adhesive resins partially compensates for their inherently slower cure rate. Hardwood phenolic resins require about 30% longer press times for adequate cure.

Notwithstanding their good adhesive capability, phenolic resins for hardwood gluing carry one distinct disadvantage: they do not prepress as well as softwood phenolic resins. Their lower condensation and longer flow are not assets. Prepressing is important because it is done to minimize face veneer losses and to reduce precure times on hot platens. Therefore, other means can be employed to obtain the required tackiness. Additives and adhesive formulations for this purpose are available. To increase the prepress capacity of low-alkali, long-flow phenolic resins suitable for hardwood gluing, small amounts of starch or polyvinyl alcohol can be added to the resin glue mix just before use.

Finished resin viscosity is increased to allow for the thickening effect of these additives without reducing the average molecular weight of the phenolic resin. Water-soluble thickeners (such as hydroxyethyl cellulose, polyethylene glycols, and maleic anhydride copolymers) contribute to prepress tack. However, they cause a large increase in resin viscosity and the amount added must be small. Consequently, the benefit to prepressing is limited. Most animal- and vegetable-based thickeners, such as gum arabic, are subject to hydrolysis in alkaline phenolic resins, and lose effectiveness in a matter of hours.

1. Additives

A number of additives and modifiers contribute useful properties to phenolic resins used for wood gluing. Multipurpose additives are the aminoresins and urea-formaldehyde and melamine-formaldehyde polymers. These include not only urea and melamine plywood adhesive resins, but also dimethylol urea, trimethylol melamine, and hexamethylol melamine. Added in amounts from 5 to 15% of phenolic resin solids, they improve resin tack and prepressing, increase long assembly time tolerance, shorten pressing times, and enhance resin functionality. This results in stronger bonds on wood veneers.

As long as they are used in limited quantities, they have a negligible effect on long-term phenolic bond durability. They appear to be well protected from hydrolytic degradation by the cured phenolic polymer network. The dispersion of the aminoresin molecules in the alkaline medium of the phenolic resin inhibits their curing reaction, which is acid catalized. It causes them to function as methylolated cross-linking units for the phenol-formaldehyde polymer.

Formaldehyde in liquid solution or solid form, and formaldehyde-generating compounds, are also phenolic resin additives which improve functionality and decrease curing times. Paraformaldehyde is the most frequently used, but hexamethylenetetramine, formaldehyde/sodium-bisulfite complexes, tris(hydroxymethyl)nitromethane, and glyoxal are also used. Significant effects are obtained when 3–5% is added, based on phenolic resin solids. Further reduction in curing time are possible if 1–2% resorcinol is added, or resorcinol-acting natural extractives such as wattle-tannin extract. This should be mixed with paraformaldehyde and added to the liquid phenolic resins glue mix.

Formaldehyde additions overcome the effect of phenolic extractives in certain hardwood species, which prevent proper cure or adhesion of phenol-formaldehyde resins. Free formaldehyde appears to react rapidly with these phenolic extractives before they can interfere with the phenolic resin curing mechanism. With certain wood species which are rich in extractives, this technique has been used to increase bond durability from interior-grade to true exterior-grade performance.

Natural phenolic compounds are used as both replacements for substantial portions of synthetic phenol in plywood adhesive resins, and as glue mix additives to improve performance; 4–6% is added, based on phenolic resin solids. They bring about improvements in assembly time tolerance and flow with no significant change in adhesion. Glue mix additions of wattle extract with additional formaldehyde produce faster hot-pressing cycles. However, some assembly-time tolerance and pot life has to be sacrificed in the process, but full exterior-grade durability is retained.

The lignin residues from wood pulp production are also a class of extractives which are currently receiving attention as phenolic resin additives. Substituted phenols, such as cresols and xylenols, have been used as glue mix additives for phenolic adhesive resins to improve assembly-time tolerance. They are also used as solvents to remove oleoresinous deposits on the surface of pitchy softwood veneers. They can be used as flow promotors in phenolic hardwood adhesives. The amount added should not exceed 3–4% of the phenolic resin solids content to avoid interference with the rate of resin curing.

Complexing additives commonly include the soluble salts of boron, chromium, zinc, cobalt, lead, manganese, magnesium, and other. Some of these compounds have been successful in reducing pressing times and improving prepress performance [23] when added to phenolic resin adhesives. Borax is widely used in North America to shorten the prepress cycles of phenolic plywood glues for softwoods. However, these compounds tend to increase the molecular weight of a phenolic resin by complexing several molecules together through their phenolic hydroxyl groups. The gain in the resin molecular weight and prepress tack is sometimes accompanied by a reduction in assembly-time tolerance and the loss of the B-stage melt-flow behavior. In hardwood gluing, this is sometimes not advantageous, and the addition of complexing salts should be approached with caution.

Mixed borate salts are very effective as a treatment for the preservation of wood products against fungi and most insects. The application of these borate-based treatments to green hardwood veneer inhibits bonding with phenolic adhesives after veneer drying. The boron salts, which become localized in high concentration on the veneer surface, tend to gel the phenolic resin before it can reach the wood surface and bond to it. However, very dilute aqueous solutions of borates (i.e., 0.25%) applied to softwood veneers in their green state decrease their thermal degradation during high temperature drying and preserve their reactive sites for bonding with phenolic adhesives. Unfortunately, a borax solution of such a low concentration does not make green veneers mold- or insect-resistant.

D. Formulation of Plywood Glue Mixes

The guiding principles for the preparation of plywood adhesives glue mixes are:

1. To maintain phenolic solids in the mixed glue as high as possible (preferably in the range 30–40%).
2. To incorporate a cellulosic filler, such as a nut shell flour of 200 mesh or finer, in about 20–40% of phenolic solids. Nonabrasive inorganic fillers may also be satisfactory.
3. Alternatively, to add about half this amount of unrefined starchy material, such as wheat flour.
4. To add no alkali, or at the most only 1–2% to disperse and stabilize the starchy material.

5. To add only enough water to produce a glue viscosity which can be handled by the gluing equipment. The preferred viscosity range is 1500–2500 cp, measured at 25°C.
6. To ensure proper wetting of the veneer by the glue film, a surface active agent should be added (about 0.1–0.25% on resin solids).

Examples of glue mixes incorporating these principles are the following:

Material	Compositions		
40–45% solids PF resin	100	100	100
300-mesh coconut (or walnut) shell flour	12	14	10
Industrial wheat flour	6	–	–
50% sodium hydroxide	2	–	–
Surfactant	0.1	0.1	0.1
Water	10	5	–
Total parts	130.1	120.1	110.1
Phenolic solids	31–35%	33–37%	36–41%

These glue mixes may still need adjustment as regards resin solids and viscosity, to allow for the behavior of other filler materials in the alkaline phenolic resin. It is important to keep glue viscosity in a satisfactory range, neither thin enough to permit excessive flow and absorption into veneer surfaces, nor thick enough to cause application problems or an unduly short pot life.

When a starch filler is used, the material must first be added to water alone and mixed briefly but thoroughly to avoid lumps. Then sodium hydroxide must be added and mixed in. About half the resin is then added, followed by the shell flour, the remaining resin, and the surfactant. After each addition, the composition should be mixed until it is homogenous and lump-free.

The small amount of surfactant performs two important functions. It improves the wetting of the veneer by the glue film, and displaces air from the veneer surface, which maximizes adhesive contact. It limits the depth to which liquid glues are absorbed, thus concentrating adhesive solids more effectively on the glue line. Defoamers are sometimes used; 0.5–1% on resin solids is sufficient.

E. Plywood Manufacturing Variables

The factors of particular importance relative to plywood assembly and pressing are: adhesive application, assembly method, assembly time, prepressing, pressing temperature, platen pressure, and panel construction.

1. *Adhesive Application*

Although curtain coaters, extruders, and spray applicators are popular in some countries, the double-roller spreader is the most frequently used plywood adhesive applicator worldwide. Glue spreads must be uniform and controllable, in a range suitable for wood veneer, namely a single glue line weight of about 150–250 g/m^2. The exact weight depends on the veneer thickness and the properties of the species. Therefore, veneers such as those which are 1.6 mm thick cannot be spread over 150 g/cm^2 without bleeding or blistering. Thick veneers, such as those of 4.2 mm thickness, require a glue spread of about 220 g/cm^2 to balance the absorption and heating rate. This is one reason for not combining thick and thin veneers inside the same panel construction.

2. *Assembly Method*

The assembly method for plywood involves the alternate lay-up of veneers spread with glue and those without glue. The sheets of veneer should be handled with reasonable care so they are not cracked or broken in the process. The pieces of veneer forming each panel should be referenced first and then placed on top of each other to avoid glue films being wiped off, which occurs if the plies are slid over each other. The handling of spread, unpressed panels should be minimized for the same reason.

3. *Assembly Time*

Assembly-time manipulation is an important step in the process of gluing dense and difficult hardwood species successfully. This involves lenthening the assembly time to allow the liquid glue lines to lose water gradually, and become tacky, high-solids films before pressure is applied.

Minimum top panel assembly times of 15–30 min are recommended for hardwood veneers. A corresponding bottom panel assembly time of 45 min to 1 hr or more is recommended. Except for prepressing the panel assemblies to make easier press loading possible and to minimize degradation, the glue lines should be allowed to dry out completely on most dense hardwood veneers. The inherent flow in properly formulated phenolic wood adhesives provides transfer, wetting, and full adhesion in the hot press.

4. *Prepressing*

Because of their high concentration of resin solids, most phenolic wood adhesives are inherently tacky and develop satisfactory prepress bonds, although somewhat slowly. A top panel stand time of 10–15 min before prepressing is recommended for this reason. Starchy materials in the resin or in the glue mix improve prepressing and shorten the required stand time. A prepressing cycle of 10 min at about 5–7 kg/cm^2 is sufficient to develop temporary bonds between most wood veneers. These cold-press bonds generally hold for 1 hr or more. It is possible, to lay up and successfully manufacture plywood panels without prepressing.

5. *Pressing Temperatures*

The main reason for listing hot-press temperature among the critical operating conditions for wood gluing is to emphasize the essential difference in heat input requirements between urea-formaldehyde and phenol-formaldehyde wood adhesives. Urea adhesives are catalyzed on application with a separate addition of acid-reactive material and also by the acidic nature of wood itself. The acid environment on the glue line is capable of taking UF adhesive films to full cure even at ambient temperatures if given sufficient time under pressure. Hot pressing accelerates this curing mechanism. Therefore, urea adhesives can be successfully cured at moderate temperatures, even if the heat rise on the innermost glue lines of a plywood panel is relatively slow. Urea adhesives can also be cured rapidly at high temperatures, thus providing an alternate fast pressing schedule. However, press temperatures above 150°C begin to cause hydrolysis of the cross-linked polymer at a significant rate and should be used with caution.

Alkaline phenol-formaldehyde adhesives do not bond satisfactorily under prolonged pressure an ambient temperature. They simply dry out and leave films of friable adhesive solids because no low-temperature cure mechanism is present. Phenolic adhesives have a definite minimum energy input requirement to cure. For highly advanced and functional softwood plywood resins, this minimum heat input level is 1 min at 100°C on the innermost glue lines of a plywood panel. For typical hardwood phenolic adhesives, the minimum heat input level for satisfactory cure is about 3 min at 100°C, or even longer, depending on the exact resin variables of functionality and flow [60].

Phenolic plywood adhesives require a much higher minimum heat input than do urea resin adhesives in order to cure. The temperature rise on the interior glue lines of a panel must also be more rapid. Some water must be present on the glue lines to permit liquefaction, flow, and transfer before curing. This is important for softwood phenolic adhesives, and to a lesser extent, for hardwood adhesives. Loss of water and drying out occurs more rapidly at intermediate pressing temperatures (where the result is a gradual rise in temperature on inner glue lines) for high-heat-activated phenolic cure. Under these conditions, phenolic glue films dry out and develop inferior, water-sensitive bonds.

Because both factors are related to heat, plywood bonded with phenolic adhesives should be hot pressed at the highest practical temperatures possible for equipment performance and without causing too many steam blisters. (There is no set upper limit to phenolic resin curing temperatures as there is with urea resins.) If blistering becomes excessive, it is recommended that the veneer moisture content be lowered before decreasing the press temperature. Curing temperatures for phenolic adhesives should remain above 125°C except for thin panels. Another practical benefit which can be derived from this minimum hot-press temperature limit is that the steam pressure required to maintain 125°C is generally sufficient to prevent waterlogging of the press platens. This in turn eliminates cold spots and insufficiently cured glue line areas. These tend to occur all too frequently at lower press temperatures when steam presses are used.

6. *Platen Pressure*

High-density veneers need extra pressure to make adequate contact on each glue line in a panel assembly. The minimum platen pressures for proper bonding for dense hardwoods and medium-density softwood are 14 and 12 kg/cm^2, respectively. For rough veneer surfaces or extremely dense wood species, a pressure of 16 kg/cm^2 is the minimum required. The loss of panel thickness due to densification during hot pressing is no higher for heavy hardwoods than it is for softwoods pressed at 12 kg/cm^2 (i.e., 4–6%). High veneer moisture contents cause a rapid increase in press densification and permanent loss of panel thickness. Veneer moisture content should therefore be controlled within fairly close limits.

The hot press should reach full pressure in 1 min or less, preferably 30 sec, to minimize the effects of precure on outer glue lines. This is especially true for very thin, decorative face veneers. Hydraulic booster pumps are available to correct this deficiency and are worth the cost of their installation in terms of the consistent improvement in adhesion obtained.

7. *Panel Construction*

It is not possible or practical to specify the optimum thickness to which veneers should be peeled for the best drying and gluing practice. However, extremes of thick and thin peel should be avoided, except for decorative application (where the panels have a thick core and very thin, decorative faces). The preferred thickness for easy drying, minimum handling losses, and maximized strength should be between 2 and 4 mm. Above that range, veneer tends to become rough and deeply cracked, drying times become too long, and chances of overdrying increase. If the veneer is less than 2 mm thick, handling losses increase substantially unless the veneer has been taped green. Thin plies promote excessive blistering on adjacent glue lines when they are used in the interior of a panel assembly.

Except for these practical limitations, phenolic adhesives can handle almost any normal plywood lay-up to about 50 mm total thickness. However, the thicker the panel, the higher the press temperatures must be to ensure that the heat rise on the center glue lines is enough to produce proper cure instead of dry-out. A temperature of 165°C is not too high for 50-mm panels.

If panel strength in terms of flexure, tension, or shear is a serious design consideration, a uniform lay-up of moderately thin plies (i.e., 2.5–3.5 mm) is suggested, even if one or two extra glue lines have to be added. Where very high strength in one direction is desired, as in truck or container floors, the ply can be applied in a single direction. There should be two cross-plies for splitting resistance and dimensional stability. The cross-ply should be about 1.5 mm thin to minimize the depth of lathe checks on the loose side, therefore avoiding early failure due to rolling shear. Parallel ply constructions of this type present no bonding problems for a properly formulated phenolic adhesive.

Very thin, decorative hardwood veneer faces should not be bonded with phenolic adhesives because of their dard color when cured and the risk of bleed-through. If a decorative grade of plywood needs an exterior-grade adhesive, dried adhesive carrier film should be used. Unfortunately, this is expensive. Liquid phenolic adhesives can be used successfully if the glue mix contains a starchy filler. Glue spreads should be carefully controlled to minimum levels consistent with good adhesion and be allowed a top panel stand time of 15–30 min before hot pressing is carried out.

F. Wood-Related Factors

The principal wood-related factors affecting the bonding of wood veneer with phenolic adhesives are: density and porosity, extractive content, drying conditions, freshness, surface smoothness, thickness control, and surface cleanliness.

1. Density and Porosity

The effect of density on adhesion is more physical than chemical. Dense, nonporous wood surfaces confine the liquefied adhesive to the immediate glue line for longer time under pressure. This causes the adhesive solids to move too much to areas of minimum contact or pressure, which results in poor bonding. Liquid glue lines on lower-density woods lose water into the wood more quickly and become viscous, high-solids films before heat and pressure are applied. In this form, they flow less and present a more concentrated layer of adhesive solids. This produces stronger, more uniform phenolic adhesive bonds.

Conversely, excessively thin glue mixes used on softwoods show not only a fast loss of water into the veneer, but also overpenetration of the whole adhesive solution into it. Far longer assembly times are needed on very dense wood surfaces to immobilize liquid glue lines and to obtain a tacky, concentrated adhesive layer. That is why paper glue lines, such as dried adhesive carrier film, work so well on hardwoods. They are largely free of solvents and the excessive mobility that solvents cause. Although the phenolic solids liquefy again under heat and pressure before proceeding to cure, the film remains uniform because of its high viscosity and the presence of the paper carrier, which acts as a continuous layer of filler.

Dense hardwoods lessen bond durability in another way. On exposure to water or weathering, they develop stronger mechanical separating forces on the glue line than do softer species. Because the extent of wood failure on dense wood surfaces tends to be shallow, bonds are more easily ruptured by these strong cross-grain expansive or warping forces.

Open-grained dense hardwoods are easier to glue than close-grained hardwoods, but they pose one or two problems. The open resin canals at the surface siphon away considerable quantities of liquid or liquefied adhesive as transfer and flow occur. Unless extra resin is applied, not enough is left to bond the heavily lignified ridges of fiber between the canals. Assuming that glue spreads are sufficient and a strong bond is obtained, the wood itself tends to shear next to the glue line through the lathe check area of an adjacent ply on exposure to water or weathering. Wood failure is almost 100%, but physical bond strength is low. With continued weathering, the glue lines in plywood made with these diffuse-porous species tend to deteriorate, thus limiting their useful life. In protected service, however, they remain satisfactorily strong.

2. *Extractive Content*

Most wood species contain various combinations and appreciable quantities of natural latex or oleoresinous materials for wound healing, phenolic substances for heartwood protection, and polysaccharide growth or nutrient compounds in their sap. These substances, which become concentrated on the surfaces of wood veneer through drying, present both physical and chemical barriers to bonding. They form a surface layer which prevents the adhesive from reaching the wood structure. Some or these materials are chemical plasticizers for phenolic resins and promote excessive flow. Others react directly with phenolic polymers to accelerate or retard cure rate. Collectively, they can seriously interfere with normal adhesive behavior and bond development.

3. *Drying Conditions*

The conditions of temperature, time, and humidity under which hardwood veneers are dried can have a profound effect on subsequent efforts to bond them into plywood. High temperatures, low humidity, and long heat exposure all promote thermochemical modifications of the wood surfaces, which destroy bonding sites and reduce adhesion. This is called overdrying and is indicated by a water-repellant veneer surface which normal adhesives will neither wet nor bond. Phenolic glue films cured against overdried veneer surfaces show little transfer.

The following is suggested to preserve bonding capability:

1. Do not use dryer temperatures above 170°C. Even lower temperatures should be used for particular species which are unusually heat sensitive. If a dryer has two zones, the second or dry zone should be about 20–25°C lower than the first or wet zone. These limits are based on the use of moisture detection equipment at the dryer outfeed. If no such instruments are available and veneer is being dried by "feel," the maximum practical dryer temperature is 150°C.
2. Close the dampers on all dryer sections until steam begins to appear at the ends and door seals. By maintaining a high level of humidity in the dryer, gluability is preserved, veneer brittleness is reduced, and the extractive concentration on veneer surfaces is minimized. A partially vented dryer does not use as much energy as a fully vented dryer.
3. Adjust drying times to produce a moisture content range of about 4–8%, with 15–20% redry. This ensures that even the thinnest veneers do not become overdried through excessive heat exposure. Some veneers, rejected because of their high moisture content, can be redried gently, with all heat to the dryer turned off; or they can be interleaved with dry veneers and left for 24 hr to equalize in the pile. The moisture detection equipment is needed to segregate wet from dry veneer at the dryer output and is a necessity for efficient drying.

4. Freshness

Numerous studies have established a definite correlation between the age of the veneer at the time of gluing and the level of adhesion which can be expected. ("Age" is elapsed time between peeling or drying and actual assembly into plywood.) The older the veneer, the poorer the bond. The loss of wood failure after 6 months of storage can be as much as 15–20%.

Remembering that plywood bonds are largely chemical, two things apparently happen to veneer during a long storage time which alter or block bonding sites available for adhesion. One is a slow but steady series of reactions between atmospheric oxygen and the cellulose, lignin, and sap constituents in the wood surface. The net effect is a progressive loss of surface functionality and therefore adhesion. The other is bacterial or fungal attack, which occurs if the veneer is stored green. Moldy wood not only bonds poorly, but is also much weaker. If prolonged veneer storage is unavoidable, dry it first.

5. *Surface Smoothness*

For high-density hardwood veneers, a smooth surface is a necessity. Where there is no surface contact, there can be no adhesion. The pressure in areas where contact does occur becomes so high that glue films which are normally satisfactory are squeezed off into no-contact areas, resulting in joint starvation. Since dense hardwoods do not crush readily, the gaps are not easily eliminated. Therefore, the use of high platen pressures to gain more glue line contact tends to be self-defeating. Using alternately low- and high-density plies is more practical, but even this is not a good substitute for the proper heating and peeling of the wood log.

Good knife maintenance and lathe adjustment ensure that the veneer surfaces does not become woolly and covered with torn fibers or other wood debris. Phenolic adhesives only bond the surfaces presented. If these surfaces are covered 20 fibers deep with damaged wood cells, the adhesive may do an excellent job of regluing the loose fibers, but never quite reach the sound wood underneath. The resulting bonds tend to be weak and shallow but uniform in wood failure.

6. *Thickness Control*

Wood veneer should also be uniformly thick for efficient bonding. The roller spreaders used in plywood mills are equipped with hard-rubber covers to promote long wear and resistance to swelling by fungicidal solutions in the glue. Therefore, they do not readily accommodate roughness or variations in veneer thickness and cause a high percentage of rejection due to incomplete glue coverage. Therefore, a smooth and uniform veneer is a potentially important source of saving of both glue and wood.

Variation in veneer thickness is definitely a hazard in the hot press. Thin veneers tend to carry heavy glue spreads and make low-pressure contact with adjacent ply. Both factors contribute to excessive panel blistering under normal gluing and hot-pressing conditions. Conversely, thick veneers pick up light glue spreads and develop poor bonds due to dry-out.

7. *Surface Cleanliness*

Veneer surfaces should be kept clean to promote the development of strong adhesive bonds. The presence of oil smears, muddy footprints, and sander dust on surfaces to be glued contributes to areas of poor adhesion and subsequent panel failure. The same is true of wood chips, broken knots, and slivers carried onto the glue lines by one or more of the ply in the panel assembly. Each chip or piece of knot can cause a low-pressure area of up to several square inches on the glue line. Therefore, common sense as well as good lay-up practice dictates that debris on veneer surfaces be swept off before boards are placed in a plywood assembly.

G. General Observations on Particleboard Manufacture

In the case of the application of phenolic adhesives to the manufacture of exterior-grade particleboard, the closest attention must be focused on the application of the resin rather than on its formulation. A good phenolic resin for plywood can be successfully used for the manufacture of particleboard once the different conditions of application have been understood. The pressing time of the board varies according to the type of adhesive, its reactivity, and the moisture content of the glued particles. In many cases a light water spray is applied to the top surfaces of the board before prepressing to shorten the pressing time. The light film of water covering the surface is vaporized when it comes in contact with the hot caul sheet of the press and migrates from the surfaces toward the core of the panel, causing a faster increase in temperature and faster cure.

The water spray prevents precuring of the adhesive on the surface of the board during closure of the press before contact with the hot top caul sheet. The wood undergoes a partially irreversible plastic deformation during pressing, caused by the combined action of pressure and heat. Different products can be obtained by varying the type of pressing cycle of the board. Different pressing procedures and diagrams are available. A diagram for industrial three-layer boards may read as follows:

Maximum pressure. 23-27 kg/m^2 is reached as fast as possible after press closure (i.e., 35–50 sec; other processes use pressure as high as 35 kg/cm^2).

Contact with the gauge bars. As a rule contact is made after 60–120 sec from the start of the press closure. The higher the density, the longer the time it takes. This, however, can be reduced by increasing the moisture content of the glued mat.

Steam escape. It is expected to begin 1–3 min after making contact with the gauge bars.

Pressure decrease. After approximately 1½–2 min of maximum pressure, the pressure is slowly decreased until the final pressure on the panel is as low as 2–3 kg/cm^2. This takes place toward the end of the cycle, just before press opening.

This pressing diagram produces a board with high-density face layers, and the shortest possible pressing time, at a given temperature and a low power consumption. The main properties of panels with high-density face layers are the stiffness of the panel; better warp resistance; high dimensional stability; hard, glossy, and shockproof surfaces which need less adhesive for subsequent veneering; and narrow thickness tolerances.

Considerable variations in the properties of the final board can be obtained by varying the moisture contents of surface and core layers, and by using faster resins in the core layer and slower reacting resins in the surface layer. This can also be achieved by varying geometry and sizes of the wood chips, the density of the board, and so on.

Small variations in the manufacture and characteristics of the phenolic resin used do not affect the property of the finished particleboard as extensively as do the factors listed above. Experiments [61] on the correlation of curing and bonding properties of particleboard glued with resol-type phenolic resins by differential scanning calorimetry show that resols tend to reach two endotherm peaks, the first at 65–80°C and the second at 150–170°C. Resols used for particleboard have been shown to begin curing at lower temperatures than those for novolak resins. Resols-glued particleboard show no bond formation at 120°C. At 130°C the resol-glued panels show internal bond strengths of 0.55–0.7 MPa. The internal bond strength for the wet tests increases as the board core temperature goes over 150°C during pressing. The normal temperature for 12- to 13-mm-thick board glued with phenolic adhesive are 170–200°C. The pressing time is 5–5½ min.

The results obtained using phenol-formaldehyde adhesives for particleboard are the following:

	Swelling after 2-hr boil		Internal bond strength		
Density (g/cm^3)	Measured wet (%)	Measured dry (%)	Dry (kg/cm^2)	After 2-hr boil (kg/cm^2)	24-hr cold-water swelling (%)
±0.700	±15	2–3	10–11	5–8	9–11

H. Dry-Out Resistance

One of the more common difficulties in bonding pine veneers and chips is adhesive dry-out. Dry-out is associated with the high liquid absorbancy of pine sapwood and it appears especially during long assembly times. This problem can be overcome by using phenolic resins modified through reaction with alkylated phenols, especially 3,4-xylenol [62]. Another technique used to achieve similar results is the manipulation of synthesis procedures used in preparing a standard phenol-formaldehyde resin [62]. The dry-out resistance imparted by alkylated phenols is due to an initial semithermoplastic character in the resin. This is derived from their monomer bifunctionality and linear polymer that is consequently formed.

If a linear and essentially non-cross-linking prepolymer is prepared from phenol and formaldehyde, it can be coreacted with a nonlinear and cross-linking prepolymer to form a resin. The latter resin will have some initially semithermoplastic or dry flow character, but will be primarily a thermosetting one. The product is an alkaline novolak-resol copolymer. Evaluation of this copolymer concept has shown that many resin possess a controlled initial semithermoplastic character which improves resistance to dry-out. Good dry-out resistance is achieved without loss of press-time efficiency or broad-range bonding ability. Such resins perform noticeably better than other types of resins which are resistant to dry out.

Such a resin of the alkaline novolak-resol type can be prepared by coreacting a prepolymer, prepared by reacting formaldehyde and phenol in the molar ratio 2.6:1.0, and a prepolymer obtained by reacting formaldehyde and phenol in a molar ratio of 1:1. The two prepolymers are then mixed in 50:50 proportions by mass and coreacted.

I. Wood Laminating and Finger Jointing Adhesives

Various adhesive formulation can be used for the manufacture of laminated wood beams and finger joints for structural purposes. Only those adhesive formulations that, at some time or another, have been used extensively in industrial applications will be described. All these formulations are based totally or partially on resorcinol and the process is carried out at ambient temperatures up to 50°C.

1. *Adhesive 1: Resorcinol-Formaldehyde Adhesive*

A resorcinol-formaldehyde novolak is produced accordingly to the following schematic reaction:

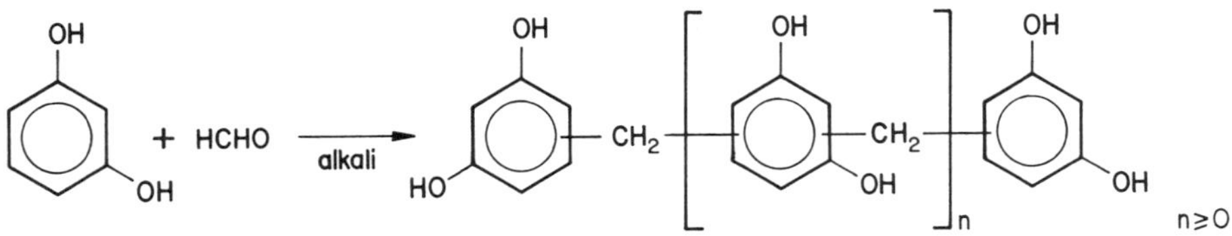

If paraformaldehyde and fillers, generally wood and nutshell flours, are added, the resin becomes capable of setting in 2–3 hr and curing in 16–24 hr at ambient temperature.

2. *Adhesive 2: Phenol-Resorcinol-Formaldehyde Adhesive and Powder Hardener*

A phenol-formaldehyde resol is prepared and resorcinol is grafted onto it according to the following schematic reaction:

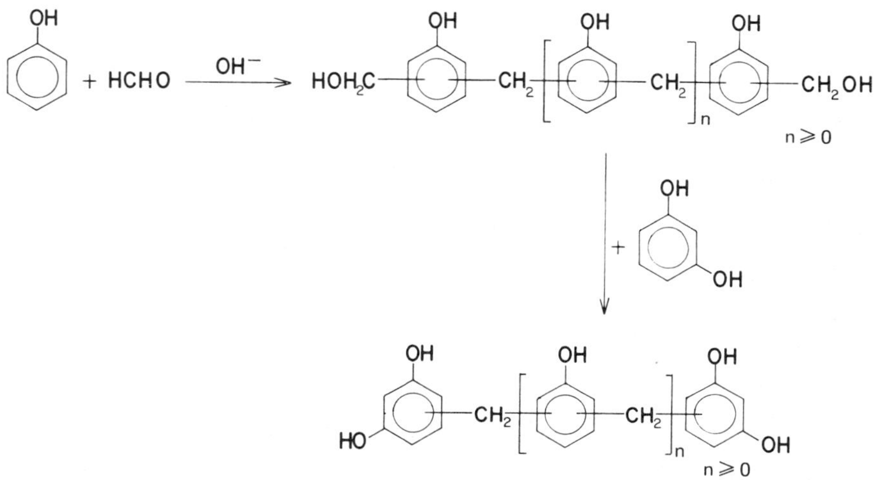

The resin produced is capable of setting in 2–3 hr and cures in 16–24 hr at ambient temperature once paraformaldehyde and flour fillers have been added.

3. *Adhesive 3: Urea-Resorcinol-Formaldehyde Adhesive*

A urea-formaldehyde resin is prepared and resorcinol is grafted onto it as terminal units. The urea-formaldehyde resin can also be purchased ready-make. The schematic reaction is as follows:

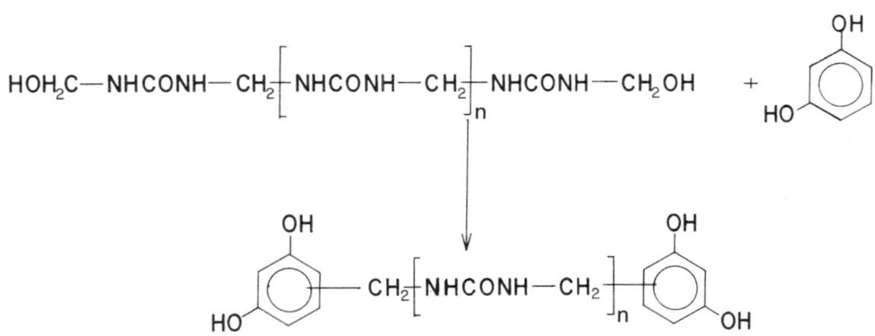

The behavior of this adhesive is identical to that of adhesive 2, although larger amounts of resorcinol are used. These adhesives have a high urea/formaldehyde ratio, are water resistant, and are capable of radio-frequency curing. They can be used also for plywood manufacture.

4. *Adhesive 4: Phenol-Resorcinol-Formaldehyde and Liquid Hardener*

A phenol-formaldehyde resol is prepared which constitutes the resin. Resorcinol chemical or resorcinol formaldehyde novolaks, in aqueous or water-alcohol solutions, are used as hardeners according to the following scheme:

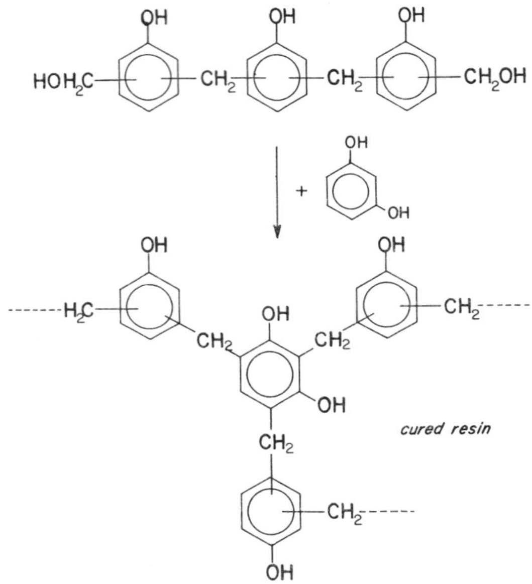

5. *Adhesive 5: Phenol-Resorcinol-Formaldehyde and Liquid Hardener*

A phenol-resorcinol-formaldehyde similar to adhesive 2 is prepared and a phenol-formaldehyde resol of the same type of adhesive 4 is used as hardener. The schematic curing reaction is the following:

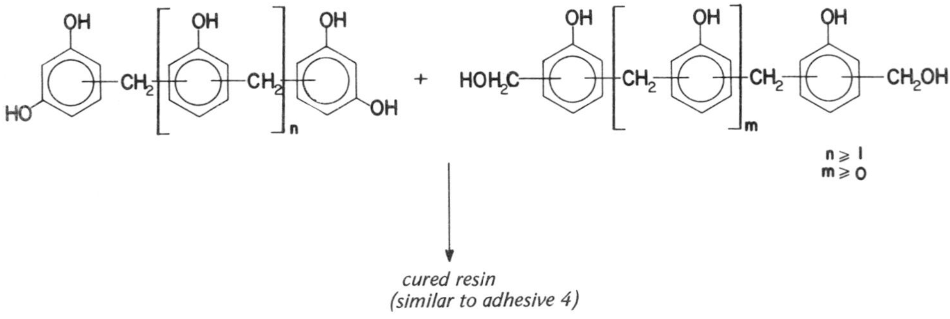

Adhesives of type 2, in which a liquid phenol-resorcinol-formaldehyde adhesive and a powder hardener are used, are currently the most commonly used industrially. Adhesives of type 1, which are by far the best of the five types presented, were used extensively up to a few years ago. They fell into disfavor because of the high price of the resorcinol chemical needed to make them. They are still used industrially when particularly difficult wood-gluing problems arise. Adhesives of types 4 and 5 have been used quite extensively in the past, but have been superseded in the last few years by adhesives of type 2, which have several handling advantages.

Adhesives of type 3 are very new and were developed as an alternative to phenol-resorcinol-formaldehyde adhesives of type 2 due to the ever-increasing price of oil-derived phenol. Adhesives of types 1, 2, and 3 are mixed before use in a mass/mass ratio of liquid adhesive resin (50–60% solid content) to powder hardener of 5:1. The hardener is generally a mixture of 10 parts paraformaldehyde and 10 fillers. It is comprised of wood flour 200 mesh or a mixture of wood flour and nutshell flour also of 200 mesh. Adhesives of types 4 and 5 have a liquid resin/liquid hardener ratio of 1:1 by mass. (This is because the hardener is also a resin.)

All properly formulated resorcinol-based adhesives must have a viscosity low enough in aqueous-alcoholic solutions to flow with ease into all the interstices of the wood surface. Wetting ability is promoted by the alcohol. The paraformaldehyde used as hardener is an addition polymer composed of a few to over 100 formaldehyde monomers. It dissolves slowly in water, by depolymerization to formaldehyde monomers. The rate of depolymerization depends on the degree of polymerization of the paraformaldehyde, the size of the particles, and the pH.

Therefore, the working life or pot life of a glue mix can be adjusted by selecting the type of paraformaldehyde and the pH correctly, as its success depends on the rate of supply of formaldehyde monomer. Fillers are added to give consistency to the glue mix, to control viscosity and thixotropic characteristics, to form a fibrous reinforcement of the adhesive film, and to lessen the cost. Wood flour is used as a filler to obtain better gap filling where rough or uneven surfaces must be bonded, or where low bonding pressures must be used. Nutshell flours, such as coconut shell flour, walnut shell flour, peach pip shell flour, macadamia nutshell flour, or even olive stone flour are used as fillers to lower absorptive qualities and to provide smooth flowing mixtures. Clays and silica smokes can be used in very small amounts to control the thixotropic consistency of the glue mix.

As the formaldehyde reacts with the resorcinol-based resin, condensation occurs, with the formation of high molecular weight polymers. There is considerable interaction between the growing resorcinol polymers and the noncrystalline hemicellulosic and lignocellulosic molecules.

The highly polar methylol groups and the phenolic hydroxy groups link to cellulose and lignin groups by hydrogen and electrostatic bonds. The growing adhesive polymers continue to interact to form colloidal particles and then a gelatinous film. This mechanism depends strongly on the moisture content of the wood, which determines the rate of water and solvent absorption.

The advantage of ambient-temperature curing is that the moisture escapes gradually from the hard film formed on curing, inducing a minimum of residual stresses on the joint, and allowing the glue line to assume the aspect of a molecularly porous solid. As a consequence, the hard film is able to transpire in the same way as wood, which minimizes checking or crazing, and allows the glued joint to survive exposure to the extremes of humidity cycles.

It is worthwhile to describe the different types of cold-setting adhesives in more detail. Adhesives of types 4 and 5 will not be discussed further, as they are now used rarely if at all.

6. *Adhesives of Type 1*

These are prepared from resorcinol by reaction with a deficient amount of formaldehyde. The molar ratio of resorcinol to formaldehyde is generally in the range 1:0.5–1:0.7. Although all resorcinol adhesives can cure efficiently under neutral conditions, it has been found commercially advantageous to cure them under mildly alkaline conditions. The resin usually has a pH of about 8.

A general recipe for these resins is as follows:

Resorcinol, 2500 parts by mass
First formalin 37%, 500 parts by mass
Second formalin 37%, 750 parts by mass
Oxalic acid crystals, 5 parts by mass

The 2500 parts of resorcinol is added to the 500 parts of formalin solution in a reactor furnished with mechanical anchor stirring, cooling and heating equipment, and a reflux condenser. The mixture is gradually and slowly heated up to 100°C. As the resorcinol goes into solution, great care must be taken that the mixture does not boil over. As the temperature reaches 100°C, the heating of the reactor is stopped and 750 parts of formalin solution is gradually added. The temperature is controlled so that it does not reach 110°C, where boiling over could occur.

The second lot of formaldehyde is usually added in increments over $1\frac{1}{4}$–3 hr, according to the size of the reaction mass. During the last addition of formaldehyde to the partially reacted mass, the rate of reaction slows down and about 5–10% of formaldehyde which has not reacted with the resorcinol remains in the reaction mass. In order to combining this "free" formaldehyde without further polymerization of the resin, 5–20 parts of oxalic acid (or acetic or other acids) is added in crystal, liquid, or solution form. The total reaction time varies from $2\frac{1}{2}$ to $3\frac{1}{2}$ hr.

While the second amount of formaldehyde is added, the reaction heating is furnished by the reaction mass itself. When the acid is added, the mass begins to cool. The reaction is then complete. If any dewatering is needed, the mixture is maintained at about 100°C during the distillation of the water and a vacuum of about 710 mm Hg is maintained on the system while the temperature gradually rises with the proceedings of the distillation. The resin can be diluted with methylated spirits or a water-methanol mixture. The resorcinol/formaldehyde molar ratio that is allowed is 1:0.5–0.3, the optimum being 0.67 formaldehyde. Instead of oxalic acid, 0.5–0.6% caustic soda can be used, and is sometimes more convenient, as the resin must be brought to slightly alkaline pH values before it can be used.

7. *Adhesives of Type 2*

The procedure for the preparation of this resin can be modified in many ways by varying catalysts, concentration, molar ratios, and condensation conditions. This results in a wide variety of adhesive products. A few recipes, from relatively easy ones to more complex ones, which may be of use in preparing adhesives, are the following:

Recipe 1

First phenol, 40 parts by mass
Water, 10 parts by mass
Methanol, 4 parts by mass
Formaldehyde, 14 parts by mass
NaOH 40%, 2.75 parts by mass
Resorcinol, 19–25 parts by mass
Second phenol, 24 parts by mass

The first lot of phenol, water, methanol, and paraformaldehyde is charged in a reaction vessel and mildly heated until the phenol is dissolved. Caustic soda solution is added and the reaction becomes exothermic. The reaction mixture is refluxed at 90–92°C for a period of 1–$3\frac{1}{2}$ hr (usually slightly over 1 hr), according to the characteristics wanted in the resin. Resorcinol is added and the temperature drops to about 60°C, followed 0–2 min later by the second lot of phenol. The mixture is again brought to reflux for 1 hr, and cooled and stored after the pH has been adjusted according to the setting time required. When the glue mix has to be prepared, paraformaldehyde and fillers are added.

Recipe 2

Resorcinol, 2250 parts by mass
Phenol, 216 parts by mass
Crude cresylic acids, 250 parts by mass

First 37% formalin solution, 500 parts by mass
Second 37% formalin solution, 750 parts by mass
Caustic soda 40%, 12–60 parts by mass

The reactions can be carried out in three different manners, giving resins with quite different characteristics:

1. As for the reaction of adhesives of type 1, by mixing resorcinol phenol, and crude cresylic acids. (This is a crude tar acid product obtained as a by-product of coal distillation and composed mainly of phenol, cresols, xylenols, and higher-boiling-point phenols). It is then reacted as for adhesives of type 1, or better, as in the following procedure.
2. By mixing phenol and crude cresylic acid and reacting them at reflux for a predetermined time with the first lot of formalin in the presence of part of the caustic soda. The resorcinol is added and the reaction proceeds as for adhesive 1.
3. The mixture of phenol and crude cresylic acids must be reacted with both lots of formaldehyde at reflux for a predetermined time in the presence of a large part or all of the caustic soda. The resorcinol is then added and the mixture refluxed again for 1–2 hr.

<u>Recipe 3</u>

Phenol, 110 parts by mass + 22 parts water
First formalin 37% solution, 49 parts by mass
H_2SO_4 10% solution, 22 parts by mass
First NaOH 40% solution, 4.5 parts by mass
Second NaOH 40% solution, 9.25 parts by mass
Second formalin 37% solution, 90–93 parts by mass
Methanol or methylated spirits, 30 parts by mass (at start of reaction)
Resorcinol, 71 parts by mass
Tannin extract, 19 parts by mass

Phenol, water, methanol, and the first amount of formalin solution are charged on the reaction vessel and mildly heated until the phenol is dissolved. H_2SO_4 is added and the temperature increased to reflux. The mixture is refluxed for 3.5-4 hr (generally ±4 hr). It is cooled to 50–60°C and the following is added: The two amounts of NaOH 40% solution (slowly) and the second amount of formalin solution. The mixture is refluxed for 4.5-4.75 hr, and then resorcinol is added. The mixture is refluxed for a further 30–50 min. Spray-dried wattle tannin extract is added immediately before or during cooling to adjust viscosity. The pH must be adjusted to 8.0–9.5 according to the pot life required. The hardener is a 50:50 mixture of paraformaldehyde 96% (usually a fast grade) and 180–200 mesh wood flour (60:40 mass proportion).

The results obtainable with phenol-resorcinol-formaldehyde cold-setting resins tested according to the British Standard BS 1204–1965 specification (carried out on beech strips) are usually and approximately the following:

Dry test		Cold soak test		6-hr boil test	
Shear strength (psi)	Wood failure (%)	Shear strength (psi)	Wood failure (%)	Shear strength (psi)	Wood failure (%)
800–850	90–100	700–800	90–100	550–700	90–100

8. *Adhesives of Type 3*

A wide variety of products can be obtained by varying the reaction conditions, especially of the urea-formaldehyde resin. Useful recipes for their preparation can be gleaned from the work of Blommers and Moult [63]. Examples of these types of resins are as follows:

a. Preparation of Resorcinol-Terminated Urea-Formaldehyde Adhesive (1): First, 220 g of resorcinol (2M), 220 g of H_2O, and 1.2 g of *para* toluenesulfonic acid are charged in a 1-liter flask equipped with a stirrer, thermometer, and condenser. The pH is approximately 1. Then 72 g of powdered dimethylol urea (0.6 M) is added slowly through an addition funnel. This takes 2½ hr, during which the temperature is not allowed to rise above 21°C. After this, the solution is stirred at room temperature (25°C) for 1 hr. The pH is adjusted to 7.1 with 50% NaOH, and the solution is poured into an evaporating dish to remove water in a forced draft oven set at 50°C.

The viscous resin is concentrated to 89.7% solids. On dilution with water, an emulsion is formed. An adhesive mix is prepared by blending 100 parts by weight of this resorcinol-terminated urea-formaldehyde resin with 5 parts of wood flour filler and 11 parts of paraformaldehyde at a pH of 8.7.

b. Preparation of Resorcinol-Terminated Urea-Formaldehyde Adhesive (2): The following are charged in a 1-liter flask equipped with stirrer, thermometer, and condenser: 110 g of resorcinol (1 M), 55 g of H_2O, 41.3 g of methanol, 13.7 g of ethyl alcohol with a small amount of methyl ethyl ketone, and 1 g of *para*-toluenesulfonic acid. Next, 72 g of powdered dimethylol urea (0.6 M) is slowly added through an addition funnel. The addition takes 5½ hr during which the temperature is not allowed to rise above 22°C. The reaction mixture is then stirred at room temperature (23°C) for 45 min. The pH is adjusted to 7.9 with 50% NaOH, and the solution is stored.

The resin solution is approximately 58% solids. On dilution with water, a sticky resin is formed. An adhesive mix is prepared by blending

66.8 parts by weight of resin with 33.2 parts of water, 7.5 parts of wood flour filler, and 11 parts of paraformaldehyde.

c. Preparation of Resorcinol-Terminated Dimethylol Methylenediurea Adhesive

1. Preparation of methylenediurea: A 4-liter flash equipped with a stirrer and condenser is charged with 1.540 g of urea, 1.540 ml of H_2O, and 250 g of 37% formalin solution. After being stirred well, the mixture is allowed to remain at room temperature for 2 days. The mixture is neutralized to pH 7 with 50% NaOH and the heavy white precipitate is collected by filtration. The impure product is dried (925 g yield) and then extracted with 1500 ml of ethanol. The alcohol-insoluble portion is dired. The yield of white powder is 471 g (m.p. > 270°C).

2. Preparation of dimethylol methylenediurea: A 300-ml resin flask equipped with stirrer, thermometer, and condenser is charged with 32.4 g (0.4 M) 37% formaldehyde (MeOH) and the pH adjusted to 8.5 with 50% NaOH. 26.4 g (0.2 M) of purified methylenediurea is added to this solution over a period of 1½ hr. The temperature is kept at 22–23°C. Next, 3 ml of ethanol and methyl ethyl ketone and 5 ml of H_2O are added. The pH is adjusted to 8 with 50% NaOH and allowed to stay at room temperature overnight. The product is now a paste.

3. Preparation of resorcinol-terminated dimethylolmethylenediurea (resorcinol/DMMDU 1:0.6): A 300-ml flask equipped with stirrer, thermometer, and condenser is charged with 36.7 g of resorcinol (0.33 M), 36.7 g of H_2O, and 0.5 g of *para*-toluenesulfonic acid. The pH is approximately 1. The pasty methylenediurea, thinned with 37 ml of ethanol, is slowly added to this solution. This addition takes 2 hr and the temperature is not allowed to go above 32°C. The milky solution is then stirred at room temperature for 45 min. The pH is adjusted to 7 with 50% NaOH and water is removed overnight in a forced-draft oven set at 45°C. The highly viscous product has a solids content of 90.5%. The percentage of free resorcinol is 7.27%.

d. Preparation of a Resorcinol-Modified Urea-Formaldehyde Chain

1. Preparation of urea-formaldehyde resin: A 300-ml flask equipped with stirrer, thermometer, and condenser is charged with 60 g of urea (1 M), and 56.7 g (0.7 M) of 37% formaldehyde (<1% MeOH). It is heated to 26°C and 10 ml of H_2O is added. The urea dissolves totally. At this point 1 drop of 10% H_2SO_4 is added to adjust the pH to 3–4. Heat is stopped and the temperature rises to 44°C over a period of 20 min. An ice bath is applied, when needed, to maintain the temperature at around 40°C for another 45 min. The milky suspension is finally cooled at room temperature and dried by evaporation at room temperature.

2. Resorcinol modification: A 300-ml resin flask equipped with stirrer, thermometer, and condenser is charged with 25 g of the resin prepared in step 1 above, and 25 g of water. Next, 1 drop of 50% NaOH (pH now 8–9) and 20 ml of water are added. A paste is formed to which 10 ml of 37% formaldehyde (the pH having been previously adjusted to 8) is added. This mixture is left at room temperature overnight. The thick paste which results is mixed with 22 g of resorcinol dissolved in 22 ml of water at a pH of 8. Heat is applied and the temperature is maintained at 50–54°C for 6 hr. During this time, the pH is checked to ensure that it is in the region 7–8. At the end of the heating period, the emulsion formed is cooled to room temperature and stored. After 3 months at room temperature the emulsion is still usable.

e. Preparation of Resorcinol-Terminated Sulfite-Modified UF Resins

1. *Sulfite modification*: A 1-liter resin flask equipped with stirrer, thermometer, condenser, and sampling tube is charged with 150 g of urea (2.5 M), 406.2 g (5.01 M) 37% formaldehyde (<1% MeOH), and 30.6 g of $Na_2S_2O_5$. The pH is adjusted to 9–10 with 10% Na_2CO_3. The mixture is heated to reflux and kept there for 30 min. It is then cooled to 53°C and 50% formic acid added until a pH of 4 is reached. Condensation is continued until a viscosity of H (on the Gardner scale) is reached. The reaction mixture (which is now an emulsion), is cooled to room temperature and the pH adjusted to 7–8 with 50% NaOH.

2. *Resorcinol termination*: A 300-ml resin flask equipped with stirrer, thermometer, and condenser is charged with 55 g of resorcinol (0.5 M), 55 g of H_2O, and 1.0 g of *para*-toluenesulfonic acid. To this solution 100 g of UF polymer from step 1 above which is of approximately 53% solids, is slowly added over 2 hr. The temperature is maintained at 19–21°C. When half the UF resin has been added, a precipitate is formed. When the addition is complete, a great deal of resin is present which has not dissolved at pH 8. When the mixture is dried overnight in a forced-draft oven set at 50°C, a white solid remains which is soluble enough to form a solution of 60% concentration in 50% dioxane.

f. Preparation of Resorcinol-Terminated Sulfite-Modified Dimethylol, Methylenediurea

1. *Sulfite modification of methylenediurea*: A 1-liter resin flask equipped with stirrer, thermometer, and condenser is charged with 132 g of purified methylenediurea (1 M), 162.2 g (2 M) of 37% formaldehyde (<1% MeOH), enough 50% NaOH to adjust the pH to 9, and 12.24 g of sodium metabisulfite. The pasty blend is stirred while being heated. In 2½ hr the temperature rises to 69°C, but the mixture remains pasty. Next, 60 ml of water is added and the heat increased. After another hour, the mixture is at 97°C and practically a clear solution. Heating is continued for another 20 min. The solution is then allowed to cool at room temperature. After 1½ hr the milky solution is at 35°C with a pH of 6–7. The pH is adjusted to 8 with 1 drop of 50% NaOH and allowed to stay like this overnight. The paste turns into a cream through the addition of 20 ml of water.

2. *Resorcinol termination at 1/0.3*: A 1-liter flask equipped with a stirrer, condenser, and thermometer is charged with 110 g of resorcinol (1 M), 110 g of water, and 1 g of *para*-toluenesulfonic acid. The

temperature is kept between 15 and 20°C. Their 116.1 g (0.3 M) of dimethylol methylenediurea polymer of step 1 is slowly added over 3 hr. The hazy solution that results is stirred at room temperature for 45 min. The pH is adjusted to 7 and the product poured into an evaporating dish for drying overnight in a forced-draft oven set at 50°C. The moderately viscous resin has a solids content of 90.5%. On dilution with water, a haze forms. The percentage of free resorcinol (dry) is 33.5%.

3. *Resorcinol termination at 1/0.6*: The same procedure described in step 2 is followed except that 232.2 g (0.6 M) of step 1 dimethylol methylenediurea polymer is added. The viscous resin that remains after drying has a solids content of 67.4%. On dilution with water, an emulsion forms. The percentage of free resorcinol (dry) is 13.3%.

g. Single-Kettle Preparation of Resorcinol-Terminated UF Resins

1. *Sulfite modification*: A 1-liter resin flask equipped with a stirrer, thermometer, and condenser is charged with 150 g of urea (2.5 M), 406.2 g (5.01 M) of 37% formaldehyde (<1% MeOH), and 30.6 g of $Na_2S_2O_5$. The pH is adjusted to 9–10 with 50% NaOH. The reaction mixture is heated to reflux and kept there for 30 min. On cooling to room temperature it becomes milky, with some precipitation.

2. *Base catalysis of 62 mol% resorcinol*: A 300-ml flask equipped with a stirrer, thermometer, and condenser is charged with 55 g of resorcinol and 40 g of water. This solution is adjusted to pH 8 with 50% NaOH. Next, 70.75 g of well-stirred sulfite-modified dimethylol urea prepared in part 1 is added. The mixture is heated to 55°C. A temperature of 50–56°C is maintained for a period of 6 hr, during which 50% NaOH is frequently added to keep the pH at 8. The final solution has 53% solids and a pH of 7.5–8.0. A sample of the reddish solution placed in a 45°C stability test oven forms a gel in 1 day.

3. *Acid catalysis of 62 mol% resorcinol*: A 2-liter flask equipped with a stirrer, thermometer, and condenser is charged with 150.5 g of urea (2.5 M) and 406.2 g of 37% formaldehyde (5.0 M). After adjusting the pH to 8 by adding 6 drops of 50% NaOH, the solution is mixed with 30.0 g of $Na_2S_2O_5$. Adjustment to pH 9–10 is made with 2 drops of 50% NaOH. The mixture is heated to 98°C and refluxing is maintained for 30 min. The heat is removed and the solution is allowed to cool at 61°C. At this point 100 ml of water is added and a cooling bath applied to drop the temperature to 42°C. When 458.4 g of resorcinol (4.2 M) is added, the temperature drops to 19°C. On adding 4.5 g of *para*-toluenesulfonic acid, the temperature begins to rise. The temperature is not allowed to go above 39°C for 5 hr. The principal range is 28–33°C. Adjustment of the pH from 1–2 to 7.2 is made with 50% NaOH. The percentage free resorcinol on a dry basis is 17.3%.

4. *Acid catalysis of 64.5 mol% resorcinol*: Case 4 is prepared as for case 3, except that 500 g of resorcinol (4.5 M) is used, together with 5.0 g of *para*-toluenesulfonic acid. The final pH is 7.4. If a sample of the reddish solution (62.9% solids) is placed in a 45°C sta-

bility oven, it changes in viscosity in 30 days from 2.7 sec to 6.2 sec. The percentage of free resorcinol (dry) is 19%.

5. *Acid catalysis of 66.7 mol% resorcinol*: Case 5 is prepared as for case 4, except that 100 ml of water, 550 g of resorcinol (5 M), and 5.5 g of *para*-toluenesulfonic acid are used. The final pH is 7.4. A sample of this reddish solution (62.2% solids) is placed in a 45°C stability test oven. It changes its viscosity in 30 days from 1.5 sec to 1.9 sec. The percentage of free resorcinol (dry) is 22%.

6. *Acid catalysis of 71.4 mol% resorcinol*: A 2-liter resin flask equipped with stirrer, thermometer, and condenser is charged with 150 g of urea, 406.2 g of 37% formaldehyde, (<1% MeOH), and 6 drops of 50% NaOH. Vigorous stirring is maintained throughout the preparation. At pH 8–9, 30.6 g of sodium metabisulfite and 2 drops of 50% NaOH are added. The mixture now has a pH of 9–10. It is heated to bring the mixture to reflux. A temperature of 98–100°C is maintained for 30 min, after which the solution is cooled to 35–40°C (1 hr). Next, 30.3 g of water and 684.8 g of resorcinol are added. The temperature is lowered to about 11°C, and then slowly increased to 15°C (20–30 min). Their 3.3 g of *para*-toluene sulfonic acid is added. A mild exotherm carries the temperature to 18°C in 10 min. Another 3.3 g of acid is added (pH 1–2), and an ice bath is used to maintain the reaction temperature below 26°C. The reaction mixture is kept between 20 and 26°C for an additional 4½ hr. At this point 9.2 g of 50% NaOH is used to adjust the pH to 7.1. The product is a clear red solution with a solids content of 60%. The percentage of free resorcinol (dry) is 27.5%.

7. *Acid catalysis of 76.9 mol% resorcinol*: Case 7 is prepared as for case 6 except that 300 ml of water, 9.6.7 g of resorcinol (8.3 M), and 9 g of *para*-toluenesulfonic acid are used. The final pH is 7.1. A sample of this reddish solution (64.5% solids), if stored for 30 days in a 45°C stability oven, changes in viscosity from 1.0 sec to 1.1 sec. The percentage of free resorcinol (dry) is 34.8%. The glue mixes, in parts by mass used [63] as an example for the final use of all these resins, are as follows:

Resin 7	100							
Resin 6		100	100	91.6				
Resin 6					100	100		
Resin 5							96.5	96.5
H_2O				8.4			3.5	3.5
Wood flour	8.0	7.4	7.4		7.4			
Walnut shell flour				18		14	13.5	13.5
Paraformaldehyde	11	11	11	11	11	11	11	9
pH	6.7	7.3	6.6	7.1	7.2	7.2	7.3	7.2

Other recipes, for adhesives of type c [64] using commercial UF resin syrups, are the following:

Resorcinol, 26.0 parts by mass
Methanol, 3.5 parts by mass
Water, 7.0 parts by mass
Urea-formaldehyde 64% syrup (commercial BASF K285), 20.0 parts by mass
Caustic soda 40% solution, 8.0 parts by mass

Process 1. Resorcinol, methanol, water, UF syrup, and caustic soda solution are mixed at ambient temperatures for 1 hr. The pH is adjusted according to the pot life required, but it is always kept alkaline.
Process 2. Resorcinol, methanol, water, UF syrup, and caustic soda solution are mixed and refluxed under stirring for 30 min at 86°C. The pH is adjusted according to the pot life required.

All of these adhesives can be hardened at ambient temperatures or by radio-frequency curing once a hardener composed of paraformaldehyde 96% powder and wood flour filler has been added to the liquid resin. The pH of the resin must be adjusted according to the pot life and hardening rate required. The results obtained with these resins, when they are well formulated, are comparable to those of adhesives of type b.

J. Fast-Setting Adhesives for Finger Jointing

Together with the more traditional finger jointing adhesives which have just been discussed, a series of ambient temperature fast-setting systems have also been developed. These eliminate the long delays caused by the use of more conventional phenol-resorcinol-formaldehyde adhesives which require lengthy periods to set. These types of phenolic adhesives are applied separately. They were first developed in the United States [65–68] to glue large components where presses were impractical. Kreibich [68] describes these separate applications or "honeymoon" systems as follows: Component A is a slow-reacting phenol-resorcinol-formaldehyde resin with a reactive hardener. Component B is a fast-reacting resin with a slow-reacting hardener. When A and B are mated, the reactive parts of the components react within minutes to form a joint which can be handled and processed further. Full curing of the slow-reacting part of the system takes place with time. Usually, component A is a traditional phenol-resorcinol-formaldehyde cold-setting adhesive at its standard pH to which paraformaldehyde hardener has been added. Flour fillers may be added or omitted from the glue mix. Component B is a phenol/*meta*-aminophenol/formaldehyde resin with a very high pH (and therefore a high reactivity), which contains no hardener or only a very slow hardener. The schematic reactions in the preparation of component B are the following:

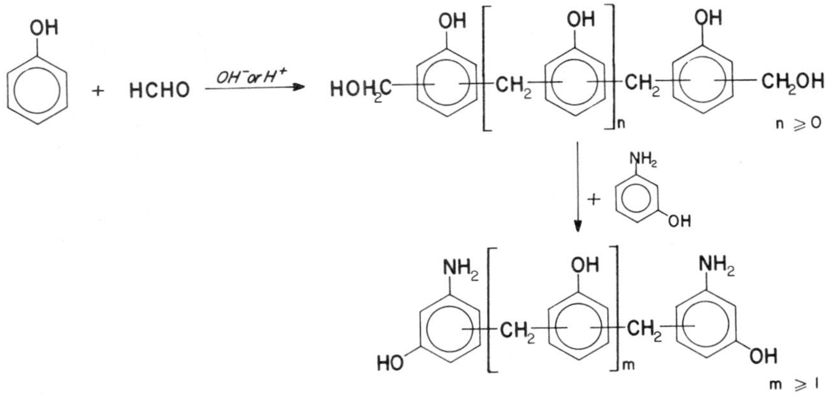

Meta-aminophenol (or *meta*-hydroxyaniline) is much faster and reacts with formaldehyde rather than resorcinol. Consequently, once the component A glue mix is spread on one finger joint profile and component B is spread on the other finger joint profile, and the two profiles are joined under pressure, the reaction of component B with the hardener of part A is very fast. Finger joints prepared with these adhesives generally reach in 30 min at 25°C the levels of strength which finger joints glued with more conventional phenolic adhesives are able to reach only after 6 hr at 40–50°C, or in 16–24 hr at 25°C.

More recently, a modification of the system described by Kreibich has been used in industry with good success. The part A of the adhesive is again a phenol-resorcinol-formaldehyde (PRF) cold-setting adhesive, with powder hardener added at its standard pH. Part B can be either the same PRF adhesive with no hardener, and the pH adjusted to 10–11, or a 50–55% tannin extract solution at a pH of 11–12, provided that the tannin is of the condensed or flavonoid type, such as wattle, quebracho, hemlock, or pine extract, with no hardener [69,70]. The results obtained with these two systems are good and the resin has not only all the advantages of the ones described above, but it is also as a result of the elimination of *meta*-aminophenol and/or the use of vegetable tannins, considerably cheaper [69,70].

REFERENCES

1. A. Pizzi, R. M. Horak, D. Ferreira, and D. G. Roux, *Cellul. Chem. Technol.* *13*:753 (1979).
2. N. J. L. Megson, *Phenolic Resin Chemistry*, Butterworths, London, 1958.
3. E. J. W. Whittaker, *Acta Crystallogr.* *6*:714 (1953).
4. H. S. Lilley and D. W. J. Osmond, *Proc. XIth Int. Chem. Congr.*, London, 1947, Abstr. 101/11P.

5. R. E. Richards and H. W. Thompson, *J. Chem. Soc.*, 458 (1947).
6. M. Szwarc, *Disc. Faraday Soc. 2*:39 (1947).
7. H. von Euler, E. Adler, J. O. Cedwall, and O. Törngren, *Ark. Kemi Mineral. Geol. 15A*(11) (1942).
8. K. Hultzsch, *Kunststoffe 32*:69 (1942).
9. E. Ziegler, *Ber. Dtsch. Chem. Ges. 74*:841 (1941).
10. H. von Euler, E. Adler, G. Eklund, and O. Törngren, *Ark. Kemi Mineral. Geol. 15B*(9) (1941).
11. E. Adler, *Ark. Kemi Mineral. Geol. 14B*(23) (1941).
12. M. S. Nickerson, *Br. Plast. 19*:345 (1947).
13. D. du T. Rossouw, A. Pizzi, and G. McGillivray, *J. Polym. Sci., Polym. Chem. Ed. 18*:3323 (1980).
14. V. N. Vorozhtov and E. N. Yurvigina, *J. Gen. Chem. USSR 1*:49 (1931).
15. E. Ziegler, *Oesterr. Chem.-Ztg. 49*:92 (1948).
16. E. Ziegler, *Mh. Chem. 78*:334 (1948).
17. C. C. Price and A. N. Sachanan, *Ind. Eng. Chem. 40*:922 (1948).
18. T. Shono, *Proc. World Eng. Congr., Tokyo, 31,* 533 (1931).
19. K. Hultzsch, *Chem Ber. 82*:16 (1949).
20. A. Zinke, *J. Appl. Chem. 1*:257 (1951).
21. H. L. Bender, *Mod. Plast. 30*:136 (1953); *31*:115 (1954).
22. D. A. Fraser, R. W. Hall, and A. J. L. Raum, *J. Appl. Chem.* 7:676 (1957).
23. A. Pizzi, *J. Appl. Polym. Sci. 24*:1247, 1257 (1979); *J. Polym. Sci., Polym. Lett. Ed. 17*:489 (1979).
24. A. E. Martell and M. Calvin, *Chemistry of Metal Chelate Compounds,* Prentice-Hall, Englewood Cliffs, N.J., 1952.
25. R. W. Kluiber, *J. Am. Chem. Soc. 82*:4839 (1960).
26. A. Pizzi and P. van der Spuy, *J. Polym. Sci. Polym. Chem. Ed. 18*:3477 (1980).
27. M. S. Akutin, A. N. Ustkachkints, B. V. Yakobson, B. A. Potapov, I. I. Zhuravleva, and G. B. Kopteva, French Patent Application No. 72-35931, publication No. 2-372-182 (1976).
28. B. W. Nordlander, *Oil Paint Drug Rep. Yearb. 130*(3):27 (1936).
29. M. M. Sprung, *J. Am. Chem. Soc. 63*:334 (1941).
30. H. von Euler, E. Adler, and B. Bergström, *Ark. Kemi Mineral. Geol. 14B*(25) (1941).
31. T. T. Jones, *J. Soc. Chem. Ind. Lond. 67*:196 (1948).
32. K. B. Goldblum, *J. Soc. Chem. Ind. Lond. 69*:102 (1950).
33. L. N. Debing, G. E. Murray, and R. J. Schatz, *Ind. Eng. Chem. 44*:672 (1943); *47*:55 (1944).
34. M. Yanagita, *J. Soc. Chem. Ind. Jap. 45*:1086, 1297 (1942); *46*:672 (1943); *47*:55 (1944).
35. J. H. Freeman and C. W. Lewis, *J. Am. Chem. Soc. 76*:2080 (1954).

36. J. I. de Jong and J. de Jonge, *Recl. Trav. Chim. Pays-Bas 72*497 (1953).
37. J. S. Fitzgerald and R. J. L. Martin, *Aust. J. Chem. 8*:194 (1955).
38. H. Kämmerer, *Macromol. Chem. 8*:72, 85, 98 (1952).
39. H. Kämmerer and M. Grossman, *Angew. Chem. 62*:263 (1953).
40. H. C. Malhotra and V. Kumar, *J. Appl. Polym. Sci. 23*:3757 (1979).
41. G. R. Sprengling and J. H. Freeman, *J. Am. Chem. Soc. 72*: 1982 (1950).
42. K. Hultzsch, *Chemie der Phenolharze,* Springer-Verlag, Berlin, 1950.
43. A. A. K. Whitehouse, E. G. K. Pritchett, and G. Barnett, *Phenolic Resins*, Jliffe, London, 1967.
44. A. A. Vanscheidt and N. N. Kuznetsova, *Zh. Prikl. Khim. 30*(13):1830 (1957).
45. R. Houwink, *Trans. Faraday Soc. 32*:122 (1936).
46. E. G. K. Pritchett, *Chem. Ind. (Rev.) 81* (1948); 295 (1949).
47. Y. E. Doroshenko, V. V. Korshak, and V. A. Sergeev, *Sov. Plast. (Engl. Transl.)*, 13 (Aug. 1966).
48. R. Houwink, *J. Soc. Chem. Ind. Lond. 55*:247 (1936).
49. R. A. V. Raff and B. H. Silverman, *Can. Chem. 29*:857 (1951).
50. A. R. Ingram, *Can Chem. 29*:863 (1951).
51. C. T. Liu and T. Naratsuka, *Mozukai Gakkaishi 15*:79 (1969).
52. P. H. Rhodes, *Mod. Plast. 24*(12):145 (1947).
53. G. G. Marra, *For. Prod. J. 6*:97 (1956).
54. W. E. St. Clair and R. H. Moult, U.S. Patent No. 3,133,034 (1945).
55. G. T. Tiedeman and M. R. Sanclemente, *J. Appl. Polym. Sci. 17*:1813 (1973).
56. G. T. Tiedeman, M. R. Sanclemente, and H. A. Smith. *J. Appl. Polym. Sci. 17*:1819 (1973).
57. A. A. Vanscheidt and O. N. Simonova, *Plast. Massy Sb. Stat.* 129 (1939).
58. R. T. Conley and J. F. Bieron, *J. Appl. Polym. Sci. 8*:103 (1963).
59. H. G. K. Pritchett, *Trans. Plast. Inst. Lond. 18*:30 (1950).
60. A. L. Lambuth, IUFRO Conf. Wood Gluing Working Party s5.04.07, Merida, Venezuela, Oct. 1977.
61. S. Chow and P. R. Steiner, *For. Prod. J. 29*:11 (1979).
62. C. M. Chen and J. T. Rice, *For. Prod. J. 26*(6):17 (1976).
63. E. A. Blommers and R. H. Moult, U.S. Patent No. 4,032,515 (1977).
64. A. Pizzi, *J. Appl. Polym. Sci. 23*:2777 (1979).
65. G. F. Baxter and R. E. Kreibich, *For. Prod. J. 23*(1):17 (1973).
66. R. W. Caster, *For. Prod. J. 23*(1):26 (1973).

67. H. Ericson, *Papper Och Trä 1*:19 (1975).
68. R. E. Kreibich, *Adhes. Age 17*, 26 (1974).
69. A. Pizzi, D. du T. Rossouw, W. Knuffel, and M. Singmin, *Holzforsch. Holzverwert. 32*, 6, 140 (1980).
70. F. A. Cameron and A. Pizzi, *CSIR Spec. Rep. Hout 207*, Pretoria, South Africa, 1981.

4

Tannin-Based Wood Adhesives

A. PIZZI / *National Timber Research Institute, Council for Scientific and Industrial Research, Pretoria, South Africa*

I. INTRODUCTION

The word "tannin" has been used loosely to define two different classes of chemical compounds of mainly phenolic nature: hydrolizable tannins and condensed tannins. The former, including chestnut, myrobolans (*Terminalia* and *Phyllantus* tree species), and dividivi (*Caesalpina coraria*) extracts, are mixtures of simple phenols such as pyrogallol and ellagic acid and of esters of a sugar, mainly glucose, with gallic and digallic acids.

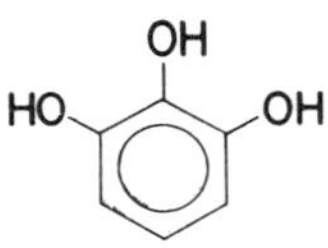

I

Pyrogallol

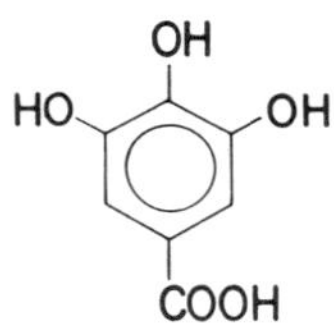

II

Gallic acid

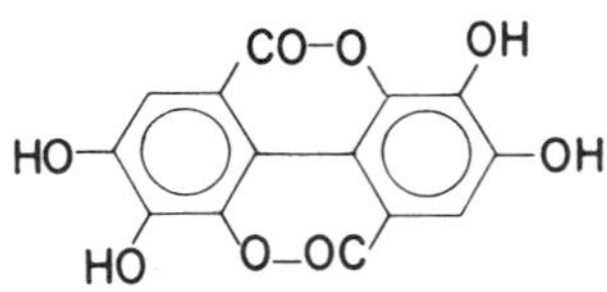

III

Ellagic acid

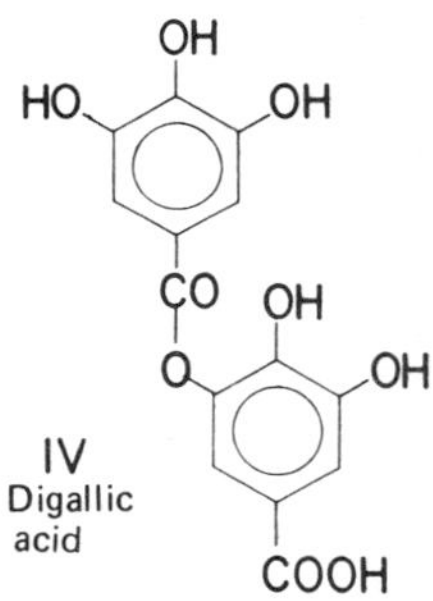

IV

m - Digallic acid

They can and have been used, sucessfully, as partial substitutes of phenol in the manufacture of phenol-formaldehyde resins [1, 2]. Their chemical behavior is analogous to that of simple phenols of low reactivity toward formaldehyde and their moderate use as phenol substitutes in the above-mentioned resins does not present difficulties. Their lack of macromolecular structure in their natural state, the low level of phenol substitution they allow, their low nucleophilicity, and limited worldwide production somewhat decrease their chemical and economical interest.

Condensed tannins, on the other hand, constituting more that 90% of the total world production of commercial tannins (±350,000 tons per year), are both chemically and economically more interesting for the preparation of adhesives and resins. Condensed tannins and their flavonoid precursors are known for their wide distribution in nature and particularly for their substantial concentration in the wood and bark of various trees. These include various *Acacia* (wattle of mimosa bark extract), *Schinopsis* (quebracho wood extract), *Tsuga* (hemlock bark extract), and *Rhus* (sumach extract) species, from which commercial tannin extracts are manufactured, and various *Pinus* (pine bark extract) species not yet commercially exploited. Where bark and wood of trees were found to be particularly rich sources of condensed tannins, commercial development ensued through large-scale afforestation and/or industrial extraction, mainly for use in leather tanning. The production of tannins for leather manufacture reached its peak immediately after World War II and has since progressively declined. Thus this progressive decline of their traditional market coupled with the increased price and decreased availability of synthetic phenolic materials due to the advent of the energy crisis stimulated fundamental and applied research on the use of such tannins as a source of condensed phenolics. The purpose of this chapter is to bring together and review critically the work published on the various fundamental and applied aspects of these polymers and to show how the fundamental knowledge of a chemical subject can bring interesting and extensive industrial applications in unusual and unthought-of fields.

II. CHEMISTRY OF CONDENSED TANNINS

A. General

Condensed tannins, consisting of flavonoid units which have undergone varying degrees of condensation, are invariably associated with their immediate precursors (flavan-3-ols, flavan-3,4-diols), other flavonoid analogs [3–5], carbohydrates, and traces of amino and imino-acids [6]. Monoflavonoids and nitrogen-containing acids are present in concentrations which are too low to influence the chemical and physical characteristics of the extract as a whole. However, the simple carbohydrates (hexoses, pentoses, and disaccharides) and complex glucuronates (hydrocolloid gums) are often present in sufficient quantities to decrease and increase viscosity, respectively, and excessive variation in their percentages would alter the physical properties of the natural extract independently of contributions related to the degree of condensation of the tannins.

B. Monoflavonoids

Monoflavonoids, commonly known as "phenolic nontannins," represent the most studied group in the commercially important tannin extracts because of their relative simplicity. They comprise flavan-3, 4-diols (leucoanthocyanidins), flavan-3-ols (catechins), dihydroflavonoids (flavonols), flavonones, chalcones, and coumaran-3-ones, thus representing most of the known classes of flavonoid analogs [3–5, 7–9]. Typical are those of black wattle (mimosa) bark extract (*Acacia mearnsii*), where the four possible combinations of resorcinol and phloroglucinol (A rings) with catechol and pyrogallol (B rings) coexist, although these flavonoids constitute a minor percentage (3%) of the total phenolics [3].

Among the groups of monoflavonoids noted above, only flavan-3,4-diols and certain flavan-3-ols do appear to participate in tannin formation. Each of the four combinations of phenolic substitution is to be found in black wattle tannins [10]. The main polyphenolic pattern is represented by flavonoid analogs based on resorcinol A rings and pyrogallol B rings (XX). These constitute about 70% of the tannins. The secondary but parallel pattern is based on resorcinol A rings and

A. Flavan-3,4-diols

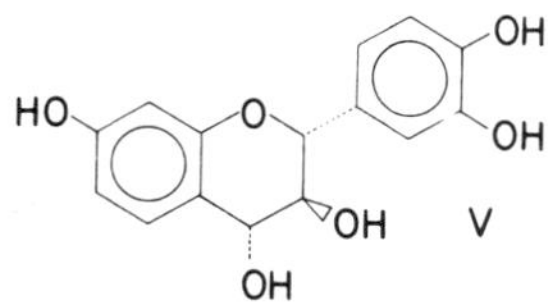

V

(+) - leucofisetinidin

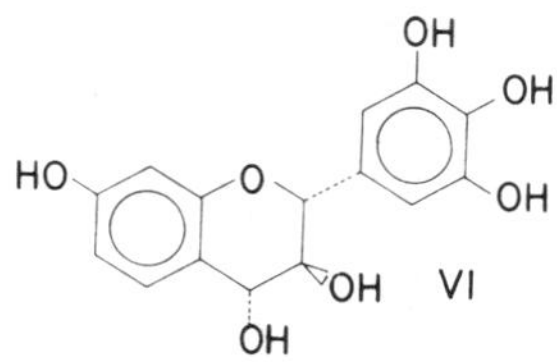

VI

(+) - leucorobinetidin

B. Flavan-3-ols

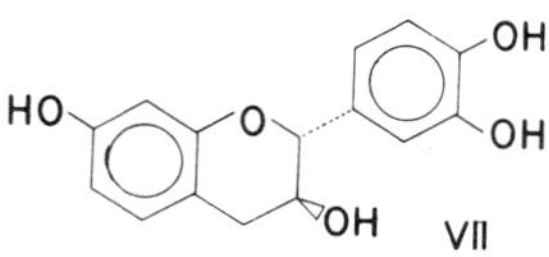

VII

(−) - fisentinidol

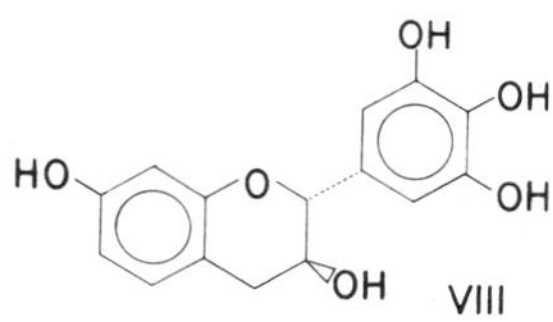

VIII

(−) - robinetinidol

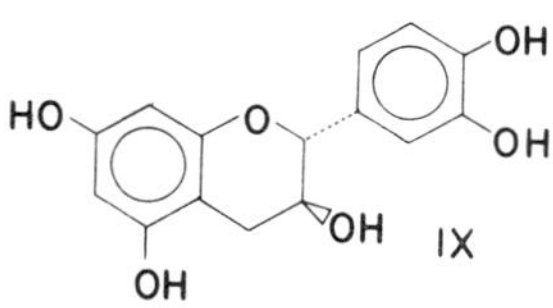

IX

(+) - catechin

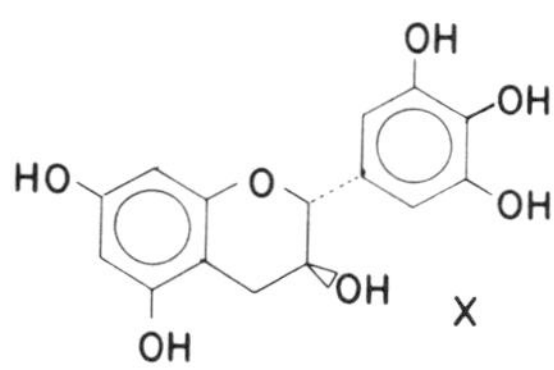

X

(+) - gallocatechin

C. Dihydroflavonols

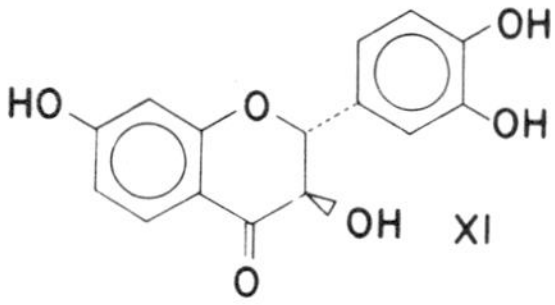

XI

(+) - fustin

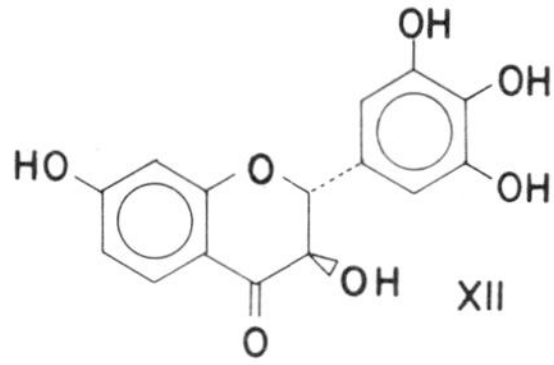

XII

(+) - dihydrorobinetin

D. Flavonols

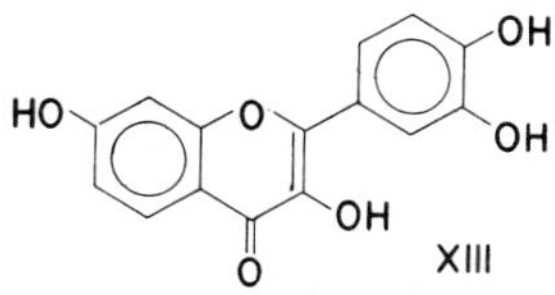

XIII

fisetin

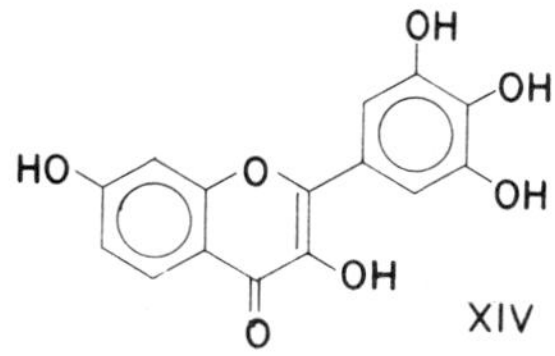

XIV

robinetin

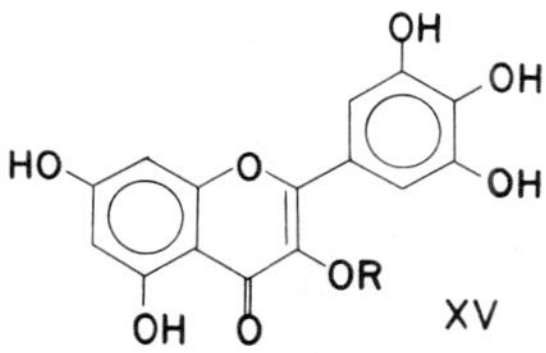

XV

quercetrin (R = rhamnose)

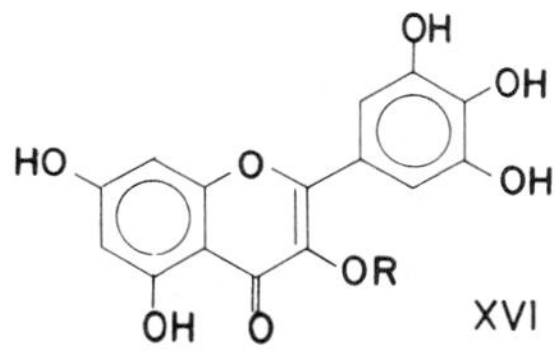

XVI

myricitrin (R = rhamnose)

E. Flavanones

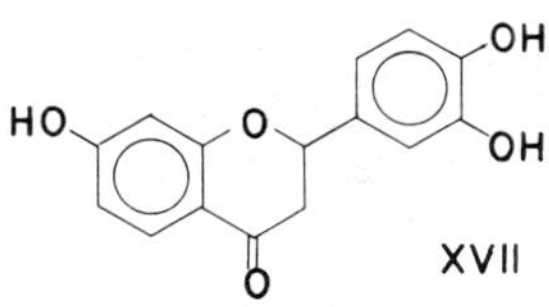

XVII

(±) - butin

F. Chalcones

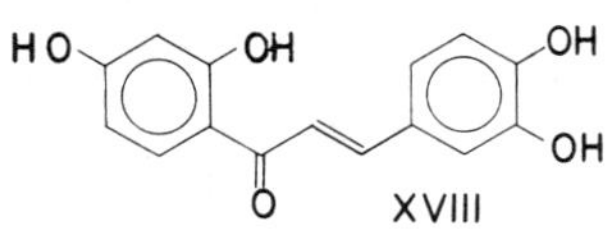

XVIII

butein

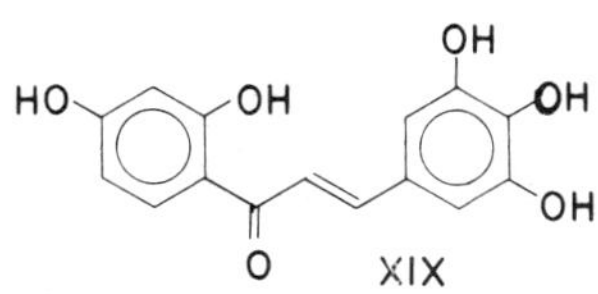

XIX

robtein

catechol B rings (XXI). These tannins represent about 25% of the total bark tannin fraction. Superimposed on these two predominant tannin flavonoid mixtures are two minor groups of the analogs arising from photosynthetic processes occurring in the leaves and immature bark. These are based on phloroglucinol-pyrogallol (XXII) and phloroglucinol-catechol (XXIII) flavonoids. These four patterns constitute 65–80% of mimosa bark extract.

The remaining parts of the wattle bark extract are the "nontannins" [10]. They may be subdivided into carbohydrates, hydrocolloid gums,

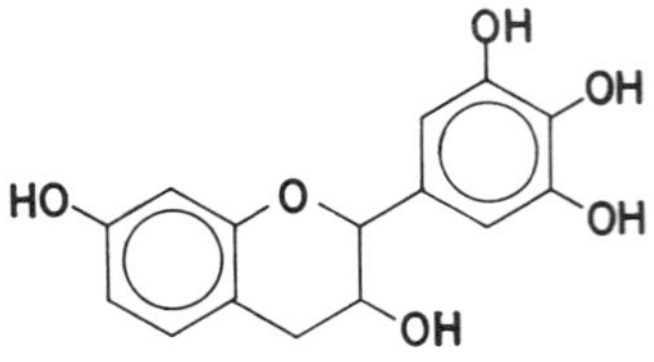

XX

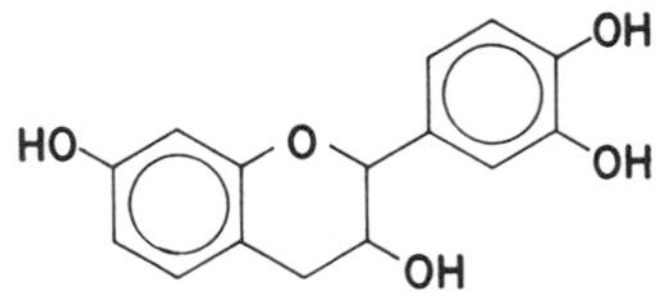

XXI

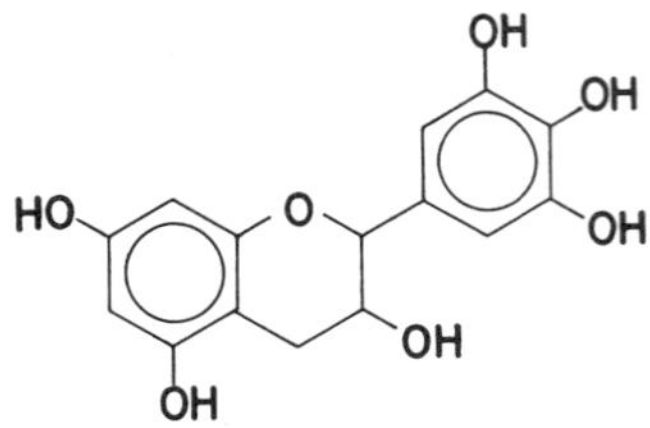

XXII

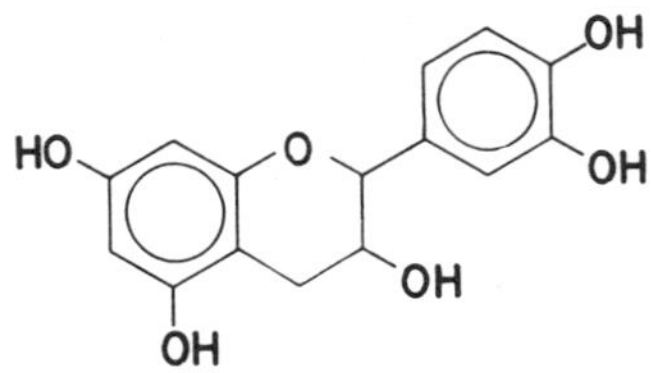

XXIII

and amino and imino acid fractions [10]. The carbohydrates (1)-pinitol and sucrose predominate, with glucose in a lower proportion [10]. The hydrocolloid gums vary in concentration from 3 to 6% and contribute significantly to the viscosity of the extract in spite of their low concentration [10]. The nitrogen compounds of wattle bark extract are mainly the imino acids (-)-L-pipecolic acid, (-)-L-4-hydroxy-*trans*-pipecolic acid, and L-proline, together with lesser quantities of the amino acids arginine, alanine, aspartic acid, glutammic acid, and serine. The nitrogen compounds constitute about 3% of the extract.

Similar flavonoid A- and B-ring relationships, although slightly different and less surely determined, also exist in quebracho (*Schinopsis lorentzii* and *balansae*) wood extract [4,5,11].

The two main substitution patterns of wattle extract are also present in quebracho extract, in which the apparent absence of the flavonols quercetrin and myricetrin [7,11,12] and the flavan-3-ols catechin and gallocatechin [7,11,12] is the most obvious and important difference. In brief, no phloroglucinolic A-ring pattern, or more probably a much lower quantity of it, is present in the quebracho extract, as can be deduced by the failure in isolating leucocyanidin but in the success of attaining cyanidin chloride [7,11,12]. Similar patterns to wattle and quebracho are followed by hemlock and Douglas fir bark extract. Completely different patterns and relationships do instead exist in the case of pine tannins. With the exception of *Pinus ponderosa*, whose principal flavonoid pattern is just about identical to that of wattle and quebracho tannins [13], the other pine species studied (*Radiata, Eliotae, Taeda, Aleppensis, Sylvestris, Patula, Pinaster,* etc.) [14–16] present instead only two main patterns. One pattern is represented by flavonoid analogs based on phloroglucinol A rings and catechol B rings

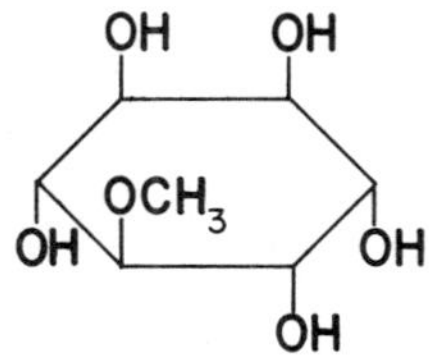

XXIV
(+)Pinitol

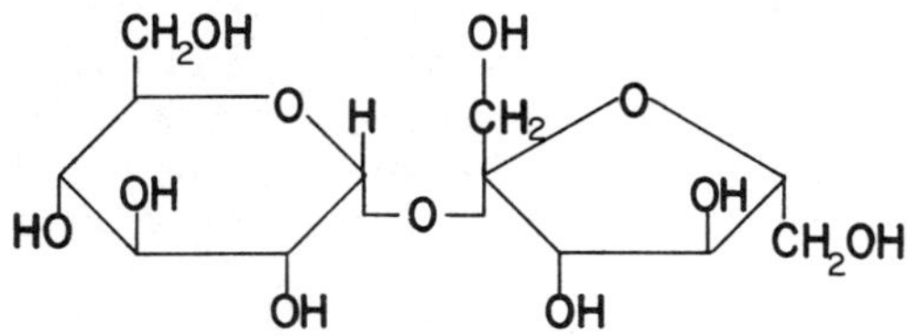

XXV
Sucrose

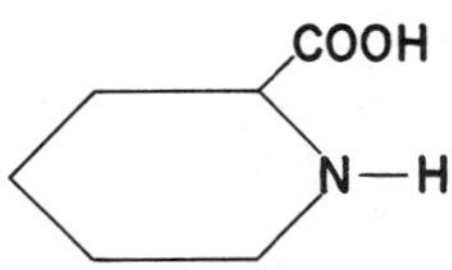

XXVI
(+)L-Pipecolic acid

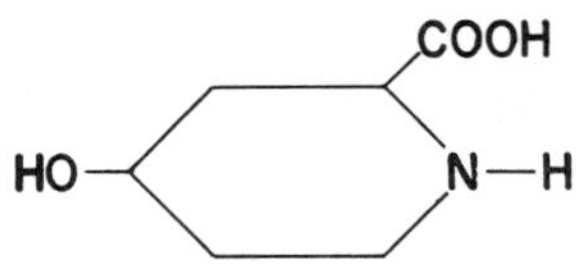

XXVII
(+)L-4-Hydroxy-trans-pipecolic acid

XXVIII

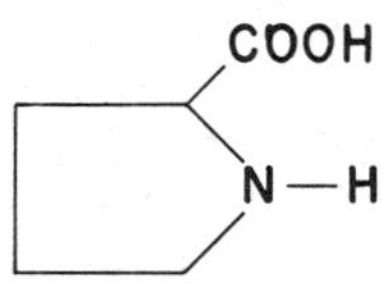

L-Proline

(XXIII) [14,15]. The other pattern, present in much lower proportion, is represented by phloroglucinol A rings and phenol B rings (XXIX) [14,15].

The A rings then possess only the phloroglucinol type of structure, with very important consequences in the use of these tannins for adhesives. Resorcinol-type A rings as well as pyrogallol-type B rings are

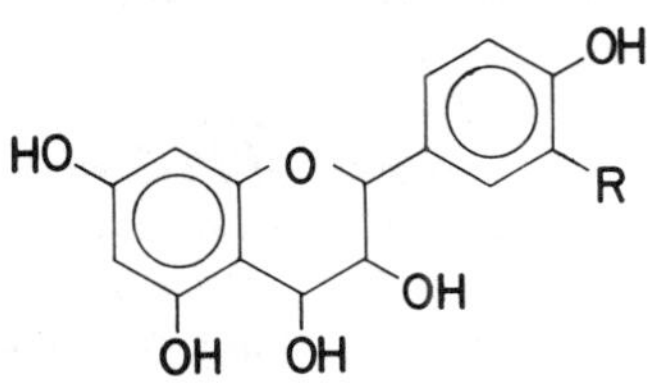

XXIX R = H
XXX R = OH

completely absent. The main leucoanthocyanidins in these extracts are leucocyanidin (XXX) and afzelechin (XXIX).

C. Biflavonoids

Of the monoflavonoids presented, only flavan-3,4-diols and flavan-3-ols apparently participate in tannin formation. This is logical, as all the other flavonoids present carbonyl groups at the 4-position (etherocyclic rings). These groups eliminate the possibility of autocondensation to biflavonoids and higher polyflavonoids by both reducing the nucleophilic character of the A rings and occupying one of the positions through which natural condensation occurs. *Meta*-disubstitution or trisubstitution with hydroxyl or heterocyclic oxygen groups on the A rings of flavan-3,4-diols promote strong nucleophilic centers at the 6- and 8-positions as well as the formation at the 4-position of benzylcarbonium ions stabilized by delocalization of the charge on the vicinal aromatic ring. Both conditions enhance the possibility of flavonoid autocondensation. Consequently, (+)-catechin (IX) and (+)-gallocatechin (X) offer the strongest nucleophilic centers while (+)-leucofisetinidin (V) and (+)-leucorobinetidin (VI) provide potential benzylcarbonium ions for electrophilic substitution. Logically, attack on the catechins should be at both of the two available centers (6- and 8-position) on the A ring and not only on the 8-position as sustained by a few authors [7]. The proved fact that the 8-position is slightly more reactive than the 6-position [7,17] does not exclude the latter one from the possibility of reaction, as proved by the types of biflavonoids isolated. Some of the anticipated products based on steric considerations are in accord with those found among wattle tannins: namely, (+)-leucofisetinidin-(+)-catechin, (+)-leucorobinetidin-(+)-catechin, and (+)-leucorobinetidin-(+)-gallocatechin.

Both types of biflavonoids, the 4,6- and 4,8-linked must be present. Similar biflavonoid units which correspond to (-)-leucofisetinidin-(+)-catechin have been isolated from the heartwood of *S. balansae* (quebracho).

Related biflavonoids were isolated by Nisi and Panizzi (from *Eucalyptus camaldulensis*) [18], Krishnamoorty and Seshadri (*Mirica nagi*) [19], and delle Monache et al. (*Ouratea* spp.) [20]. Drewes, et al. [21,22] characterized the sterochemistry of two groups of crystalline leucofisetinidins (XXXI–XXXIII) from the wood of the black wattle tree (*Acacia mearnsii*) and also determined the stereochemistry of a homologous series of all-*trans*-5-deoxyleucoanthocyanidin-(+)-catechin from the bark extract of the same tree (XXXIV–XXXVI).

The C- and F-ring heterocyclic systems were shown to have half-chain conformations for most individual flavonoid moieties, exceptions being twisted boat conformations where 2,3-*trans*-3, 4-*cis* relative configurations pertained in the upper units (C rings). The x-ray structure

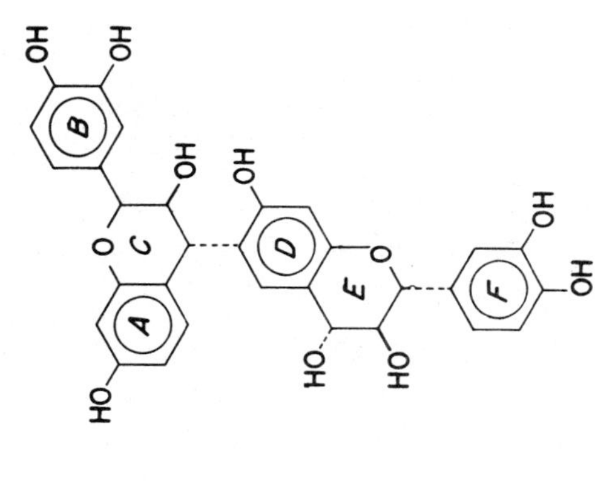

XXXI

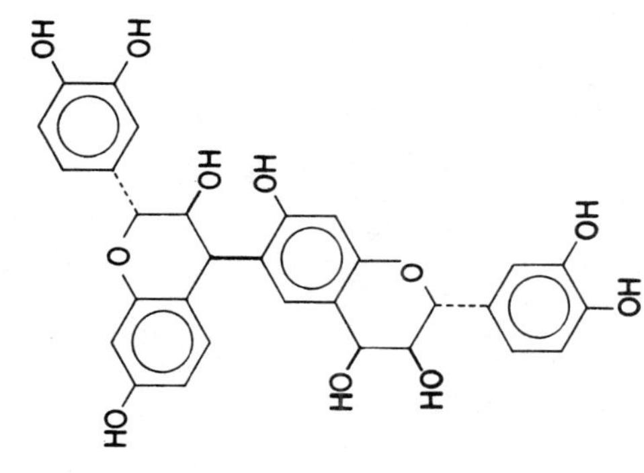

XXXII

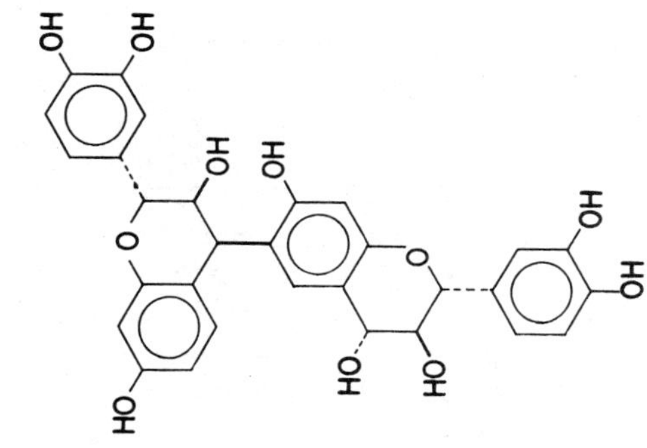

XXXIII

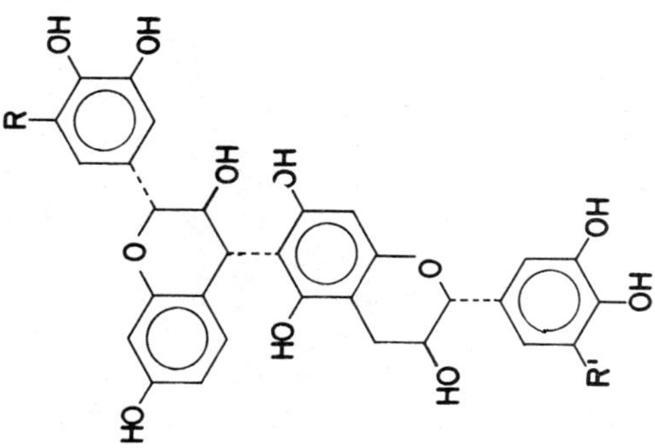

XXXIV R = R' = H
XXXV R = OH, R' = H
XXXVI R = R' = OH

of 8-bromotetra-O-methyl-(+)-catechin confirms the 2R, 3S absolute configuration of the parent compound (+)-catechin [17]. In the heterocyclic ring both the heterocyclic oxygen atom and the 4-carbon lie marginally above the mean plane of the adjacent aromatic A ring with the 2-carbon and 3-carbon more definitely higher and lower, respectively, of the same mean plane [17]. The conformation of the etherocyclic ring is midway between a C-2 sofa and a C-2, C-3 half-chair conformation in the crystal [17]. The phenyl and hydroxy groups at positions 2 and 3 in the heterocyclic ring of the 8-bromoderivative are in the *trans* diequatorial position [17]. Confirmation of the 4,6-coupling for bileucofisetinidins (XXXI–XXXIII) and the all-*trans*-5-deoxyleucoanthyocyanidins (XXXIV–XXXVI) was obtained by chemical shift data [23]. The biflavonoids (XXXIV–XXXVI) from the *A. mearnsii* bark extract are also accompanied by their apparent precursors, (+)-leucofisetinidin, (+)-leucorobinetidin, (+)-catechin, and (+)-gallocatechin and by higher oligomeric analogs [23]. Combined, these constitute the tannins of "mimosa" or wattle extract of commerce, the mixture having a number of average molecular weight of 1250 [24]. Weinges et al. [25, 26] isolated also 4,8-linked leucocyanidins from various sources (XXXVII), while Hemingway and McGraw [14] isolated and identified similarly 4,8-linked biflavonoids from the bark of loblolly pine. The etherocyclic ring of both units of these 4,8-linked biflavonoids were also found to be in the half-chair conformation.

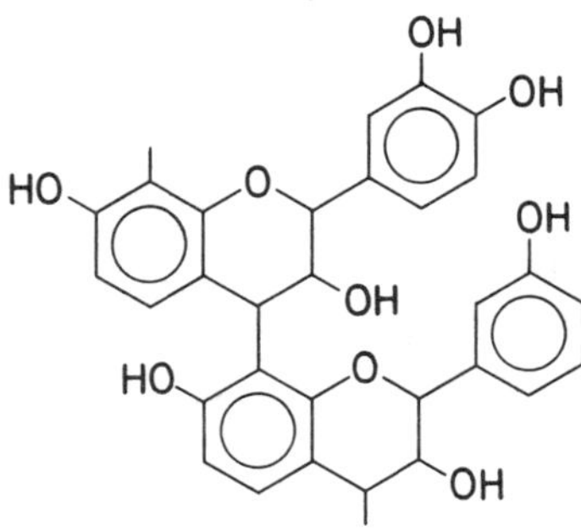

XXXVII

D. Triflavonoids and Tetraflavonoids Condensed Tannins

Roux et al. [7] indicated that the principle of condensation based on 4,6-links between resorcinol units, following initial 4,8-links between resorcinol and lower terminal phloroglucinolic units, appear to be a general flavonoid autocondensation pattern. While the 4,6-links between resorcinolic units are beyond any doubt, considerable doubt still exists about:

1. The positioning of the phloroglucinolic flavonoid unit as the tannins "lower" terminal unit

2. The 4,8-links being the only phloroglucinolic condensation pattern possible, as proposed.

This was reputedly shown by the isolated triflavonoid condensed tannin from the heartwood of the mopane tree (*Colophospermum mopane*) [27,28] and a tetraflavonoid unit from the karree tree (*Rhus lancea*) [22].

The recurrent 4,6-linkages between the "upper" units was demonstrated by nuclear magnetic resonance (NMR) spectrometry [22] and by chemical degradation, while the 4,8-linkage to terminal units is based on the interpretation of the shifts of all methoxyl NMR resonances during progressive C_6D_6 addition to $CHCl_3$ solutions of the methyl ethers of the tannins [7]. The same author, at different stages, advances two different structures [7,23,29] for the mode of linkage of the lower terminal phloroglucinolic flavonoid unit. Structures XXXIX, 4,8-linked, and XL, 4,6-linked, have both been proposed for the same tetraflavonoid, indicating the uncertainty of such an interpretation [7,23,29].

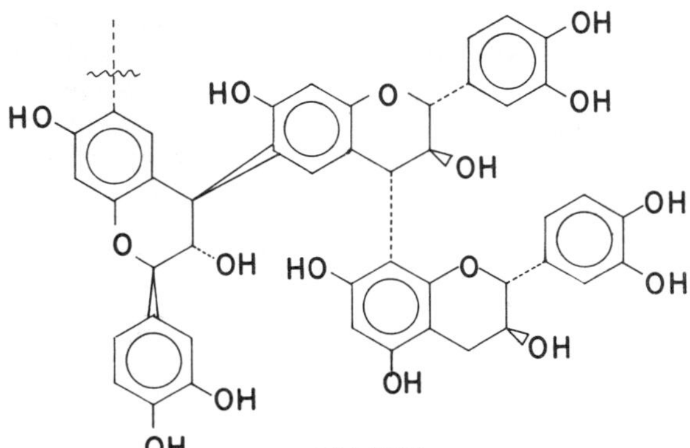

XXXVIII

Commercially extracted wattle bark extract (*A. mearnsii*) show a similar and continuing system of condensation at the triflavonoid level. Descrepancies [30] in the experimental behavior and reactivity toward formaldehyde of tannin-based adhesives, especially wattle tannin adhesives, from what expected from any of the proposed structures XXXIX and XL brought to the realization that the phloroglucinolic unit is not available for reaction and should not, in all probability, present any free position (nor the 6- or 8-position) available to formaldehyde attack. If all the structures present should present a lower phloroglucinolic terminal unit free to formaldehyde attack, then the initial condensation, on addition of formaldehyde, between tannin macromolecules should occur through these phloroglucinolic units, as phloroglucinolic units are between 10 and 15 times more reactive toward formaldehyde than the corresponding resorcinolic flavonoids [31]. Thus a very fast and

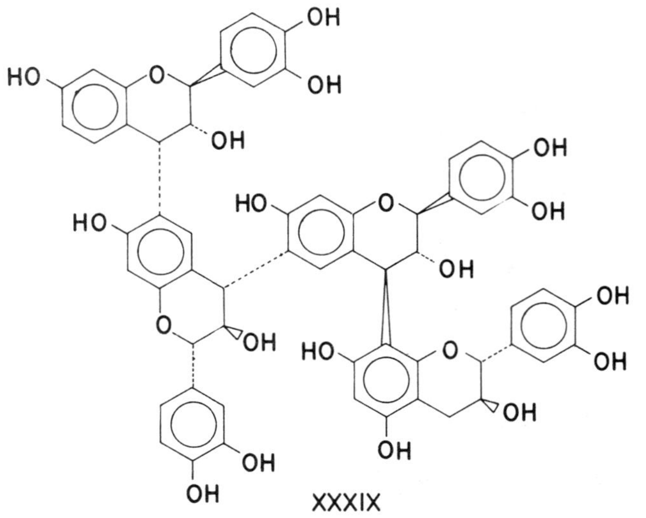

XXXIX

considerable increase in the viscosity followed by a somewhat slower rate of viscosity increase of tannin-formaldehyde mixtures should be noticeable. This is not so [30] and viscosity increase graphs in function of time are smooth exponential curves. There are no indications of sudden changes of rate of viscosity increase at the beginning of the

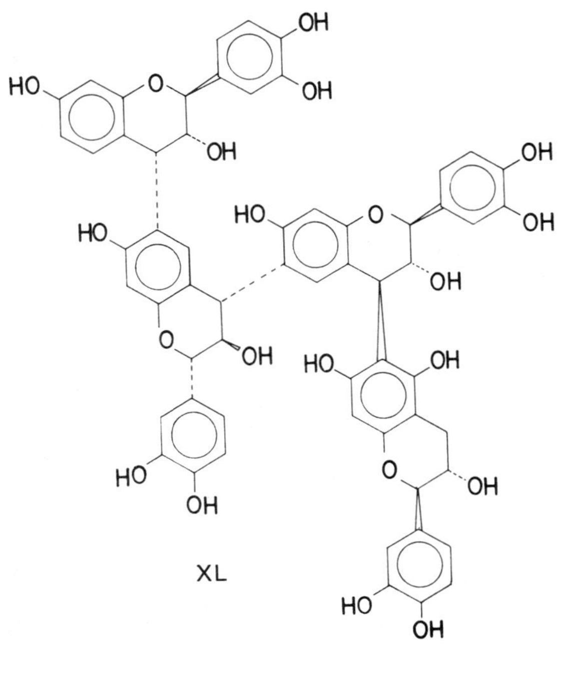

XL

curve due to a change of the species reacting with formaldehyde. Quite independently from such applied considerations, Roux [32] and coworkers realized by the use of high resolution and progressively high temperature NMR spectra of triflavonoids obtained by novel synthetic ways [33] that the most probable structure of triflavonoids and position of the phloroglucinolic unit are as follows:

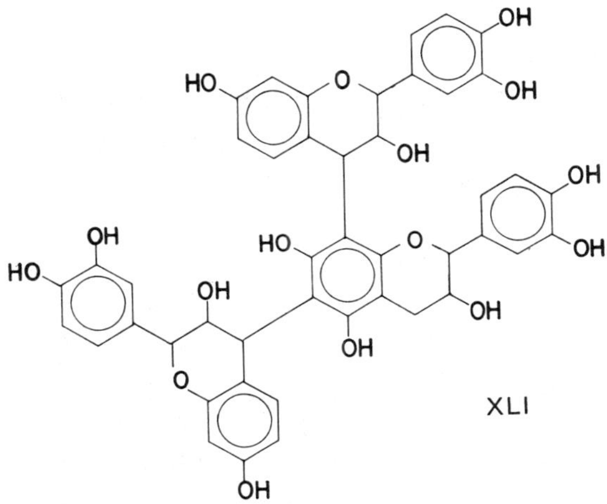

XLI

From structure XLI it is quite evident that no position highly reactive with formaldehyde is available on the phloroglucinolic unit, as both the 6- and 8-positions are blocked by other flavonoids. Hence in phloroglucinolic flavonoid units, all three 4,6-, 4,8-, and 6,8-linkages are probably allowed. This finding may bring to the realization that all the three types of linkages, through positions 4, 6, and 8 of the phloroglucinolic unit, may be present simultaneously, with the formation of branched rather than linear polymeric tannins. It is not possible to conclude with the data available at this stage if the tannins macromolecules are linear polymers, as assumed up to now, or branched polymers. A few structural possibilities are open, such as:

1. Branched tannins when phloroglucinolic units are linked with the flavonoids at the 4-, 6-, and 8-positions.
2. Linear condensed tannins in which a polymeric tannin of type XXXIX or XL has the phloroglucinolic unit no longer terminal but with the 6- and 8-position, respectively, blocked by a further resorcinolic flavonoid unit, which now becomes the tannin lower terminal unit.
3. Linear condensed tannins in which the 6- and 8-positions of the phloroglucinolic flavonoid unit function as links between two flavonoid chains composed of 4,6-linked resorcinolic units. This is valid for tannins such as those present in commercial wattle and quebracho extracts. As regard pine extracts, whose

flavonoids are only phloroglucinolic in nature, the accepted continuing pattern of condensation is 4,8-linked (XLII).

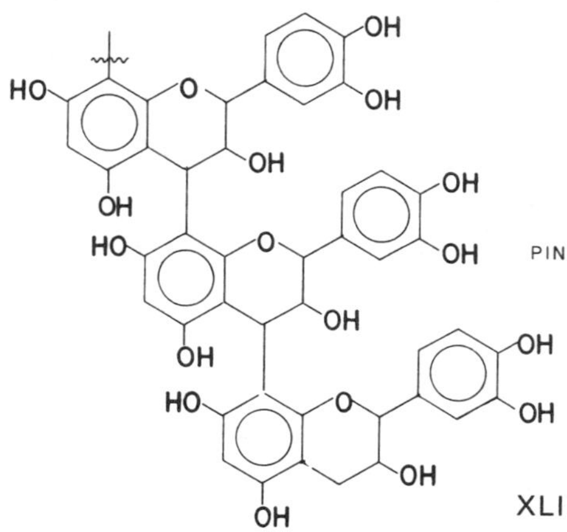

XLII

To conclude, the 8- and 6-positions provide the nucleophilic function, while the electrophile is presumably represented by a 4-carbonium ion generated from the flavan-3,4-diols. In this way the principle of self-condensation may apparently be continued, resulting in units as large as 11 or 20 linked flavonoids, respectively, for wattle tannins (number average mass 1250) [34] and for polyflavonoids of higher average mass range as quebracho (average 1784) [34] and pine (average ±4300) [15].

E. Methods for the Analysis of Phenolic Materials Content in Tanning Extract

Various methods of analysis are available for the determination of tannin content. These methods can generally be grouped into two broad classes:

1. Methods aimed at the determination of tanning material content in the extract. These methods were devised to determine which percentage of the extract would participate in leather tanning. They may also be used to give an indication of the amount of phenolic material that can react with formaldehyde present in the extract. Their main drawback, as regard adhesives, is in the incapability of detecting and determining the approximate 3% of monoflavonoids, or phenolic "nontannins," present in the extract, which do not contribute to tanning capacity but which do definitely react with formaldehyde and contribute to adhesives preparation.

2. Methods aimed at the determination of phenolic material present in the extract that can be reacted with formaldehyde. These methods

were devised particularly for tanning extracts used in adhesives preparation and are all based on the determination of some of the products of reaction of the flavonoids with formaldehyde. Their main defect is that the phenolic material content is expressed as an absolute number. They are excellent in comparing different extracts as regard their relative phenolic content, hence their suitability for adhesive manufacture; the absolute numbers obtained, though, bears no relationship with the real percentage of phenolic material in the extract.

Accepted methods of the first type comprise the hide-powder method [35], the refractometric method [36], and various visible, ultraviolet [36,37], and infrared spectrometric methods. Accepted methods of the second type comprise the Stiasny/Orth method [47,48] and its modifications and the Lemme [38], sodium bisulfite back titration method. While an in-depth discussion of all the methods noted above is beyond the scope of this review, a brief discussion of some of the more important or interesting ones is necessary:

1. Hide-powder method [39,40]. This is the oldest accepted analytical method for the determination of tannins in a tanning extract. It is based on the reproduction of the leather tanning process in which the material to be tanned is standardized powdered hide. The amount of tannins absorbed and fixed by contact of predetermined amounts of tanning extract and hide powder is expressed as a percentage of the original tanning extract. The method is widely used by leather tanners, extract producers, and leather standard and research institutions. All the extract-producing factories around the world deliver every industrial batch of tanning extract they produce with the analyzed percentage of tannins obtained with the hide-powder method.

2. Molybdate ions spectrophotometric methods [37]. These methods are based on the examination by visible/ultraviolet (UV) spectrophotometry of the intensely yellow-orange complexes formed by solutions of molybdate ions with *o*-dihydroxybenzene and 1,2,3-trihydroxybenzene derivatives [41–43]. Mono- and polyflavonoids such as condensed tannins, which contain *p*-dihydroxybenzenic (catechol) and 1,2,3-trihydroxybenzenic (pyrogallol) B rings, also form yellow-orange complexes with molybdate ions in aqueous solutions [37,43]. An *o*-dihydroxyl group is essential for complexation since five-membered chelates are formed, consisting of an *o*-dihydroxo group and a central molybdate ion [44]. The absorbance values at 400 nm of solutions containing an excess of sodium molybdate or ammonium heptamolybdate and varying amounts of tannin extract at a buffered pH in the range 4.0–5.6 vary. The amount of condensed tannin present in the extract is obtained by the linear relationship existing between tannin amounts and absorbance values [37]. This method also has the advantage of being able to detect monoflavonoids, or phenolic nontannins, which are overlooked by the hide-powder method. This and related UV methods [35] are useful for the determination of the phenolic material content of extracts used for adhesives.

3. Infrared spectrometric determination of A rings [45]. This method is based on an extension of a method developed by Chow and Steiner [46] by using the 1140/1120-cm absorbance ratio, in the infrared region, to determine the resorcinol content of phenolic resins. It is based on the linear relationship between the variation of the infrared 1490/1450-cm^{-1} absorbance ratio and the amount of A rings contained in a tanning extract. Of the two bands used to calculate the ratio the 1490 cm, one is characteristic of resorcinolic and phloroglucinolic materials only, while the other is general of all the extract components. As the ratio of A rings in a wattle tanning flavonoid is constant at 33% for resorcinolic flavonoids and 43% for phloroglucinolic flavonoids, it is possible to calculate the percentage of phenolic material in the extract. The regression equation for such a system and the limits of its application are as follows:

correlation coefficient $r^2 = 0.950$
$y = 0.02375x - 0.00275$ for $25\% \geqslant \times \geqslant 100\%$

4. Modified Stiasny method [47—50]. This method is based on the gravimetric determination of the reaction products precipitated during the reaction of tannins with formaldehyde in the presence of HCl. The tanning extract solution is treated at reflux, in an acid environment, with a molar excess of formaldehyde and the fractions of the extractives that are capable of reacting with formaldehyde precipitate out of solution; they are then filtered, dried, and weighed, and the results are reported as a "formaldehyde precipitation number." The method, although not giving the exact percentage of phenolic material in the extract, is widely used, as it has the advantage of giving a comparative measure of the amount of tannin being capable of reacting with formaldehyde under the conditions of formation of a thermosetting phenolic-type resin.

III. REACTIVITY OF TANNINS AS MACROMOLECULES

Considering their macromolecular nature, condensed tannins exhibit unique reactions as well as reactions normally expected of flavan-3-ols units. Knowledge of the more useful of these reactions is important in the industrial application of tannin extracts to adhesives.

A. Reactivity and Orientation of Electrophilic Substitutions of Flavonoids [7]

The relative accessibility and/or reactivity of flavonoid units has been examined by selective bromination with pyridine hydrobromide perbromide using units of the phloroglucinol and resorcinol series. (+)-Tetra-O-methyl catechin (XLIII) is brominated preferentially in the

8-position [7]. Only when this position is filled does substitution commence at the 6-position. The B ring is particularly unreactive, but with excess brominating agent a low degree of substitution at the 6-position does occur. The bromination sequence 8 > 6 >> 6' for (+)-catechin tetramethyl ether is shown in structure XLIII.

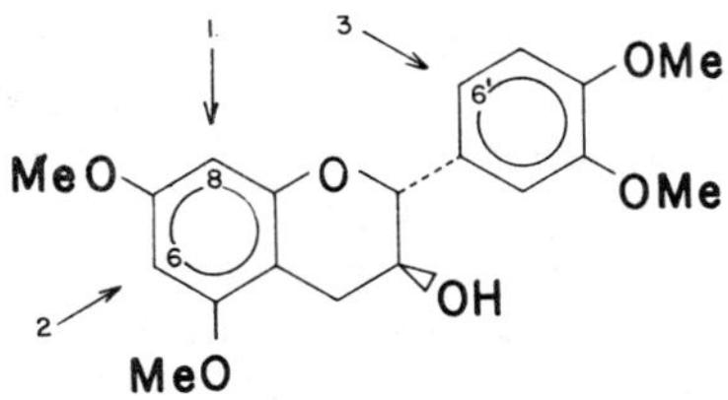

XLIII

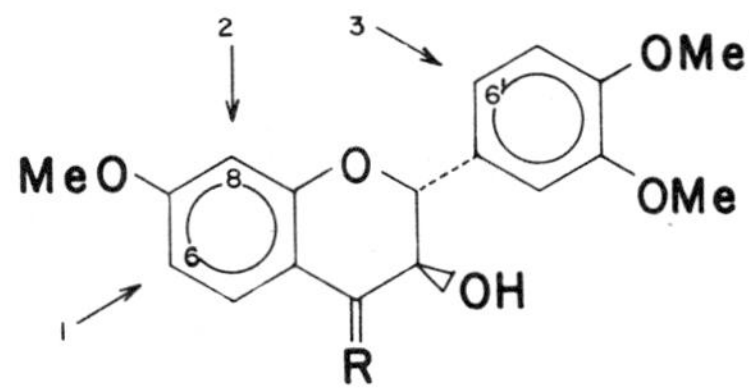

XLIV

However, for the resorcinol equivalent, (-)-tri-O-methylfisetinidol [(XLIV), R = H_2] and for (+)-tri-O-methylfustin [(XLIV), R = O] the substitution sequence for the A ring is modified to 6 > 8 >> 8' as indicated [7]. Preferential 8- and 6-substitution in phloroglucinol- and resorcinol-type flavonoids, respectively, must presumably be related to the greater accessibility of these positions in each instance. In particular, by using resorcinol as a simple model compound it has been shown [30], as expected, that di-, tri-, and tetramers formed by the acid or alkaline-catalyzed reaction of resorcinol with formaldehyde do not favor the position *ortho* to both hydroxy groups for formaldehyde attack. These preferential substitution effects which appear to be obeyed at the biflavonoid range, and confirmed by reaction of simple phenolic model compound, find further parallels over the tri- and tetraflavonoid ranges.

B. A- and B-Ring Reactions with Aldehydes and Their Kinetics

Tannins, being phenolic in nature, undergo the same well-known reaction of phenols with formaldehyde either base- or acid-catalyzed, weakly basic base-catalyzed reactions being predominantly used in industrial applications. Increasingly alkaline conditions lead to progressive activation of the phenol as nucleophile, especially above pH 3, where phenateions are formed. The nucleophilic centers on the A rings of any flavonoid unit tend to be more reactive than those found on the B ring. This is due to the vicinal hydroxyl substituents, which merely cause general activation in the B ring without any localized effects as on those found in the A ring. Formaldehyde reacts with tannins to produce polymerization through methylene bridge linkages to reactive positions of the flavonoid molecules, mainly the A rings. In condensed tannin molecules the A rings of the constituent flavonoid units retain only one highly reactive nucleophilic center, the remainder accommodating the interflavonoid bonds. Resorcinolic A rings (wattle) show reactivity toward formaldehyde comparable, though slightly lower, to that of resorcinol [51]. Phloroglucinolic A rings (pine) behave instead as phloroglucinol [31]. Pyrogallol or catechol B rings are by comparison unreactive, and may only be activated by anion formation at relatively high pH [7]. Hence, the B rings do not participate in the reaction except at high pH values (pH 10) where the reactivity toward formaldehyde of the A rings is so high that the tannin-formaldehyde adhesives prepared have unacceptably short pot lives [51]. In general, tannin adhesives practice only the A rings are used to cross-link the network. However, because of their size and shape, the tannin molecules become immobile at a low level of condensation with formaldehyde, so that the available reactive sites are too far apart for further methylene bridge formation [51]. The result is incomplete polymerization that leads to the weakness and brittleness that are characteristic of many tannin formaldehyde adhesives. There are indications, though, that at least pyrogallol B rings are capable of a limited degree of condensation with formaldehyde, even in the presence of an excess of the more reactive resorcinol and in mildly acid or alkaline reaction conditions [30], as shown by the formation of pyrogallol-formaldehyde dimers and resorcinol-pyrogallol-formaldehyde dimers and trimers, with limited pyrogallol participation, in model compound reactions [30] carried out at ambient temperature. The latter constitutes a rethinking and indicates that notwithstanding the possibility of a very limited participation of the B rings to a cross-linked tannin-formaldehyde network, such a network is still weak.

Bridging agents with longer molecules [51–53], such as phenolic and aminoplastic resins, have been used to solve this problem by helping to bridge distances too large for interflavonoid methylene bridges. The latter trend is evident by the type of industrial tannin adhesive formulations exposed in the application part of this chapter. Hillis and Urbach [54,55] have shown that while catechol and the catecholic B rings do not react with formaldehyde at pH lower than 10, through their lack of formaldehyde consumption, the addition of zinc acetate to the reaction mixture induces the B rings to react with formaldehyde at lower pH values, the optimum being in the pH range 4.5–5.5, as shown by the higher amount of formaldehyde consumed [54,55]. This finding implies that in the presence of zinc acetate further cross-linking of the tannin-formaldehyde network could be achieved through B-ring participation to the reaction. This could eliminate the need of increasing cross-linking by addition of synthetic phenolic and aminoplastic fortifiers. Pizzi [56] has shown that at the low levels of addition of zinc acetate which are economical (namely 5–10% on resin solids), there is an improvement in strength given by a higher degree of cross-linking, but not enough to give the same performance of fortified tannin resins. The main effect of bivalent metallic ions was found to be as accelerators

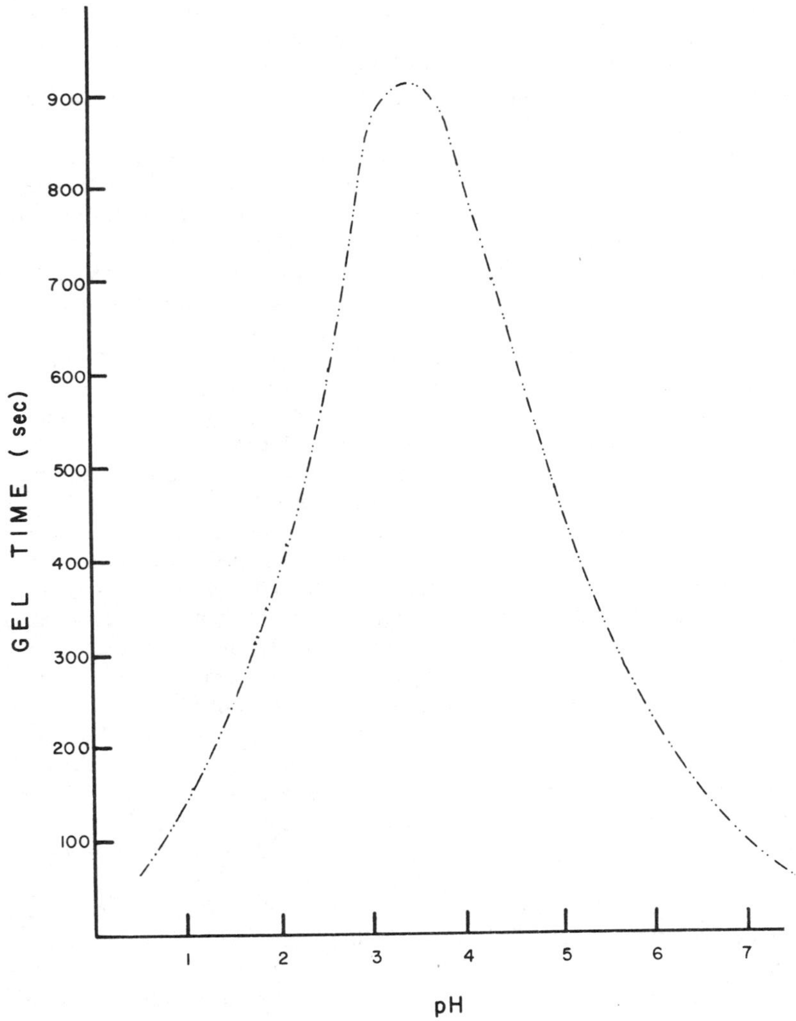

Fig. 4.1. Wattle tannin extract-formaldehyde gel times versus pH. [From A. Pizzi, *J. Macromol. Sci., Rev. Macromol. Chem. C* 2(18): 247–315 (1980).]

of the reaction of flavonoids with formaldehyde, as described in Chapter 3. As regards the pH dependence of the reaction with formaldehyde, it is generally accepted that the reaction rate of wattle tannins with formaldehyde is slowest in the pH range 4.0–4.5 [57]; for pine tannins, between 3.3 and 3.9. The quantity of formaldehyde that reacts with the polyphenols in this pH range has been found to be minimal [57].

At neutral pH, rapid reaction with formaldehyde at the 6- and 8-positions of the monomeric unit occurs, accompanied by a much slower reaction of positions 2' and 6' on the pyrogallolic or catecholic B ring. The dependence of the gel time of tannins with formaldehyde at different pH values is shown in Figure 4.1.

Formaldehyde is generally the aldehyde used in the preparation, setting, and curing of tannin adhesives. It is normally added to the tannin extract solution at the required pH both as liquid formalin solution and in its polymeric form of paraformaldehyde, which is capable of fairly rapid depolymerization under alkaline conditions. Hexamethylenetetramine (hexamine) may also be added to resins due to its formaldehyde-releasing action under heat. Hexamine is, however, unstable in acid medium [58] but becomes more stable with increased pH values [59]. Hence under alkaline conditions the liberation of formaldehyde might not be as rapid and as efficient as described. Under acid conditions hexamine decomposes to produce 6 mol of formaldehyde and 4 mol of ammonia and is soluble in water up to 50% by weight. At an alkaline pH only 3 mol of formaldehyde is liberated, accompanied by the formation of trimethylamine. Since, under alkaline conditions, hexamine liberates formaldehyde only when heated, mixtures of hexamine and wattle solution exhibit an indefinite pot life [30] at room temperature. It has been fairly widely reported with a few notable exceptions [30,60] that bonds formed with hexamine as hardener are not as boil resistant [61] as those formed by formalin or paraformaldehyde. This leads to the notion that the ammonia released is responsible for the degree of loss in weather resistance [61]. A few authors [30,60] have instead conclusively shown that hexamine can give bonds as good or better as paraformaldehyde, in alkaline environment only; this may be ascribed to the formation of trimethylamine rather than ammonia with less damage to the bonds formed. The reaction of formaldehyde with tannins may be controlled by addition of alcohols to the system. Under these circumstances some of the formaldehyde is stabilized by the formation of hemiacetals [e.g., $CH_2(OH)(OCH_3)$ if methanol is used] [31, 62]. When the adhesive is cured at an elevated temperature, the alcohol is driven off at a fairly constant rate and formaldehyde is progressively released from the hemiacetal. This ensures that less formaldehyde is volatilized when the reactants reach curing temperature, and also that the pot life of the adhesive is extended. In view of the fact that the methylene linkages may be too short for the optimum cross-linking, other aldehydes which also have bifunctional characters have been substituted for formaldehyde. Of these, one of the most frequently used is furfuraldehyde. This has been found to be unsuitable because of its slow reaction with phenols [63], but Pizzi and Scharfetter [51,64] have shown that furfuraldehyde is an efficient cross-linking agent and excellent plasticizer for tannin adhesives when coupled with formaldehyde. Glutaraldehyde has been shown to react with tannins to produce a slow-forming precipitate [65], whereas precipitates with formaldehyde forms much faster. The reaction kinetics of formaldehyde, acetaldehyde, propionaldehyde, *n*-butyraldehyde, isobutyraldehyde, and furfuraldehyde with both resorcinolic and phloroglucinolic-type condensed tannins have been investigated [15]. The same experiments were repeated on resorcinol, phloroglucinol, and catechol used as simple model compounds [15].

The tannins have also been found capable of reactions with glyoxal and benzaldehyde by means of high-temperature gel-time measurements of tannin solutions with the two aldehydes [15,30]. Discrepancies in the speed of reaction of resorcinolic and phloroglucinolic tannins with various aldehydes have been observed. In the case of phloroglucinolic tannins, the rate of reaction with different aldehydes is as follows:

$$\mathrm{HCHO} \gg \mathrm{CH_3CHOCH_3CH_2CH_2CHO} > \mathrm{CH_3CH_2CHO}$$

$$> \begin{matrix}\mathrm{CH_3}\diagdown \\ \mathrm{CH_3}\diagup\end{matrix}\mathrm{CH\text{-}CHO} > \text{furfural}$$

For resorcinolic flavonoids instead, the scale of reactivity is as follows:

$$\mathrm{HCHO} \gg \text{furfural} > \mathrm{CH_3CHO} > \mathrm{CH_3CH_2CH_2CHO} > \mathrm{CH_3CH_2CHO}$$

$$> \begin{matrix}\mathrm{CH_3}\diagdown \\ \mathrm{CH_3}\diagup\end{matrix}\mathrm{CH\text{-}CHO}$$

The obvious difference between the two scales of reactivity is the position of the furfural. The high reactivity of furfural with phenols and resorcinolic flavonoids is severely limited by steric hindrance reasons in the case of phloroglucinolic flavonoids. In the resorcinolic flavonoids the aldehyde attack is carried out at the available 8-position, which is considerably less sterically hindered than the 6-position of phloroglucinolic flavonoid units.

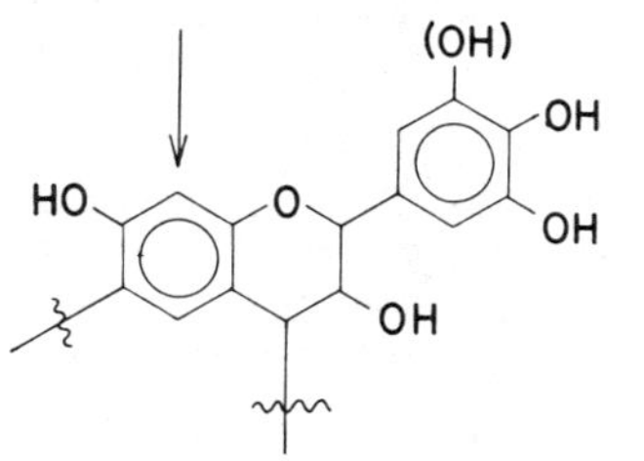

XLV

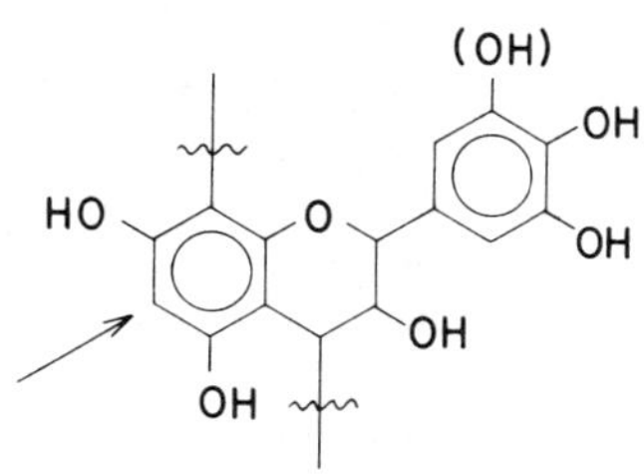

XLVI

Similar behavior is observed when phloroglucinol and resorcinol are used as model compounds of the reaction. The relative position of furfural in the reactivity scale with these two phenols is also steric hindrance dependent, being higher in the scale in the case of resorcinol than in the case of phloroglucinol [15].

The reaction of tannins with aldehydes follows, under alkaline conditions, a second-order kinetic reaction in which the rate or reaction is:

$$\text{Rate} = k_1[\text{aldehyde}][\text{tannin-OH}] + k_2[\text{aldehyde}][\text{tannin-O}^-] \qquad (1)$$

$$\text{Tannin-OH} \underset{}{\overset{\text{pH 8}}{\rightleftarrows}} \text{tannin-O}^-$$

The reaction is mainly a k_2 reaction in the case of pH values higher than 8.2, where the k_1 term of equation (1) is negligible, as the reaction is determined by the reactivity of the tannate formed. It is mainly a k_1 reaction at pH values lower than 7.9, where the concentration of the k_2 term of the equation is negligible. Consequently, equation (1) can be split in equations (2) and (3) according to the alkaline pH range used.

$$\text{Rate} = k_1[\text{aldehyde}][\text{tannin-OH}] \quad \text{pH} < 7.9 \qquad (2)$$

$$\text{Rate} = k_2[\text{aldehyde}][\text{tannin-O}^-] \quad \text{pH} > 8.2 \qquad (3)$$

The reaction of the aldehydes with the tannin/aldehyde condensates formed during the reaction also follows a second-order kinetic:

$$\text{Rate} = [\text{aldehyde}] \sum_n k_n [\text{(CH}_2\text{tannin)}_n] \qquad (4)$$

This is valid, with different k, for both the condensates in the tannin or the tannate form according to the pH range used. The initial part of the total kinetic curve under alkaline conditions (Fig. 4.2) is the sum of equations (1) + (4). The tail part of the curve is instead represented by equation (4) only. In equation (1), terms independent of the concentration of the hydroxymethylated tannins obtained after the initial formaldehyde attack (i.e., $k[\text{HCHO}][\text{tannin-CH}_2\text{OH}]$) can be excluded, at the flavonoid unit level, as no other positions available for formaldehyde attack are available on the flavonoid A ring. They cannot be completely excluded at the flavonoid polymer level, although their contribution is negligible, as the tannins resorcinolic and phloroglucinolic A rings hydroxybenzyl alcohols are too unstable and reactive [30] to allow further reaction with formaldehyde before condensing with other tannin A rings.

The total reaction rate equation can then be summarized as follows:

$$\text{Rate} = k_{1+2}[\text{aldehyde}][\text{tannin-OH}]^a[\text{tannin-O}^-]^{1-a} + [\text{aldehyde}] \times \sum_n k_n [\text{(CH}_2\text{-tannin)}_n] \qquad (5)$$

where [tannin-OH] is the concentration of the tannin form and $[\text{tannin-O}]^{1-a}$ is the concentration of the tannate. At 25°C and at pH 9.0 the value of the kinetic constants for the reaction of wattle flavonoids and formaldehyde are $k_{1+2} = 4.61 \times 10^{-2}\ \text{sec}^{-1}$ and $\sum_n k_n = 3.30 \times 10^{-3}\ \text{sec}^{-1}$ for a linked wattle flavonoid unit reactive points/formaldehyde molar ratio of 1:1 (for wattle tannin, 1.25 average reactive points for linked flavonoid unit). In acid conditions, the reaction is also of second order. The situation is very complex, as a great variety of species contribute to the total kinetic curve.

In the case of formaldehyde:

$$\begin{aligned}\text{Rate} = {} & k_1[\text{HCHO}][\text{tannin-OH}] + k_2[\text{CH}_2\text{OH}^+][\text{tannin-OH}] \\ & + k_3[\text{HCHO}][\text{tannin-OH}_2^+] + k_4[\text{CH}_2\text{OH}^+][\text{tannin-OH}_2^+] \\ & + k_5[\text{HCHO released}] + [\text{HCHO}] \sum_n k_{n1} [\leftarrow\text{CH}_2\text{-tannin}\rightarrow_{n1}] \\ & + [\text{CH}_2\text{OH}^+] \sum_n 2\, k_{n2} [\leftarrow\text{CH}_2\text{-tannin}\rightarrow_{n2}] \end{aligned} \tag{6}$$

Some of these terms are in turn negligible as according to the different conditions of acid pH values used they contribute in varying proportions to the rate of reaction. From the kinetic curves in acid conditions (Fig. 4.2) it is possible to see the contribution of the k_5 term by the upsurge of the curve indicating formaldehyde released rather than consumed. At all practical effects the equation (6), at mildly acid pH values, by combining k_1 and k_2 terms together and disregarding the k_3, k_4, and k_{n2} terms, which are negligible at pH 5–6, can be expressed as

$$\text{Rate} = k_1[\text{HCHO}][\text{tannin}] + k_5[\text{HCHO released}] + [\text{HCHO}] \sum_n k_n[\leftarrow\text{CH}_2\text{-tannin}\rightarrow_n] \tag{7}$$

where the values of the kinetic constants at 25°C and pH 4.9 for a linked wattle flavonoid unit reactive points/formaldehyde molar ratio of 1:1 are $k_1 = 6.94 \times 10^{-3}\ \text{sec}^{-1}$, $\sum_n k_n = 4.3 \times 10^{-4}\ \text{sec}^{-1}$, and $k_5 = -1.9 \times 10^{-4}\ \text{sec}^{-1}$.

The curves shown in Figure 4.2 for the concentration of free aldehydes as a function of the reaction time indicate clearly that two competitive reactions are present:

1. The reaction of the aldehyde with tannin and with low molecular weight tannin-aldehyde condensates, which are responsible for the aldehyde consumption.
2. The liberation of aldehyde, available again for reaction. The latter reaction is probably due to the passage of unstable $—\text{CR}_2—\text{O}—\text{CR}_2—$ ether bridges initially formed to $—\text{CR}_2—$ linked compounds.

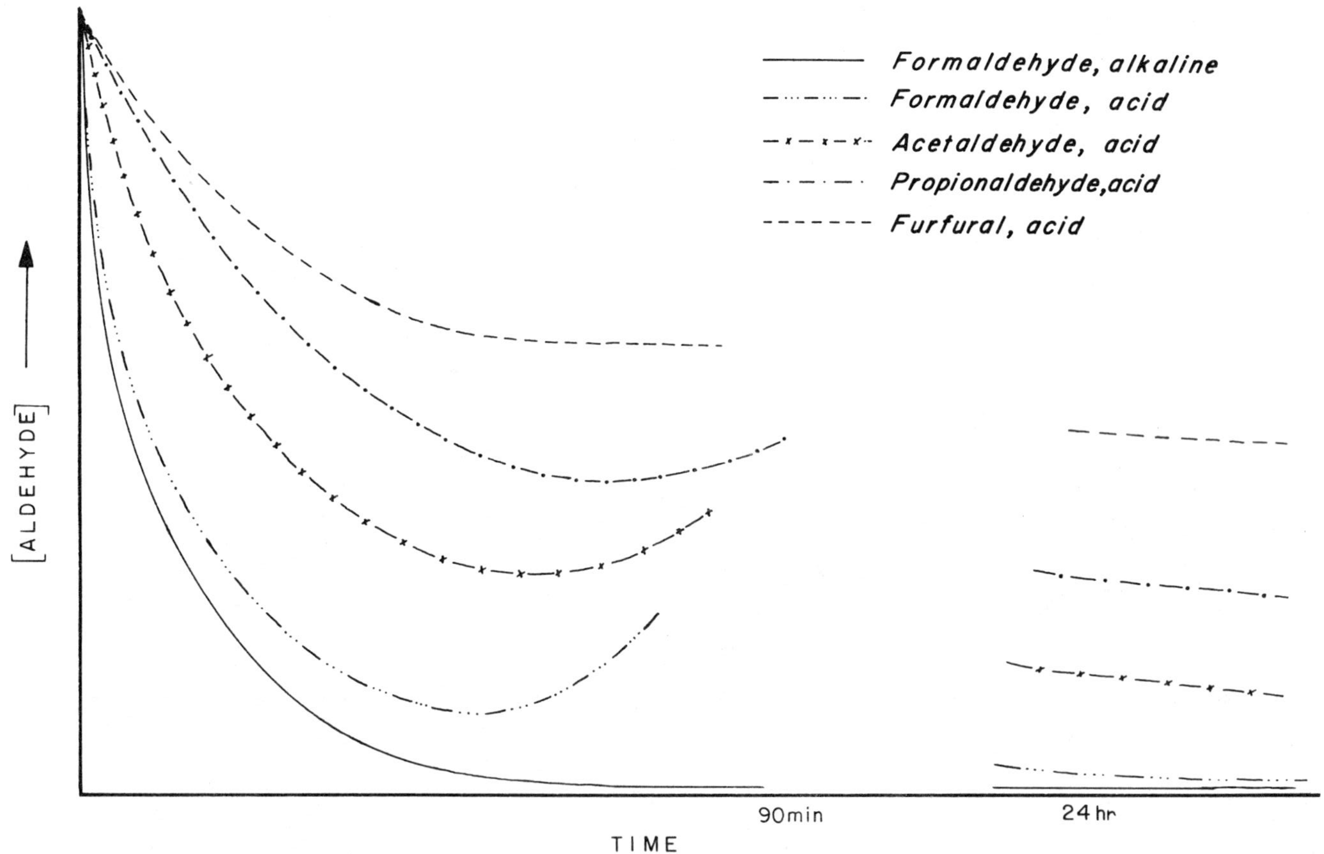

Fig- 4.2. Schematic representation of wattle tannin extract total kinetic curves. [From A. Pizzi, *J. Macromol. Sci., Rev. Macromol. Chem. C* 2(18):247–315 (1980).]

It is interesting to note that while $—CH_2—O—CH_2—$ ether bridged compounds have been isolated for the phenol/formaldehyde [66] reaction, their existence for fast-reacting phenols, such as resorcinol and phloroglucinol, has been postulated, but they have not been isolated, as these two phenols have always been considered too reactive with formaldehyde. An analysis of the phloroglucinol-formaldehyde and resorcinol-formaldehyde condensates formed [30, 51] showed that no difference exists between the products obtained under acid and alkaline conditions. The kinetic curves shown in Figure 4.2 are probably the first indication that such ethers are indeed formed by fast-reacting phenols such as resorcinol, phloroglucinol, and the flavonoids A rings [15]. The rates at which they form and disappear could explain the difficulty of their isolation as well as the observation that the final condensation products obtained under acid and alkaline conditions are identical. Furthermore, while the ethers formed during the phenol-formaldehyde reaction decompose readily at temperatures higher than 130°C, the higher reactivity of resorcinol, phloroglucinol, and flavonoid A rings nuclei toward formaldehyde favors both formation and indeed decomposition of their methylene ethers at much lower temperatures (i.e., 25°C), the latter as consequence of the ether's tendency, always present, to eliminate formaldehyde in order to form a methylene bridge.

Increasing the aldehyde molecular weight, hence its steric hindrance, should cause a decrease in the amount of ether formed as well as a decrease in its rate of formation [15]. This is observed in Figure 4.2, where the steepness of the aldehyde release curve decreases passing from formaldehyde (the steepest) to more sterically limited aldehydes. Stability of the ethers formed by different aldehydes also have a bearing, of course, especially in the case of furfural, where the ether formed should be most stable. It is not possible, of course, to exclude the possibility that the aldehyde released during the course of the reaction could be due to other causes, and this matter may still be open to discussion.

To all practical effects, mainly two aldehydes, formaldehyde and furfural, have been used to date in the preparation of tannin adhesives. Other aldehydes are sometimes used in a supportive role of formaldehyde, when particular effects or physical characteristics of the adhesive are needed. For example, substitution of 10–30% formaldehyde, on an aldehyde group molar basis, with *n*-butyraldehyde, which due to its saturated hydrocarbon chain is somewhat water repellant, can improve the water resistance of cured tannin-formaldehyde networks by a structural modification rather than "cosmetic" addition of water repellents such as waxes [15]. Tannin adhesives prepared and/or set and/or cured with other adhesives only, or with mixtures of formaldehyde and high proportions of other aldehydes, can give cured bonds weaker than those obtained with formaldehyde alone or its mixtures with furfural [15, 30].

The k_1 values for a wattle flavonoid/formaldehyde ratio of 1:1 are higher at pH 9 than at pH 4.9, indicating, correctly, that the reaction is faster at higher pH values. As a consequence, it is then possible to correlate the rate of reaction, or better the kinetic constants, to the gel times of tannin-aldehyde networks. The qualitative balance of different terms in rate of reaction equations can be easily deduced by the diagram of Scheme I, representing the gel time proportional to the total rate of reaction in function of the pH. One curve represents the reactivity toward phenols of the different chemical species given by formaldehyde. The other curve represents the reactivity toward formaldehyde of the different phenolic (or tannin) species. The third curve is the resultant effect that can be compared to the experimental gel-time curves obtained (Fig. 4.1). Scheme I is only qualitative.

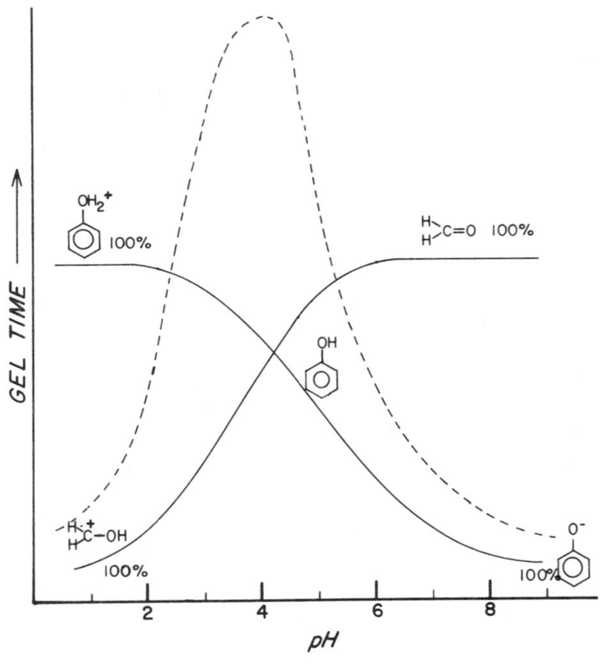

------ GEL TIME CURVE, TOTAL

——— MOLECULAR SPECIES GEL TIME CURVES

Scheme I

To conclude, the reaction rate is also dependent on the concentration of the catalyst, generally NaOH, hence from the pH. For this reason the k values must be given for well-defined pH values. In the case of acid-base-catalyzed reaction, as flavonoid-formaldehyde reactions are, the kinetic constant of a reaction can be expressed as

$$k = k_0 + k_{H^+}[H^+] + k_{OH^-}[OH^-]$$

As the kinetic constant given at pH 4.9 and 9.0 are, respectively, obtained with no addition of OH^- catalyst (pH 4.9) and with addition of OH^- catalyst (pH 9.0), we have that

$$k_{pH4.9} = k_0 \begin{cases} \text{tannin} & = k_{0'} = 6.94 \times 10^{-3} \\ \text{tannin-HCHO condensates} & = k_{0''} = 4.8 \times 10^{-4} \end{cases}$$

$$k_{pH^{9.0}} = k_0 + k_{OH^-}[OH^-]$$

As $k_{pH}{}^{9.0}$ and OH^- at pH 9.0 are known values and $k_0 = k_{pH4.9}$ is also known, it is possible to calculate k_{OH^-}.

$$k_{OH^-} = \frac{k_{pH9} - k_0}{[OH^-]} = \begin{cases} \text{tannin} & = k_{OH} = 7.43 \times 10^{-1} \\ \text{tannin-HCHO condensates} & = k_{2OH} = 5.35 \times 10^{-2} \end{cases}$$

As a consequence, the reaction rate can be expressed from equation (5) taking into account the effect of the OH^- catalyst, as

$$\text{Rate} = (k_{0'} , k_{1OH}[OH^-])[\text{aldehyde}][\text{tannin-OH}]^a[\text{tannin-O}^-]^{1-a}$$

$$+ [HCHO] \times \sum_n (k_0'' + k_{2OH}[OH^-])_n [(CH_2\text{-tannin})_n]$$

C. Metal Ions Catalysis

Phenol-formaldehyde resins may be prepared by carrying out phenol-formaldehyde condensation under such conditions that a high proportion of 2- and 2,6-substituted phenolic nuclei are produced [67,68]. The term "high *ortho* substitution" refers to phenolic resins prepared under such conditions. It has been reported that if the reaction is carried out within the pH range 4—7, in the presence of salts of bivalent metals, an *ortho*-directing effect occurs, resulting in the production of high-*ortho* phenolic resins [67]. The bivalent metallic ions also accelerate both the reaction of formaldehyde with the phenolic nuclei and the condensation of methylol phenols with other phenolic nuclei in the preparation of resins [67]. Conversely, trivalent metallic ions do not show this catalytic effect at all but tend to slow down the reaction [69]. The type of mechanism followed by the reaction is one that involves the formation of a chelate ring with the formation of complexes of type XLVIII [69].

Pizzi [69] substantiated the possibility of such a mechanism by isolating and characterizing the chromium-resorcinol-formaldehyde and chromium-phenol-formaldehyde complexes XLVIII and XLIX. As in the

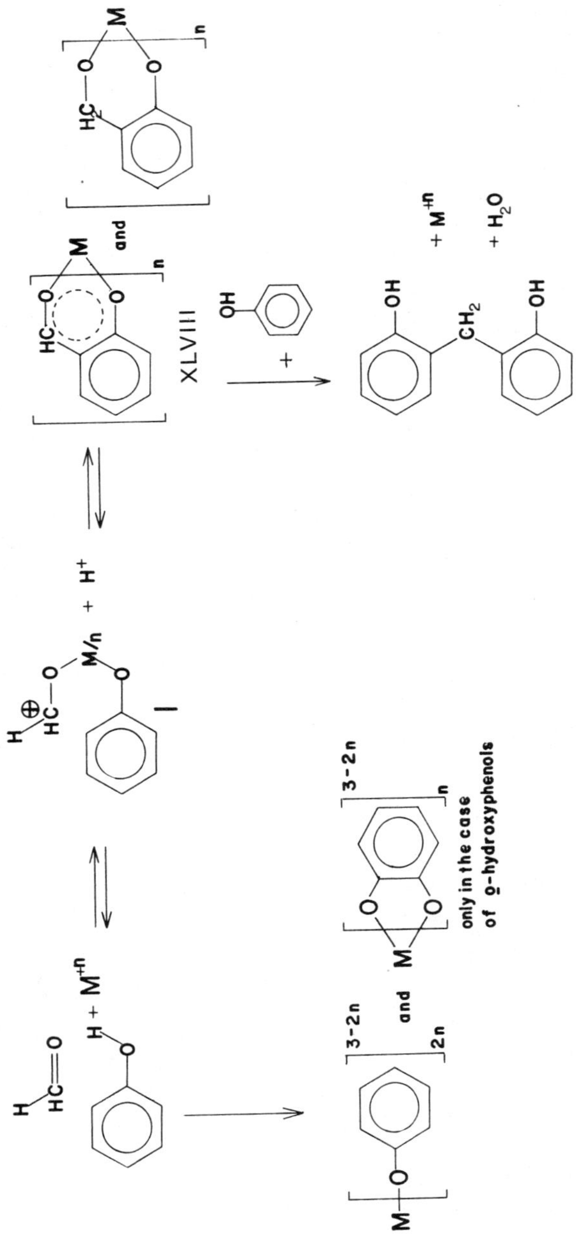
XLVIII
and
only in the case
of o-hydroxyphenols

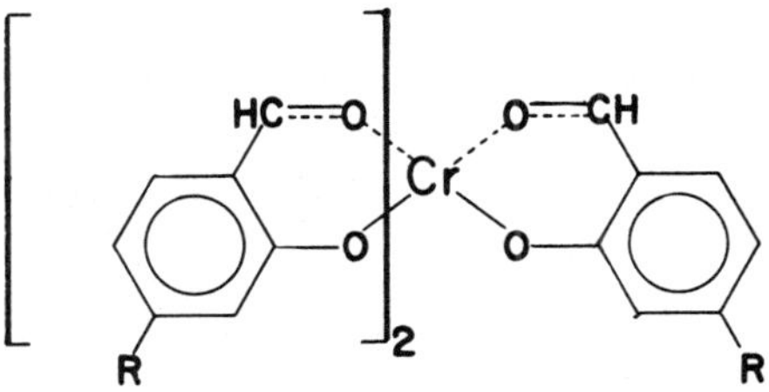

XLVIII, R = —OH

XLIX, R = —H

case of the equivalent acetylacetonates [70, 71], the nature of the bonding between metal and ligands in these chelates is at least partly covalent and their stability may be ascribed to their partly aromatic character. In the phenol-formaldehyde reactions investigated, the metal catalyst does not change its valence state but interacts with molecules or ions containing electron donor groups and accelerate reactions in a manner similar to hydrogen ions [71]. The acid catalysis due to the metal ion differs only in degree from that of the hydrogen ion [71]. The effect of the metal is stronger than that of hydrogen ions, since because of higher charge and greater covalence, its interaction with donor groups is often much greater [71]. This is important, as in the preparation of phenol-formaldehyde adhesives for wood very acid pH values cannot be used in order to avoid wood deterioration. Metal ion catalysts have, then, at a higher pH, the same catalytic effect on the preparation and setting of the resins which is shown by higher hydrogen ions concentration, hence lower pH values. This allows setting of phenolic resin adhesives in milder acid conditions with no lengthening of setting time and without any wood deterioration. The gel times with formaldehyde obtained for resorcinol and catechol indicate that the metal catalysis effect is probably valid for most phenols having a free *ortho* position to the hydroxy group [69]. Most bivalent metallic ions accelerate phenol/formaldehyde reactions. The extent of this effect is directly proportional to the quantity of metallic ion present [69]. Bivalent metals, first will not inhibit the reaction, as the complexes they form are unstable and their rate of exchange in solution are high, and second, they accelerate the reaction in a manner similar to hydrogen ions, although to a higher degree, due to their stronger charge and greater covalence. The formation of the identified stable complexes of types XLVIII and XLIX generally slows down or inhibits the reaction from proceeding to phenol-formaldehyde resins when trivalent metals are used. An exception is Fe^{III}, as shown by the effect of ferric acetate on resorcinol-formaldehyde gel time [69]. This indicates that Fe^{III} is not capable of giving stable complexes of type XLVIII or that the rate of metal exchange between complex and solution is high. This finding is consistent with the scale of rate of metal exchange given by Kluiber [72] for metal acetylacetonates in which a few metals are grouped in the following order of decreasing rate of exchange:

$$Cu^{II} > Fe^{III} \gg Al^{III} > Cr^{III}, Co^{III}$$

The order of the metals in the scale does correspond to the gel times obtained in the case of the resorcinol-formaldehyde condensation [69]. It is consequently better to express the extent of the accelerating or retarding effect of the metal ions in the order of their metal rate of exchange in solution rather than in the order of their stability [69]. The order of the metals in a scale of their decreasing stability does not bear any relationship to the gel times obtained for phenol-formaldehyde condensations [67]. The scale of the extent of the metal ions accelerating effect is then the following:

$$Pb^{II}, Zn^{II}, Cd^{II}, Ni^{II} > Mn^{II}, Mg^{II}, Cu^{II}, Co^{II} > Mn^{III}, Fe^{III}$$

$$\gg Be^{II}, Al^{III} > Cr^{III}, Co^{III}$$

An interesting structural aspect is that both types of complexes [(L) and (LI)] are formed.

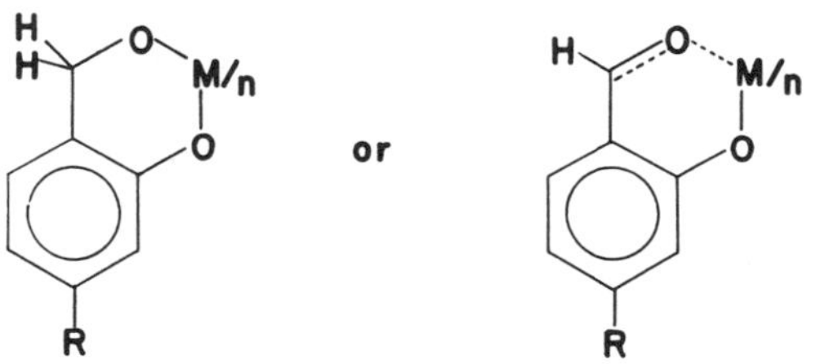

L LI

R = –H, –OH

The metal ions effect on phenol-formaldehyde reactions can be applied to condensed tannins of the flavonoid type with some degree of success. Hillis and Urbach [54, 55] have shown that while catechol and the catecholic B rings of the flavonoid unit do not react with formaldehyde at pH lower than 10, the addition of zinc acetate to the reaction mixture induces the B rings to react with formaldehyde at lower pH values, the optimum being in the pH range 4.5–5.5, as shown by the higher amount of formaldehyde consumed. This finding implies that in the presence of zinc acetate further cross-linking of the tannin-formaldehyde network could be achieved through B ring participation to the reaction. This can eliminate the need of increasing cross-linking by addition of synthetic phenolic or aminoplastic fortifiers. Although Hillis and Urbach [54, 55] use an equal amount in mass of zinc acetate and flavonoid tannin to increase B-ring cross-linking, the catalytic effect on the rate of reaction is also present when much smaller quantities of metal salts are used [69, 73]. The accelerating effect on phenol-formaldehyde reactions caused by the presence of metal ions is maintained in the case of the phenolic compound being a tannin [69].

However, the retarding effect caused by certain trivalent metal ions (Cr^{III}, Al^{III}, Co^{III}) in the case of the reactions of simple phenols with formaldehyde is not maintained when using tannins as the phenolic compound [69]. The latter finding can be explained only in light of the higher steric hindrance of the tannin polymer due to its close tridimensional structure. The metallic ion can probably still form a complex with formaldehyde and the tannin phenolic nuclei, by virtue of its strong positive charge. The high steric hindrance of the tannin, though, causes strong stresses on the partly aromatic etherocyclic ring containing the metal, with consequent instability of the ring and finally ring opening. Once the metal-containing ring is open, the methylol phenolic compound formed, not blocked any longer by a slow-exchange rate metal, is free to proceed in the reaction and to form methylene bridges. In this way the retarding effect of the metal ion,

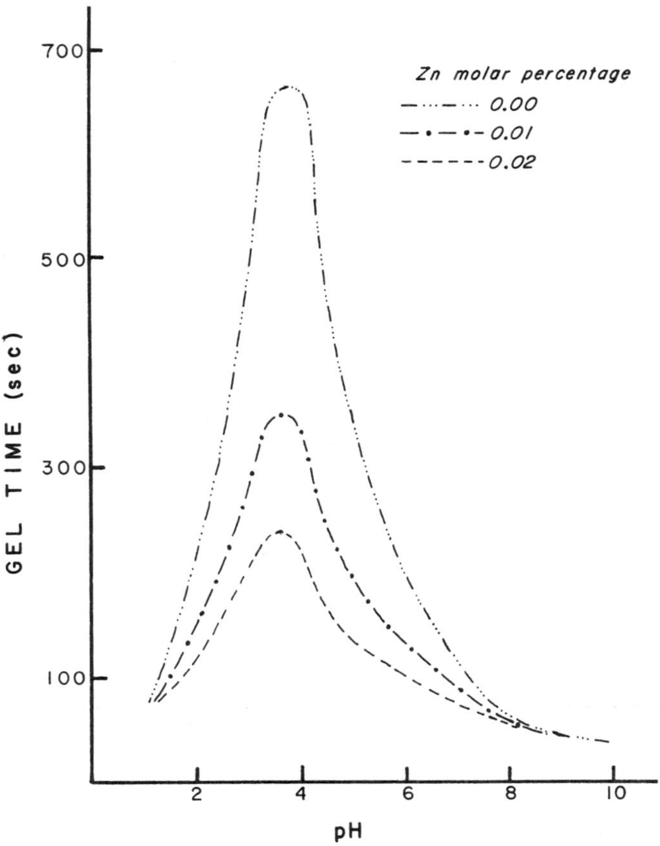

Fig. 4.3. Dependance of wattle tannin-HCHO gel times on amount of metal ion. [From A. Pizzi, *J. Macromol. Sci., Rev. Macromol. Chem.* C 2(18):247–315 (1980).]

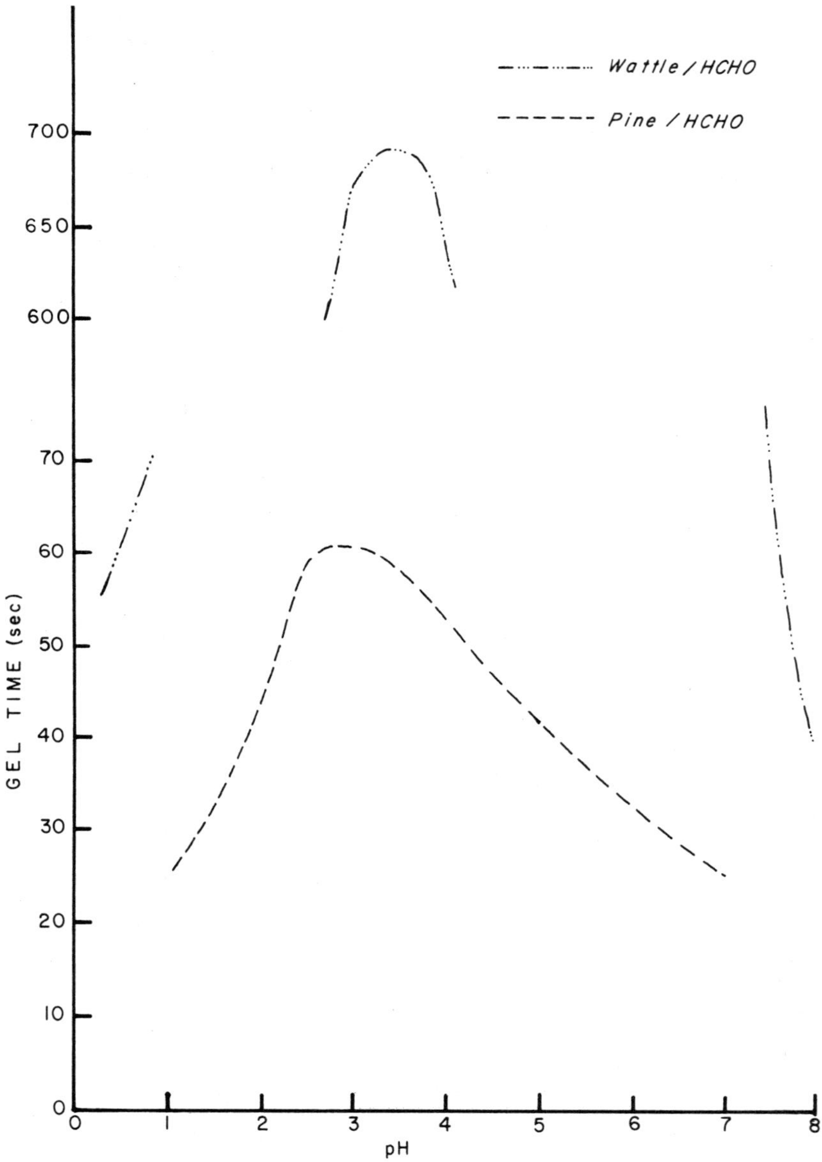

Fig. 4.4. Effect of tannin structure on gel time (96°C). [From A. Pizzi, *J. Macromol. Sci., Rev. Macromol. Chem.* C *2*(18):247–315 (1980).]

due to the formation of a stable ring, is lost, while its accelerating effect due to its strong charge is not. Consequently, all bivalent and trivalent metals function, to a different extent, as accelerators of the flavonoid tannins-formaldehyde reaction.

Proof that tannin steric hindrance is the cause of this peculiar behavior has been obtained by the reaction rates with different metal ions of formaldehyde with sulfited tannins [69]. In these the tannin steric hindrance has been considerably reduced by opening of the etherocyclic ring [7,69,74]. The sulfited tannin will be faster reacting with formaldehyde than unsulfited tannin as the flavonoid A ring now behaves more as a resorcinol nuclei than a 3-methoxyphenolic one [69]. However, at the faster gel times obtained, it is possible to observe that bivalent metals will accelerate the reaction still further, while trivalent ones will again slow it down [62], in line with their observed behavior in the simple phenol-formaldehyde reaction [69]. The metal ion accelerating effect has different intensity at different pH ranges (wattle, Fig. 4.3) [69] and with different tannins [73].

It is noticeable from Figures 4.3 and 4.4 that the accelerating effect of the metal ions will produce a flatter curve with little difference in gel time at low and high pH but with considerable difference around the pH of minimum reactivity. The lack of difference in gel time at low and high pH between tannins with or without metallic ions can be ascribed to the "drowning" of the metal ion effect by the enhanced reactivity, due to the pH, of the tannin-formaldehyde system [69]. The addition of small percentages (±5% on tannin extract) of metal ions have been found to give small increases in flavonoid B-ring cross-linking [56]; consequently, fortification of the tannin-formaldehyde network cannot always be completely eliminated [56,69]. High amounts of metal salts will considerably increase B-ring cross-linking [54,55], but the quantities of metal ions needed render this possibility uneconomical. The effect of the metal ions is then mainly as accelerators of the reaction, resulting in more complete resin curing, hence better cross-linking, when using comparable curing times. Similar effects to those obtained by using metallic acetates [56,67,69] are observed when using other salts of the same metals [67,69] in or out of conjunction with sodium acetate. The effects, though, are sometime less marked [69].

As a general rule of the influence of the anion on the effectivity of the metal as an accelerator of the reaction, the following scale applies in order of decreasing accelerating effect for Zn^{2+}:

$$Zn(CH_3COO)_2 > ZnCl_2 + CH_3COONa > ZnCl_2$$

The anion effect could be explained by the relative affinities of the metal for the anion used and the complex it forms. It should therefore be more difficult to have Zn^{2+} released from a strong acid than a weak acid salt.

It might be expected, for instance, that the Fe^{III} complex would contain the structure

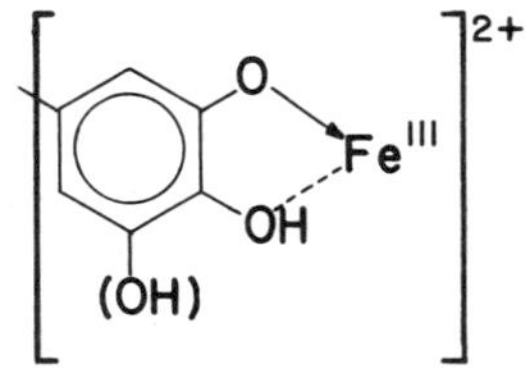

in which only one of the hydroxyl groups is ionized. However, despite the high pK_a value for the hydroxyl group of the C_4 position of the singly ionized anion, the greater stability of the doubly ionized chelate leads to the formation of an *ortho*-diphenol complex of the type [75, 76]

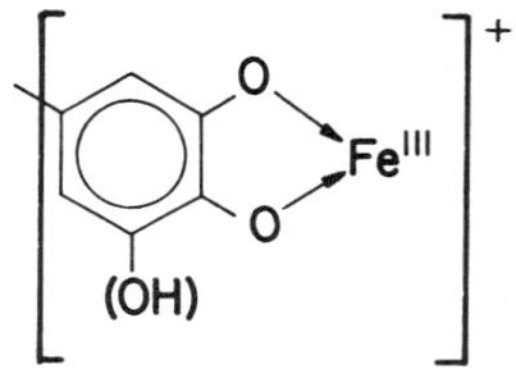

The violet Fe^{III} tannate formed between Fe^{3+} (hexa-aqueous) and mimosa (wattle) tannin under acid conditions is the bis chelate [75, 76]:

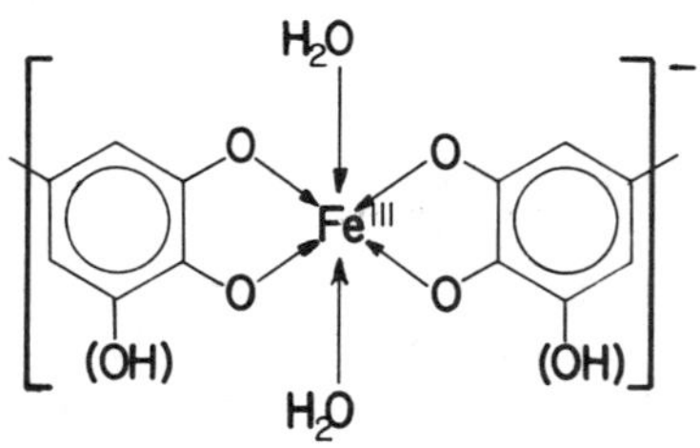

The tris chelate is not likely to form, except under alkaline conditions. The formation of the metal tannate complexes in the absence of formaldehyde does not inhibit the accelerating effect of the same metals [77] on the tannin-formaldehyde condensation once formaldehyde is added to the system [77], indicating that a stability advantage exists in favor of the six-membered etherocyclic ring metal-tannin-formaldehyde complexes over the five-membered ring metal-tannate complexes.

D. Hydrolysis and Acid and Alkaline Autocondensation

When heated in the presence of strong mineral acids, tannins are subject to two competing reactions. One is degradative leading to anthocyanidins and catechin formation as illustrated for a typical biflavonoid

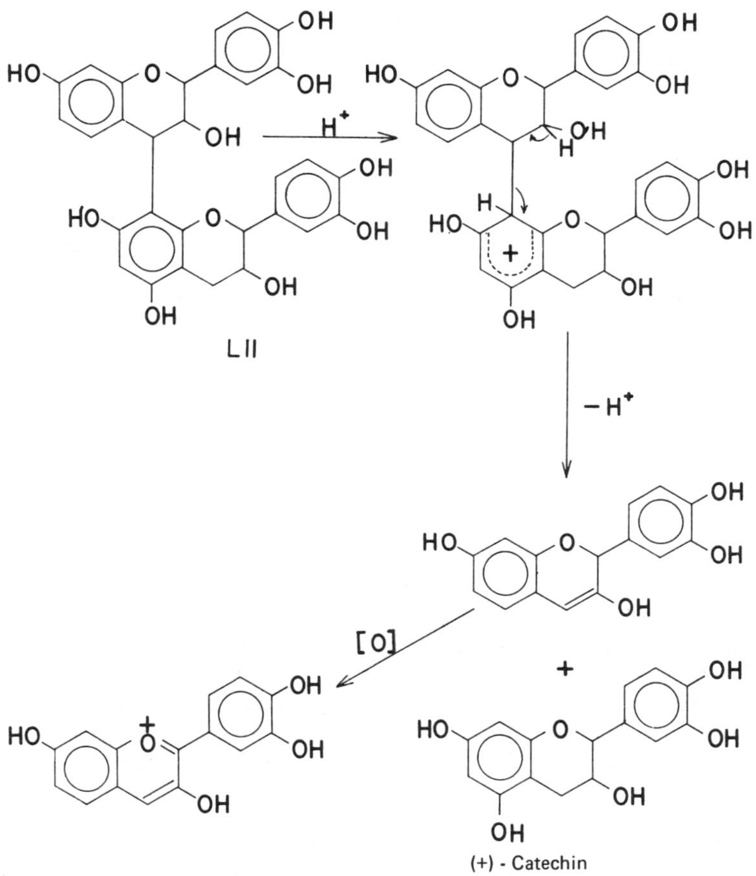

(LII) [7], whereas the second one is condensative as a result of hydrolysis of heterocyclic rings (*p*-hydroxybenzyl ether links) [7]. The *p*-hydroxybenzylcarbonium ions (LIII) created condense randomly with nucleophilic centers on other tannin units to form "phlobaphenes" or "tanner's red" [78–81]. Other modes of condensation (e.g., free radical coupling of B-ring catechol units) cannot be excluded in the presence of atmospheric oxygen.

Predominantly alcoholic conditions (e.g., 80–100% EtOH) lead to a preference for hydrolysis and anthocyanidin formation, although self-condensation is not excluded, while under aqueous conditions phlobaphene formation or formation of insoluble condensates predominates. In absolute alcohols (e.g., ethanol, *n*- and isopropanol), partial etherification of the alcoholic 3-hydroxy groups occurs in the presence of mineral acid, leading to both the anthocyanidin and its 3-O-alkyl derivative in increased yields [82]. Hydrolysis of the interflavonoid link under these conditions occurs far more readily (e.g., the polyleucocyanidin tannins of mangrove extract) [7]. Partial autocondensation in

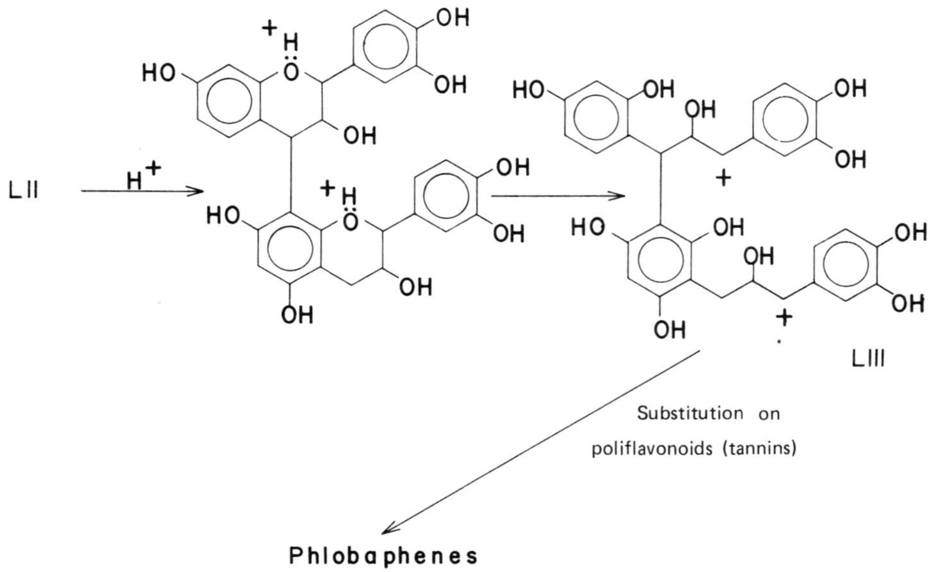

strongly alkaline conditions does also take place as observed by considerable viscosity increases at high pH values. Theoretically [30], this is thought to be due to reactions of the flavonoids C-4 of the lower terminal units with C-6 and/or C-8 of flavonoid units in other tannin polymers, according to which of the latter two is readily available. This is due to the reactivity of hydroxy-group-carrying C-4.

E. Sulfitation

Sulfitation of tannins is one of the oldest and most useful reactions in flavonoid chemistry and can be particularly useful in the preparation of tannin-based adhesives [83].

Sulfitation generally affords tannins of lower viscosity and increases solubility. Both effects are due to:

1. The elimination of the etherocyclic ether group, which is water repellent (it is well known that tannin extract water solutions are not true solutions but hydrocolloids suspensions, in which part of the tannin molecule keeps the tannin in solution while other parts tend to push the tannin out of solution) [84].
2. The introduction of a sulfonic group and another hydroxyl group, both hydrophilic.
3. The decrease in polymer rigidity, steric hindrance, and intermolecular hydrogen bonding obtained by the opening of the etherocyclic ring.
4. Acid hydrolysis of the hydrocolloid gums [88] and of the interflavonoid bond [14].

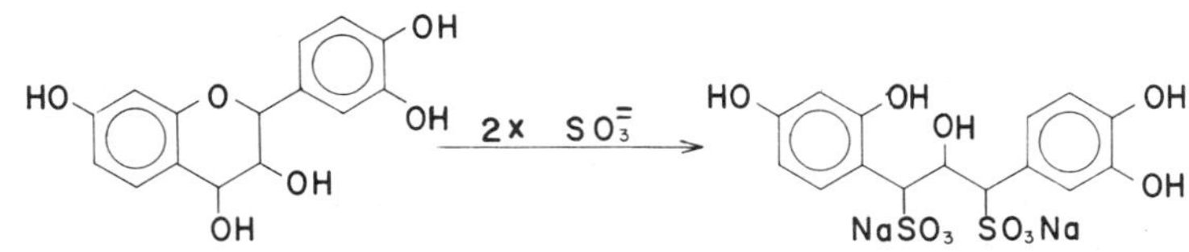

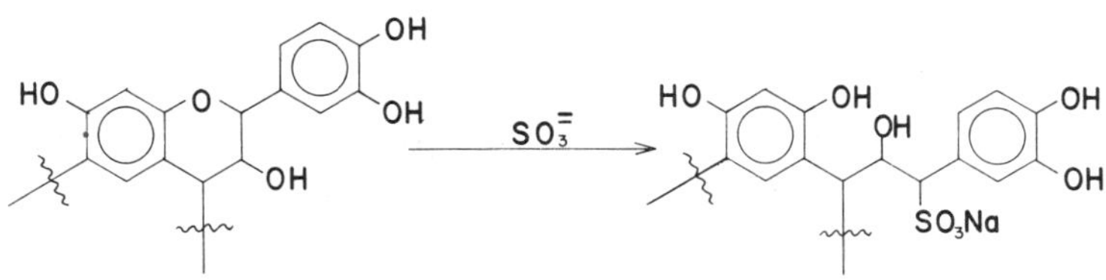

The total effect of all the modifications indicated particularly the introduction of sodium sulfonate groups, while affording the important advantages of higher concentration of tannin phenolics in adhesive applications due to enhanced solubility and decreased viscosity, and of higher moisture retention by the tannin resins, allowing slower adhesive films dry-out, hence longer assembly times [62], also represents, in certain types of adhesives, a distinct disadvantage in that sulfonate groups promote sensitivity to moisture with adhesive deterioration in the absence of adequate cross-linking [85–87]. In the case of phenolic fortified or grafted tannin resins, in which the tannin-formaldehyde network is adequately cross-linked, even high levels of sulfitations do not adversely affect the performance of the resin [62, 74]. To conclude, in sulfited tannins the flavonoid A rings (resorcinol or phloroglucinol) are more reactive toward formaldehyde, having been modified by the opening of the etherocyclic ring from the equivalent of a methoxyphenol to a hydroxyphenol and consequently to a stronger nucleophile [74]. Furthermore, they are more readily available for reaction with formaldehyde because of the generally increased mobility of the tannin molecule as a whole [74]. Both these two factors lead to better cross-linking and consequently better strength of sulfited tannin-formaldehyde networks.

IV. CHEMISTRY AND TECHNOLOGY OF INDUSTRIAL TANNIN ADHESIVE FORMULATIONS

A. General

The purity of vegetable tannin extracts varies considerably. Commercial wattle bark extract and quebracho extract normally contain 70–80% active phenolic ingredients, while pine bark yields an extract containing only 50–60% [31]. The nontannin fraction, consisting mainly of sugars and high molecular weight hydrocolloid gums, cannot participate in resin formation with formaldehyde. In order to demonstrate the effect of the presence of nontannins on the quality of adhesives, a commercial phenol-resorcinol-formaldehyde cold-setting adhesive was adulterated, separately, with various amounts of glucose and gum arabic [31]. The

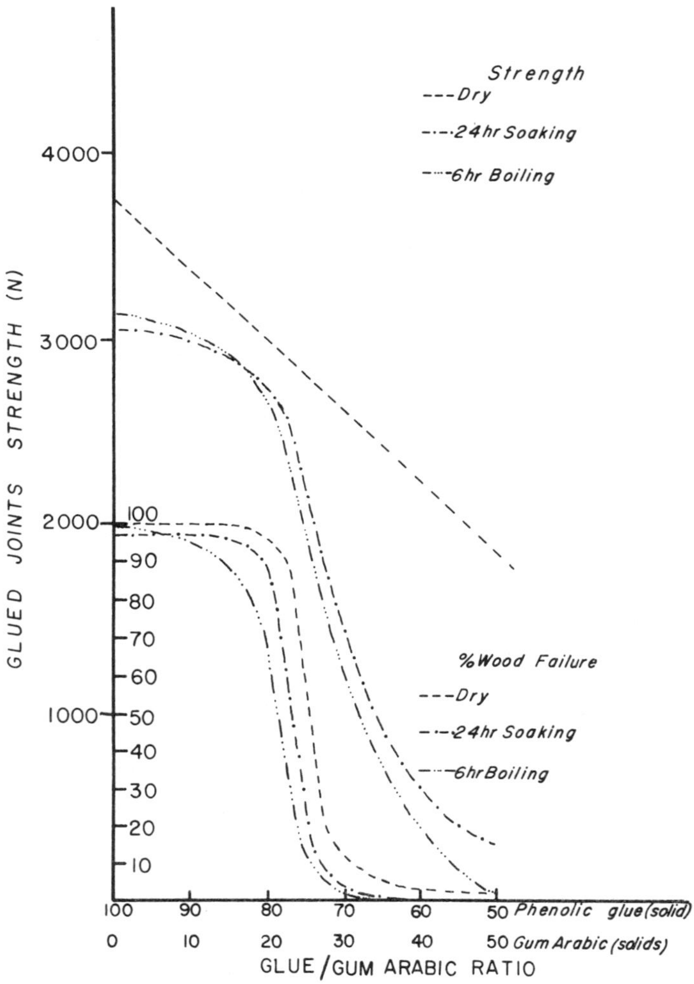

Fig. 4.5. Influence of gum arabic on glued-joint quality. [From A. Pizzi, *J. Macromol. Sci., Rev. Macromol. Chem.* C 2(18):247–315 (1980).]

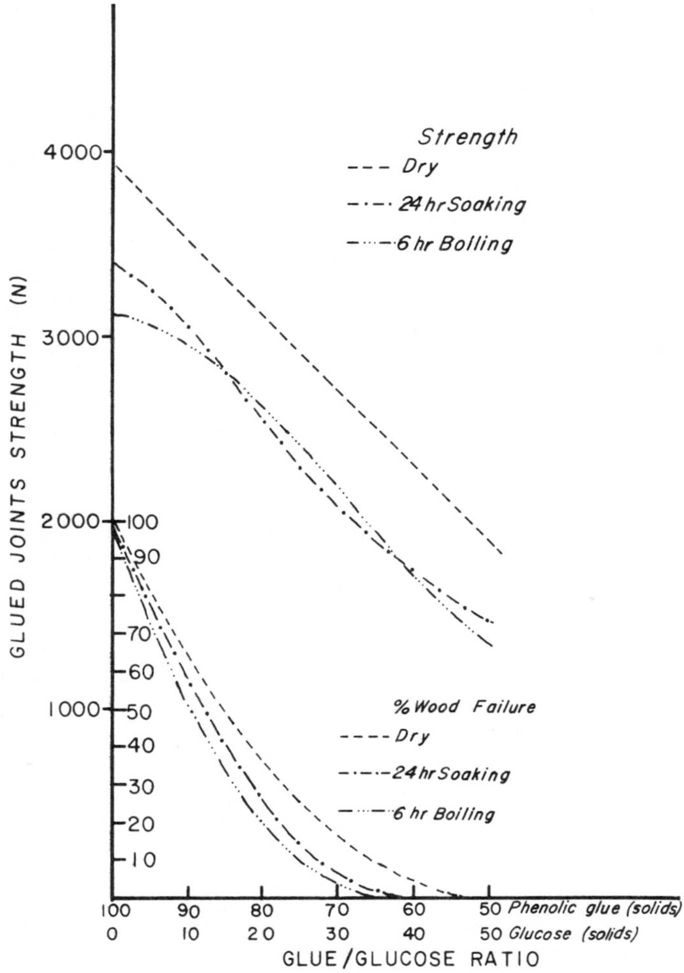

Fig. 4.6. Influence of glucose on glued-joint quality. [From A. Pizzi, *J. Macromol. Sci., Rev. Macromol. Chem.* C 2(18):247–315 (1980).]

effects on strength, wood failure, and water resistance of the glued joints are shown in Figures 4.5 and 4.6.

Sugars reduced the strength and water resistance in proportion to the amount added. Their effect is a mere dilution effect of the adhesive resin solids with consequent proportional worsening of adhesive properties. The hydrocolloid gums, instead, have a much more marked effect on both original strength and water resistance of the adhesive (Fig. 4.3) [88]. If it is assumed that the nontannins in tannin extracts have a similar influence on adhesive properties, it can be expected that unfortified wattle tannin-formaldehyde networks can achieve only 70–80% of the performance shown by synthetic adhesives, and unfortified pine adhesives merely 50–60% [31].

In many glued wood products, the demands on the glue line are so high that unmodified tannin adhesives are unsuitable. The possibility of refining extracts has proved fruitless, largely because the intimate association between the various constituents makes industrial fractionation difficult. Fortification is in many cases the most practical approach to reducing the effect of impurities. Fortification generally consists of copolymerization of the tannin with phenolic of aminoplastic resins [53]. It can be carried out during the manufacture of the adhesive resin, during glue mix assembly, just before use, or during adhesive use. Various synthetic resins have been found effective if added in sufficient quantity to reduce the nontannin fraction to below 20%.

Formaldehyde reacts with tannins and causes polymerization through methylene linkages at reactive positions on the flavonoid molecules, mainly the A rings. However, because of their size and shape, the tannin molecules become immobile at a low level of condensation with formaldehyde, so that the available reactive sites are too far apart for further methylene bridge formation. The result is incomplete cross-linking and therefore weakness. The problem becomes more acute with increasing molecular size of the tannins and is therefor more serious with pine than with wattle or quebracho tannins. Longer cross-linking agents are capable of bridging the distances, which are too long for methylene bridges formed by formaldehyde. Again, fortification, hence copolymerization of the tannins with various phenolic [51] and aminoplastic [53,45] resins, is a convenient method of obtaining longer cross-linking agents and overcoming the problem. The main resins used are of two types:

1. Phenol-formaldehyde or urea-formaldehyde resols with a medium to high methylol group content. These resins can fulfill the functions of hardener, fortifier, or both. These resins are suitable for both resorcinolic and phloroglucinolic types of condensed tannins. They can copolymerize with the tannins during resin manufacture, during use, or both.
2. Resorcinol-terminated phenol-formaldehyde [51,52] or urea-formaldehyde [45] novolaks not containing any methylol groups. They copolymerize with the tannin during use (and curing) after addition of further formaldehyde. They are suitable only as fortifiers, not as hardeners, for the resorcinolic type of condensed tannins.

In the first type of resin, copolymerization and curing are based on the condensation of the tannin with the methylol groups carried by the synthetic resin. In the second type of resins, copolymerization is instead based on the simultaneous condensation with the available formaldehyde of the flavonoids A rings and the resorcinol terminal groups of the synthetic resin. This is brought about by these two groups' comparable nucleophilicity, which makes them compete for the available formaldehyde.

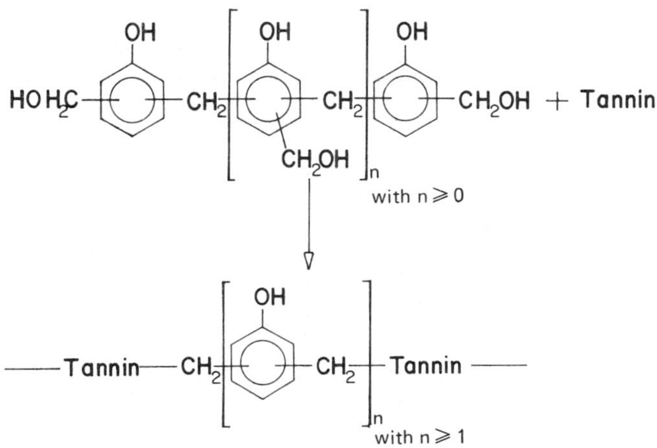

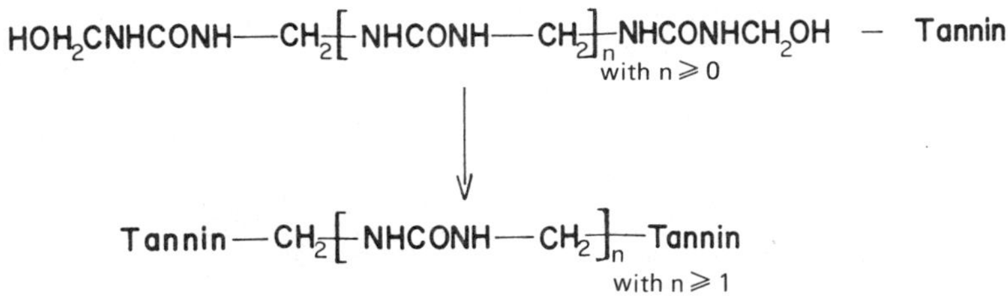

Since tannin molecules are generally large, the rate of molecular growth in relation to the rate of linkage is high, so that tannin adhesives generally tend to have short pot lives. From the point of view of both molecular size and reactivity, phloroglucinolic tannins are worse than mainly resorcinolic tannins, as shown by their relative gel times at their pH values of lowest reactivity: 960 sec for wattle extract (resorcinolic) and 65 sec for pine extract (phloroglucinolic) [31]. Reduction of the molecular size is difficult and therefore does not offer a practical solution. The only ways of lengthening adhesive pot life are to lower initial viscosity by dilution and to retard the reaction. The latter is done by:

1. Addition of alcohols to the adhesive glue mix. Alcohols form hemiacetals with formaldehyde and therefore act as retardents of the tannin-formaldehyde reaction [31]. Methanol is the most efficient alcohol in this respect and the lengthening of pot life obtainable is proportional, within limits, to the amount used. The first reaction step between formaldehyde and all hydroxyaliphatic compounds is the formation of hemiacetals:

$$HOCH_2\text{-}OH + CH_3OH \longrightarrow HO\text{-}CH_2\text{-}O\text{-}CH_3 \quad 15\ \text{kcal/mol}$$

$$HO(CH_2O)_nH + ROH \longrightarrow HO\text{-}(CH_2O)_n\text{-}OR$$

This reaction might account for the consumption of formaldehyde by tannin extracts on curing, to obtain a good cured adhesive, which is considerably higher than what theoretically calculated in the case of only A rings of the flavonoids reacting. Excluding extensive participation of the B rings from the reaction, the three to four times higher formaldehyde consumption might be due to the formation of hemiacetals as a competitive side reaction, between formaldehyde and the 20–25% of sugars and gums present in the extract.

2. Adjustments to the adhesive's pH [31].
3. The use of hexamine as hardener. Hardeners other than formaldehyde can be used to lengthen pot life. Hexamethylenetetramine gives a very long pot life at ambient temperature and fast curing at elevated temperatures [30].
4. The addition of additives other than alcohols retarding the tannin-formaldehyde reaction (e.g., sodium sulfite, which tends to react with formaldehyde).

The viscosity of bark extracts is strongly dependent on concentration. The viscosity increases very rapidly above a concentration of 50%. Compared to synthetic resins, the tannin extracts are more viscous at the concentrations normally required in adhesives. High viscosity of aqueous solutions of condensed tannins is due to the following causes, in order of importance:

1. Presence of high molecular weight hydrocolloid gums in the tannin extract [88]. The viscosity is directly proportional to the amount of gums present in the extract [88,89]. Simple sugars, instead, moderately decrease the viscosity of tannin extracts [31,88].
2. Hydrogen bonds and electrostatic tannin-tannin, tannin-gum, and gum-gum interactions [30,88]. Aqueous tannin extract solutions are not true solutions, but rather colloidal suspensions in which water access to all parts of the molecules present is very slow. As a consequence it is difficult to eliminate intermolecular hydrogen bonds by solution only [88].
3. Presence of high molecular weight tannins in the extract [88].

The high viscosity of tannin extracts solutions has also been correlated with the proportion of very high molecular weight tannins present in the extract. This effect is neither certain nor well defined and in any case appears to be quite small. In most adhesive applications, the viscosity is not critical and can be manipulated by dilution. The tannin extract solutions high viscosity can be reduced in a number of ways.

1. Acid and alkaline hydrolysis of the hydrocolloid gums reduces the extract viscosity by breaking down to sugar the high molecular weight gums [88,30].

2. The addition of small amounts (up to 3% of extract solids) of hydrogen bond breakers reduces the extracts viscosity [30,88]. Simple low molecular weight aromatic molecules such as phenol, phenyl acetate, and naphtalene are the most common hydrogen bond breakers used [30,88]. The poor solubility in water of some of these aromatic compounds appears to confirm that the viscosity reduction effect of these materials is localized in parts of the extract molecular species that are not in true solution [30,90]. There are definite indications that the hydrogen-bond-breaking action is mainly due to the aromatic rings [30,88] and that equivalent amounts of aromatic rings cause a similar decrease in viscosity [30,88]. Aromatic rings, because of their planar configuration, are easily accommodated in narrow intermolecular spaces, disrupting hydrogen bonds by both mechanical and electrical means [91]; mechanically, by simply interposing the plane of the aromatic ring between the two atoms forming a hydrogen bond; electrostatically, by the formation of an alternative hydrogen bond between the hydrogen atom and the π-electron cloud of the aromatic ring. Benzene, phenol, and pyrrole, although quite weak hydrogen-bond bases, have behaviors that correlate quite well with the criteria for well-recognized hydrogen-bond systems [31]. This effect involves a specific orientation of the proton relative to the π electrons [31]. The orientation of the covalent bond of the proton should be perpendicular to the plane of the π-electron cloud of the aromatic ring; deviations from linearity not larger than 10–15° are generally accepted. Hydrogen bond breaking through aromatic rings action must be viewed as a dhynamic rather than static situation. Hydrogen bonds among different chemical species of the system keep breaking and forming. At any given moment of time the number of possible hydrogen bonds is lower when aromatic-ring hydrogen bond breakers are present. This causes a reduction in viscosity; consequently, it is possible to state that aromatic hydrogen bond breakers act as lubricants, increasing the molecular mobility of the system. Hydrogen bond breakers not containing an aromatic ring are also known, such as urea, thiourea, sucrose, and citric, chloroacetic, and *p*-sulfonic acid. Their mechanism is different from that ascribed to the aromatic rings; it can be visualized as a process of competitive solvation, as these agents are capable of severing existing hydrogen bonds between two atoms by bonding preferentially to such atoms. Caution should be exercised in comparing the values of viscosity reduction obtained, as the decrease in extract viscosity given by some of these reagents could be due to effects other than hydrogen bond breaking.

3. Sulfitation and bisulfitation reduce considerably the viscosity of the extract by opening of the tannin etherocyclic ring and increasing

tannin molecular mobility, hence decreasing viscosity [74]. Part of the viscosity reduction is also due to hydrolysis of the gums caused by the sulfite or bisulfite ions.

Tannin adhesives have a tendency to thicken more rapidly than do synthetic resin adhesives after application to a wood surface [31,62]. This is caused mainly by the higher affinity of water for wood than for this particular type of adhesives. The thickening or drying out of the adhesive is related to:

1. The amount of fillers in the adhesive, the problem becoming more acute with heavily extended adhesives. In this case the thickening appears to occur as a result of the low-viscosity tannin solution leaving the fillers in the glue line and being drawn into the wood. This leads to starved glue lines and poor flow. The problem can consequently be identified, in this instance, as one of "apparent viscosity" [31,62].
2. The hydrophobicity of flavonoid etherocyclic rings. Opening of these rings has shown considerable lengthening of drying out times [62,90]. The easier, industrially meaningful way of opening the etherocyclic ring is by sulfitation with sodium sulfite or bisulfite. Introduction of the sulfonic group at the 2-position results in heterocyclic ring opening. Introduction of a sodium sulfonate group as well as a phenolic hydroxyl, and elimination of the ether link, help the adhesive to retain water on the glue line, extending the allowed assembly time of the glued joint and partially eliminating drying out. The etherocyclic flavonoid ring can also be opened by a route different from sulfitation. The flavonoid is treated with a strong acid in the presence of a nucleophile, such as resorcinol, more powerful and mobile than the flavonoid A rings. The acid-induced hydrolysis of the benzylether link causes the formation of a carbocation that reacts with the more powerful nucleophile present (i.e., the resorcinol). This system is also successfully employed in reducing drying out. Other methods employed are based on the addition of phenolic or aminoplastic fortifiers, which improve water retention on the glue line.
3. The use of volatile organic solvents such as methanol also causes drying-out problems. Elimination of part or most of the organic solvent also reduces drying out [62].

It must be pointed out, in connection with items 1 and 2, that the disappearance of the solvent in the glue line due to the higher affinity of water for the wood rather than the tannin, as well as the disappearance of the total adhesive solution when its real viscosity is much lower of its apparent viscosity (i.e., at higher filler content), can be eliminated easily without recurring to structural modifications of the tannin as those described in item 2. Addition of humectants, such as glycerine and ethylene glycol, is not effective.

Most effective is the addition to the adhesive glue mixes of a small amount (±3% of total adhesive solids) of carboxymethyl cellulose (the 70% degree of substitution grade is particularly effective) [77]. Carboxymethyl cellulose (CMC) holds both water and/or low-viscosity adhesive solutions on the glue line without releasing them to the wood by both absorbing available liquid and forming a spreadable thixothropic adhesive paste; furthermore, it reduces high pressure requirements in wood adhesives [77]. No deterioration of water and/or weather resistance is observed, and addition of CMC constitutes one of the best solutions of the drying-out problem. Pre-gelled starch can also be used in place of CMC and it has proved to be effective in reducing drying out, although not as conveniently and effectively as CMC. Adhesives where pre-gelled starch has been used are generally more prone to water- and weather-induced deterioration. Sodium sulfite not reacted with the tannin, but simply added to the glue mix, also improves drying-out times by holding onto the water. To conclude, opening of the flavonoid etherocyclic rings by sulfitation or other means, use of sodium sulfite unreacted with the tannin, addition of carboxymethyl cellulose or of pre-gelled starch, use of phenolic and especially aminoplastic resins as fortifiers, and elimination of part or most of the organic solvent all may lengthen the adhesive drying-out time on a thin glue line by as much as 30%–90%.

The carbohydrates present in the tanning extract do not, generally, contribute to adhesive preparation. A few instance in which there are indications of the formation of an adhesive system in which the carbohydrates probably constitute the most active part have also been reported [13,77,88]. In these systems strongly acid adhesives are produced by reaction of considerable amounts of maleic anhydride [13,88] and sometimes malic acid [13] with the carbohydrate and tannin fractions of the extract in the presence [13] or absence [13,88] of *p*-toluenesulfonic acid. The behavior of these adhesives is that of thermoplastic rather than thermosetting resin, and their dry, room-temperature strength can be in excess of 800 psi. As with many thermoplastic resins, they are not really adhesives suitable for wood gluing. Furthermore, their strongly acid character precludes their use as wood adhesives, as under such acid conditions rapid wood deterioration caused by acid hydrolysis of the cellulose intervenes [13].

By comparison of differential thermal analysis (DTA) data for tannin-formaldehyde mixtures and phenol-formaldehyde and phenol-resorcinol-formaldehyde resins, Steiner and Chow [92] gathered further evidence supporting similarities in cure characteristics between condensed tannin-formaldehyde mixtures and phenol-resorcinol-formaldehyde (PRF) rather than phenol-formaldehyde (PF) resins. Phenol-formaldehyde resins exhibit a characteristic two-endotherm thermogram, the major endotherm peak due mainly to vaporization of water in the range 115–130°C, while a much smaller peak occurs between 150 and 165°C [92]. In addition to these two endothermic peaks, PRF resins show an exothermic peak at about 180°C. This exotherm is resin reactivity dependent [77], rather than only pH dependent [92], and relates to the cure properties of the resin, indicating its room-temperature-cure capability [92]. Resorcinolic-type tannin extracts with formaldehyde at pH 5.0 exhibit similar differential thermal analysis (DTA) thermograms to PRF resins [92]. With increasing pH [92] or by addition to the extract of resorcinol at fixed pH [77], hence with a resin reactivity increase, the exotherm shifts to lower temperature while the second endotherm shifts to 170–175°C, with, in some cases, a significant increase in magnitude coupled with a corresponding decrease in first endotherm size. The presence of an exotherm at less than 100°C is desirable in terms of potential room-temperature-cure capabilities [92]. At pH values higher than 6.0, there is a significant drop in exotherm temperature, from 70°C to 49°C at pH 8.2 [92]. Addition of resorcinol to the tannin-formaldehyde mixture [92] as well as grafting of resorcinol onto the tannin [52,77] before addition of formaldehyde hardener also induces a significant drop in the exotherm temperature [77,92]. Coupling increased pH values with resorcinol additions considerably increases the ease of room-temperature cure, indicating the capability of suitably modified tannin-formaldehyde mixtures and resins to function as cold- or ambient-temperature-setting adhesives [52], behaving as synthetic PRF resins. Coupled with the exotherm peak-shift phenomenon is an increase in exotherm heat, with increasing pH and/or resorcinol addition, which is indicative of accelerating resin reactivity. The shift in size of the first and second endotherms is explained by Steiner and Chow [92] as due to some pre-gelling in the tannin-formaldehyde mixture which forms a loose network binding water to such an extent that its vaporization is delayed.

B. Standardization of Industrial Tanning Extracts

Industrial utilization sets varying standards of chemical and physical reproducibility, of which the following are all or in part currently applied to meet requirements of the tanning industry:

1. Standardized methods of afforestation
2. Harvesting of extractable material on a fixed rotation to ensure continued supplies in the case of certain tanning extracts
3. Consistent climatic and soil conditions to avoid the production of extracts of too great seasonal variation of components
4. Standardized extraction procedures (e.g., countercurrent open vat leaching at fixed times and temperature or pressure leaching under similar conditions)
5. Standardized concentration procedures (e.g., spray-drying)

Considering the macromolecular nature of condensed tannins, the quantity of formaldehyde required in their adhesive applications is well below that required for the preparation of phenol-formaldehyde resins and their setting. A number of products and processes based on tannin adhesives are detailed below.

C. Exterior-Grade Plywood Adhesives

Seven different formulations [51,53,73,93–95] and some of their combinations have been, to date, proved successful for use at the industrial level. These formulations are discussed next.

1. Resorcinolic tannins (wattle, sulfited quebracho, hemlock)
   a. System: adhesive formulations 1 [51,73,94] and 2 [51]

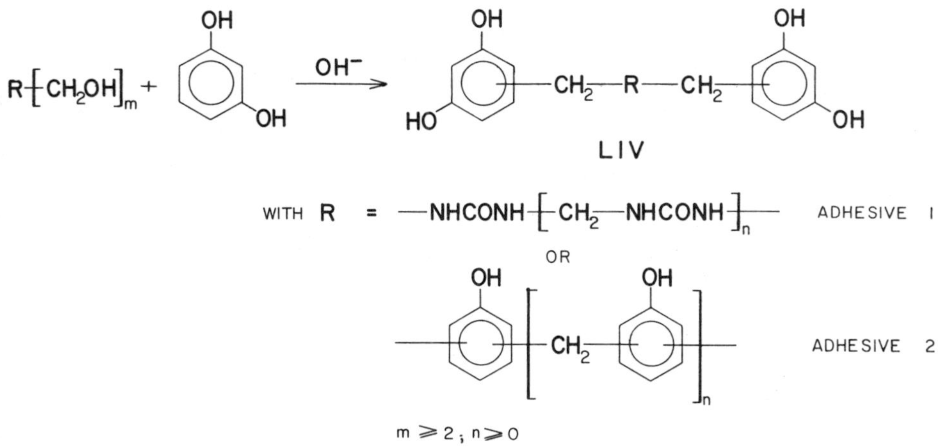

In these formulations, 10–20% (w/w) on tannin extract solids or resin LIV solids is mixed with a tannin extract solution. The mixture is an adhesive that can cure on addition of further formaldehyde (generally paraformaldehyde) and supply of heat. Curing occurs by cross-linking of resin LIV with the A rings of flavonoid polymers through simultaneous reaction with formaldehyde of the resorcinol terminals of resin LIV and the flavonoid A rings. Resin LIV is generally manufactured using commercial urea-formaldehyde (UF) resins or specially manufactured phenol-formaldehyde resins of low level of condensation. The resorcinol percentage (w/w) on resin LIV solids is generally of the order of ±20%. The pH of operation of these adhesives is in the range 5.0–7.0 for both adhesives.

   b. System 2: adhesive formulations 3 [51,53,94], 4 [51], and 5 [51]

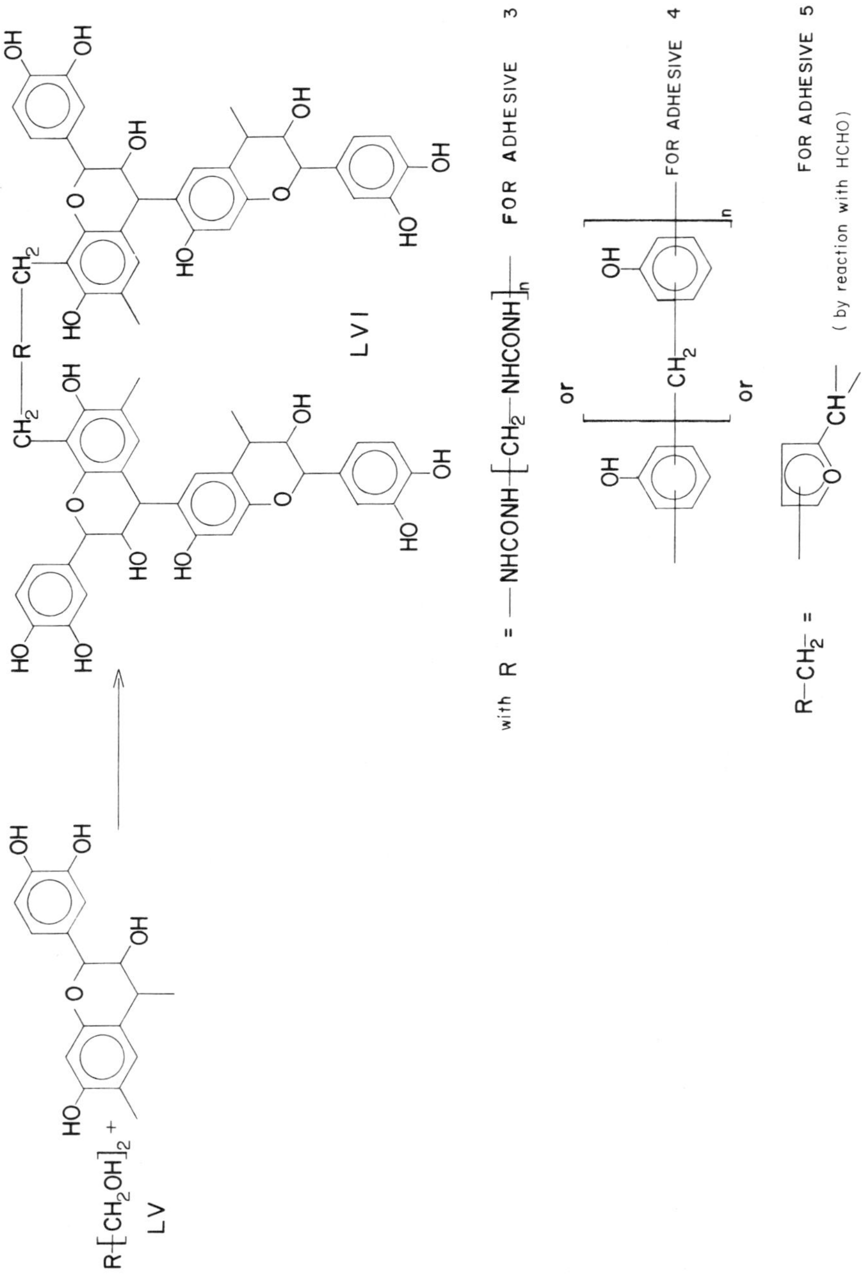
R[CH2OH]2 +
LV
LVI
with R = —NHCONH[CH2—NHCONH]n— FOR ADHESIVE 3
or
FOR ADHESIVE 4
or
R—CH2— =
FOR ADHESIVE 5
(by reaction with HCHO)

Resin LV generally constitutes 10–20% of the total resin solids of resin LVI. Resin LVI is an adhesive that can be set and cured by addition of further formaldehyde (generally paraformaldehyde) and supply of heat. Curing occurs by cross-linking of resin LVI with methylene bridges between different flavonoid A rings. Resin LVI is similar to resin LIV, system 1, in that whereas in resin LIV the reactive terminal group of the phenolic resin is resorcinol, in resin LVI it is the tannin itself. No resorcinol is used in adhesives 3 and 4 and they are very inexpensive formulations. The pH of application of adhesive 3 is generally in the range 4.5–5.5. The pH of application of adhesives 4 and 5 is in the range 5.5–7.0.

c. System 3: adhesive formulation 6 [93]

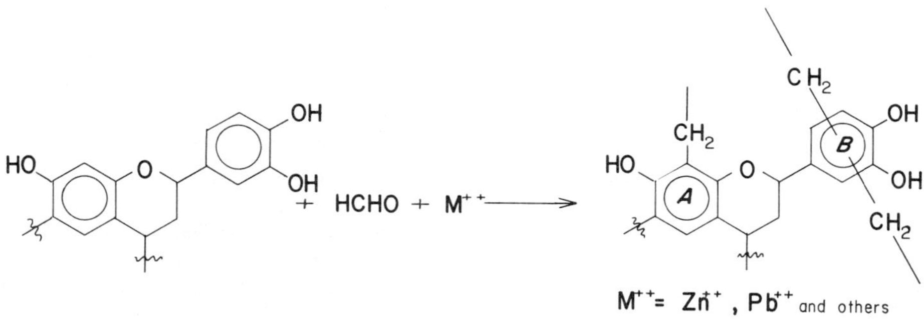

This system is based both on the accelerating effect that bivalent metallic ions have on the reaction of tannins and formaldehyde, causing better curing in shorter times in the cross-linking of flavonoid A rings, and improved strength through B-ring participation to the cross-linking of the tannin-formaldehyde network. The pH of application of this adhesive formulation is generally in the range 3.5–5.5. This system may be used by itself for easy-gluing, low-density veneers, or in conjunction with adhesive 3, system 2, for which it is particularly suited in more difficult cases.

2. Phloroglucinolic tannins (pine tannins)

a. System 4: adhesive 7 [31]

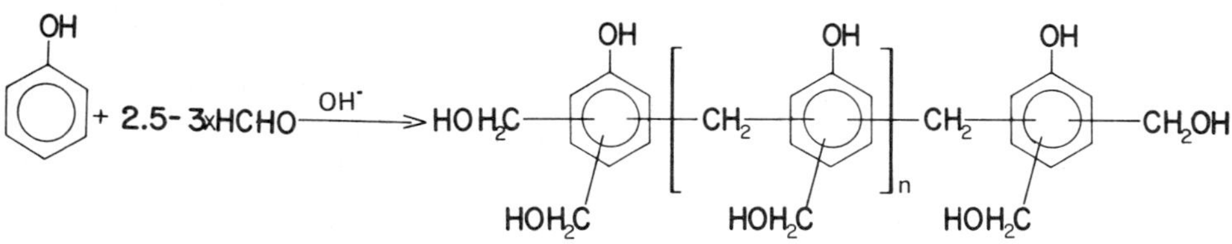

LVII

25–30% (w/w) on tannin solids of resin LVII solids is mixed with a tannin solution. No further formaldehyde is added. Lack of further formaldehyde during curing forces the tannin to cross-link with the only available reactive groups, the methylol groups of resin LVII. Resin LVII then functions as both hardener and fortifier. Resin LVII is also slower than formaldehyde in reacting with tannins, allowing, in the case of phloroglucinolic tannins, better control of the reaction.

The pH of application is in the range 3.5–5.5 for pine tannins. Of these adhesives the most extensively used in industry are adhesives 2, 3, and 6 or a combination of 3 and 6. Adhesive 2 is the oldest of the industrially applied formulations and it is now mainly superseded by formulations 3 and 6, which are the most used, or a combination of formulations 3 and 6, which present the advantages to be easily compounded by the end user from easily available materials (tannin extract, UF resins, paraformaldehyde) and have considerably longer pot lives. Interesting in adhesive 3 is the role of the urea-formaldehyde (UF) resin. UF resins are not weather- and boil-proof, and consequently resistance of tannin-UF copolymers to such conditions has been given close consideration. When small amounts of a UF resin are added to a solution of wattle extract, a reaction takes place immediately, at ambient temperature, giving a soft mass easily dissolved by fast stirring and the addition of small amounts of water. The formation of the gummy mass is due to increase molecular size due to tannin-UF copolymerization; this occurs by reaction of the $—CH_2OH$ groups of the UF resin with the resorcinolic A rings of the flavonoid unit [53,94].

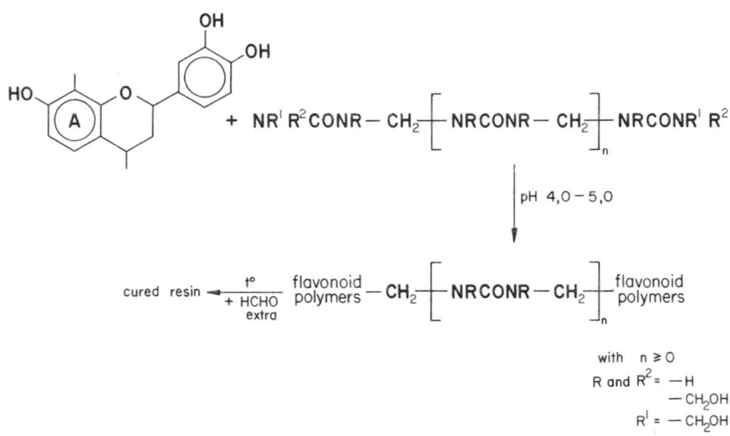

When copolymerized with tannins, small amounts of UF resins fulfill the function of fortifiers and are sufficiently shielded from the effect of cold and boiling water to prevent their normally experienced deterioration [53]. This is valid only up to 15–20% UF resin solids content on total tannin-UF adhesive solids, as for higher UF amounts, reduced water and weather resistance is observed.

In plywood applications both the use of sulfited tannins and the use of carboxymethyl cellulose are allowed. Especially the latter is used when shuttering board panels are produced. A shuttering board composite panel is similar to plywood in the fact that it is formed of layers of cross-grained veneers; the only difference is that the plywood control inner veneers is substituted with a pine block core 16–18 mm thick. This thick pine core, on hot pressing, functions as a "heat sink" and the temperature increase in the glue lines in direct contact with the core is much slower than in the case of plywood. As a consequence, the adhesive in these glue lines remains at low viscosity (the drop due to the rising temperature) for a longer period than in the case of plywood. The liquid resin is then absorbed by the pine block core before curing, leaving the glue line. The adhesives then cure in the block, giving rise to "starved" glued inner lines and consequently a weak panel. This can be avoided by addition of small amounts of carboxymethylcellulose, which forms a gel with the adhesive, avoiding absorption of both adhesive solution and solvent by the pine core block. The conditions of application of the different plywood adhesives can be summarized in general as follows:

Table 4.1. Comparative Results of Different Tannin-Based Plywood Adhesives (Three-Ply 9-mm Plywood)

	Knife test (British Standard BS 1088, 1957)		
Adhesive type	Dry	24-hr cold-water soak	72-hr boiling
1	8	8	8
2	8	8	8
3	8	8	8
4	8	7	7
3 + 5	8	8	8
6	7	7	7
7 (pine tannin)	7	6	5
3 + 6	10	9	9

Source: A. Pizzi, *J. Macromol. Sci., Rev. Macromol. Chem. C 2*(18): 247–315 (1980).

Pressure, 12–16 kg/cm^2
Prepressing, 0–10 min at 5 kg/cm^2
Glue spread, 180–260 g/m^2, single glue line
Veneer moisture content, up to 10–12%
Pressing time, variable according to veneer and plywood thickness

Results for comparable types of panels for the seven plywood adhesives are shown in Table 4.1.

D. Cold-Setting, Fast-Setting, and Radio-Frequency Laminating Adhesives

A series of different novolak-like materials have been prepared by copolymerization of urea-formaldehyde resins or phenol-formaldehyde resols with resorcinol and/or resorcinolic A rings of polyflavonoids, such as condensed tannins [52, 90, 94]. The copolymers formed have been used as cold-setting exterior-grade wood adhesives complying to the relevant international specifications [96]. The formulations used industrially are given below [97].

1. System 1: Grafting of resorcinol on a tannin-formaldehyde resol or on a tannin-UF copolymer having free hydroxybenzyl groups.

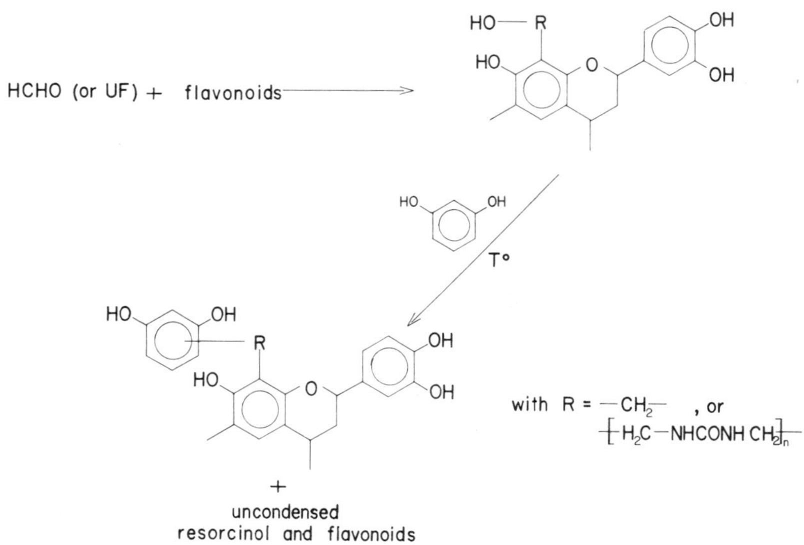

The final mixture of products of this system is an adhesive that can be set and cured at ambient temperature by addition of paraformaldehyde.

2. System 2: Simultaneous synthesis of resorcinol-formaldehyde and flavonoid-formaldehyde condensates [52]. Simultaneous synthesis of resorcinol-UF and flavonoid-UF copolymers.

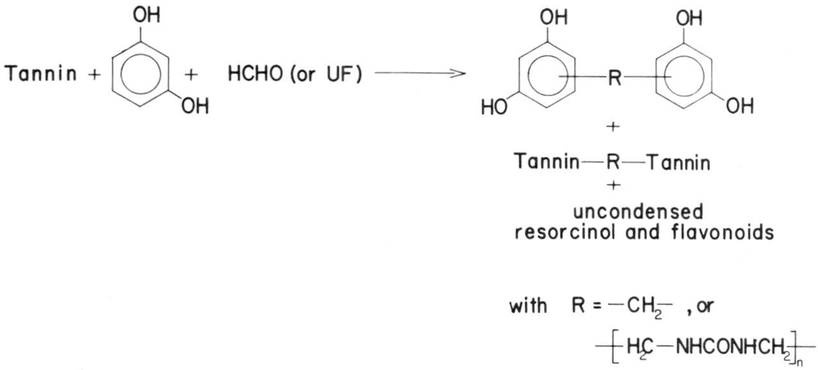

The final mixture of reaction products is an adhesive that can be set and cured at ambient temperature by addition of paraformaldehyde.

3. System 3: Synthesis of stable resorcinol-formaldehyde, phenol-resorcinol-formaldehyde, or resorcinol-urea-formaldehyde resins and subsequent addition to flavonoid polymers [52,94,98]. Resorcinol-formaldehyde, resorcinol-terminated phenol-formaldehyde, and resorcinol-terminated urea-formaldehyde resins are prepared and subsequently added to flavonoid polymers. The mixture of synthetic resin and tannins is used as an adhesive that can be set and cured at ambient temperature by addition of paraformaldehyde.

These three systems give the most commonly used cold-setting tannin-based resins. However, two other unusual systems have also found industrial application. One of them is a fast-setting finger jointing "honeymoon" adhesive, and the other comprises a series of adhesives based on the opening of the tannin etherocyclic ring. These are discussed briefly below.

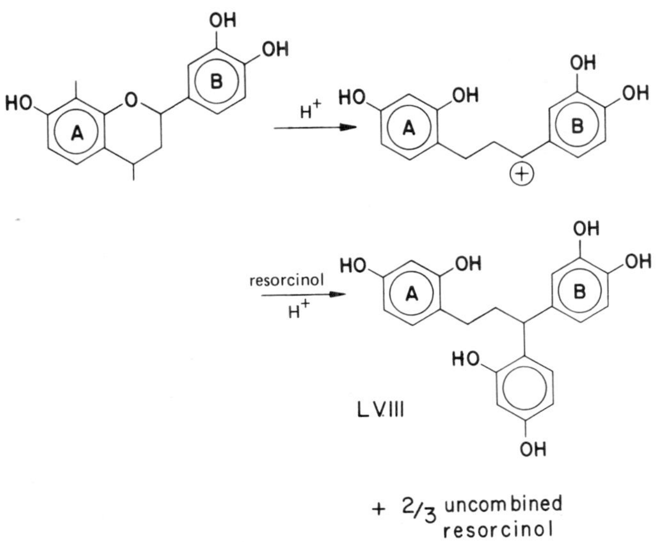

4. System 4: Acid hydrolysis of flavonoid etherocyclic ring and simultaneous grafting of resorcinol without formaldehyde [90].

The resorcinol must be present at the moment of the formation of the carbocation [78–81,90], as otherwise the carbocation would attack the next best nucleophile, the flavonoid A rings, with the formation of phlobaphenes [80], insoluble polymers unusable for adhesives. Only about one-third of the resorcinol necessary to produce a good tannin-based cold-setting adhesive can be grafted in this manner, and to improve the adhesive cohesive strength, reaction of mixture LVIII with formaldehyde is effected to obtain the following reaction:

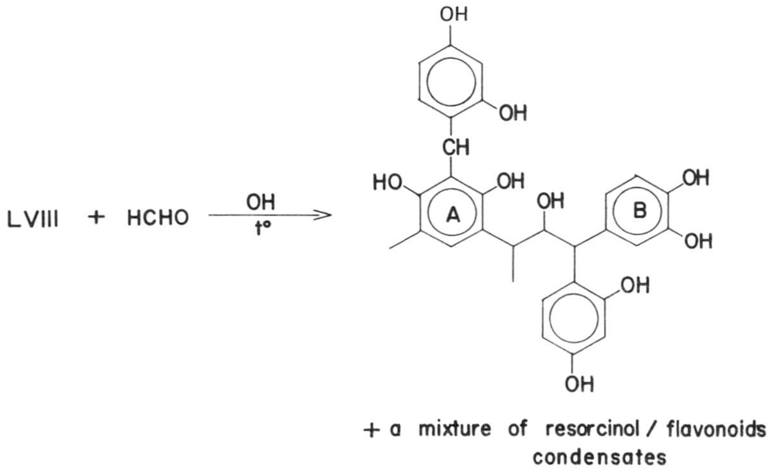

5. System 5: "Honeymoon" or "separate-application" tannin-based adhesive system for finger joints [99] and glulam [97]. Separate-application adhesives can be described according to Scheme II [100].

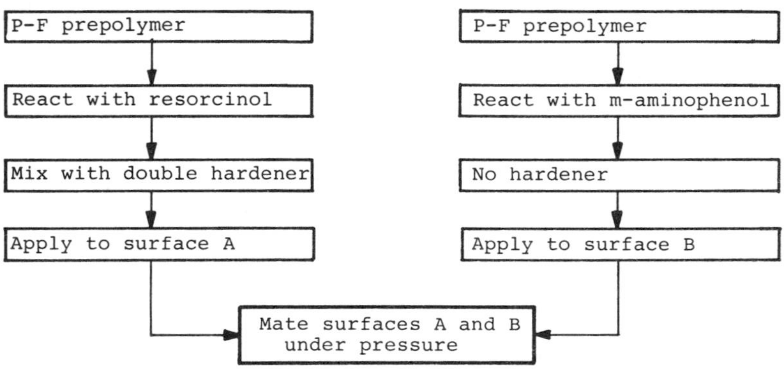

In the case of tannin-based adhesives, component A is a slow-reacting resin of the types described in systems 1 to 4, but of low pH with double the amount of hardener, while component B is a reactive resin containing no hardener. The component B resin is composed of a small amount of specially prepared *m*-hydroxyaniline-terminated phenol-formaldehyde resin mixed with a tannin solution at high pH.

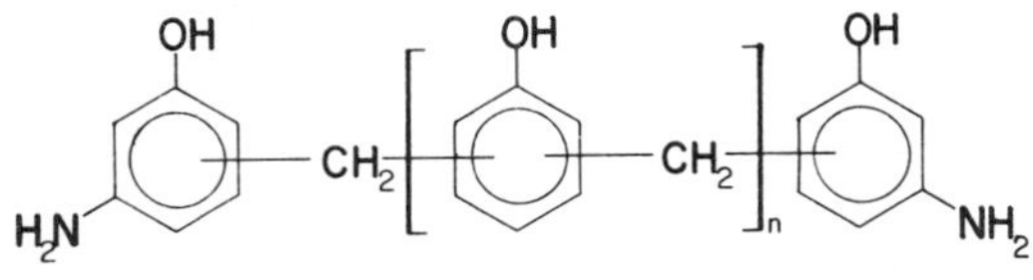

When parts A and B are mated, the reactive parts of the components react to form a joint strong enough to be handled and processed further within minutes. Fullcuring of the slow-reacting part of the system takes place with time. Development of bending strength versus time in finger-jointed boards is shown in Figure 4.7. New types of fast-setting adhesives were also prepared by using plain wattle tannin and/or pine tannin extract solutions of pH 12 as component B of the system [97]. Component A is a wattle cold-set adhesive or a phenol-resorcinol-formaldehyde adhesive. The results are also very good and the adhesive considerably cheaper [97].

Both cold-setting and thermosetting plywood adhesives can be used in radio-frequency curing. Some of the organic solvent must however be removed to reduce electrode arching [101].

The adhesives used for components A and B of the honeymoon systems are the following:

Component A	Component B
A1. Commercial wattle-resorcinol-formaldehyde cold-setting laminating adhesive + hardener + fillers (pH 7.5)	B1. Commercial wattle-formaldehyde adhesive fortified with phenol/*m*-aminophenol/formaldehyde resin; no hardener; pH 9.0
A2. Commercial phenol-resorcinol-formaldehyde cold-setting laminating adhesive + hardener + fillers (pH 8.0)	B2. Wattle-resorcinol-formaldehyde cold-setting laminating resin; no hardener; pH adjusted to 11.4
	B3. Pine tannin extract; no hardener; pH adjusted to 12.4
	B4. Wattle tannin extract; no hardener; pH adjusted to 12.6
	B5. Commercial phenol-resorcinol-formaldehyde resin; no hardener; pH adjusted to 11.4

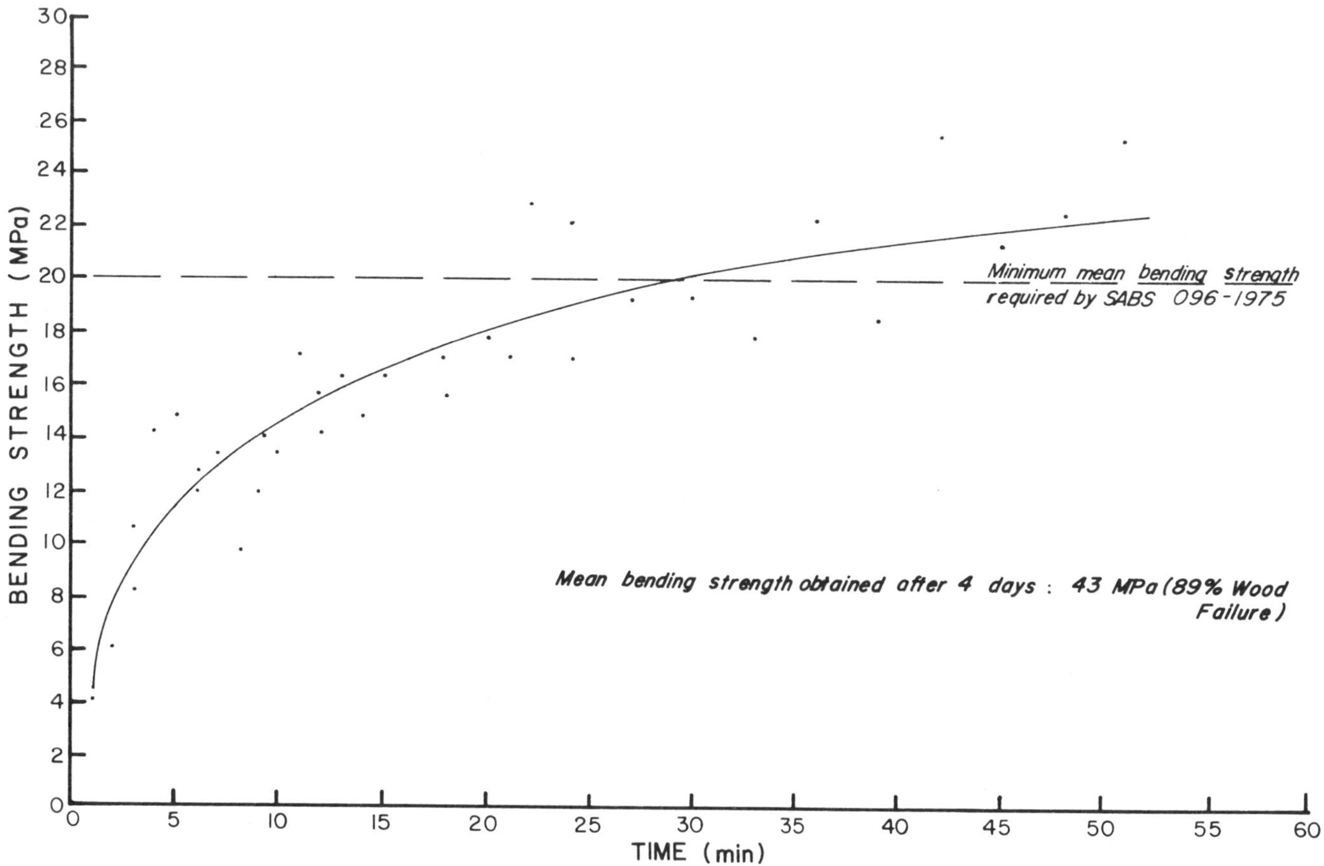

Fig. 4.7. Development of finger joint bending strength, "honeymoon" adhesive. [From A. Pizzi, *J. Macromol. Sci., Rev. Macromol. Chem C* 2(18):247–315 (1980).]

Table 4.2. Typical Shear Strength and Wood Failure Results of Tannin-Based Cold-Setting Adhesives: Close Contact Joints

Adhesive system	Dry test		24-hr cold-water soak		6-hr boiling	
	Shear strength (psi)	Wood failure (%)	Shear strength (psi)	Wood failure (%)	Shear strength (psi)	Wood failure (%)
1	799	100	614	74	570	87
2	875	100	664	100	641	98
3	830	79	695	100	684	100
4	528	36	532	25	583	74
2-sulfited	780	97	645	89	630	92
2 + 4	690	85	608	78	578	80
British Standard BS 1204–1965 requirements			500	—	325	—

Source: A. Pizzi, *J. Macromol. Sci., Rev. Macromol. Chem. C* 2(18):247–315 (1980).

All the combinations of components A and B (10 combinations) can be used as fast-setting adhesives for the manufacture of finger joints.

The comparative wood failure and strength results of the different cold-setting adhesives are shown in Table 4.2. Also shown in Table 4.2 are the results obtained using 10% sulfited tannin to improve drying-out time and to lengthen assembly time.

E. Exterior-Grade Particleboard Adhesives

Unmodified condensed tannins, when reacted with formaldehyde, give adhesives having characteristics that do not suit particleboard manufacture: namely, high viscosity, low strength, and poor water resistance. The most commonly used process to eliminate these disadvantages in the preparation of tannin-based particleboard adhesives consists of a series of subsequent acid and alkaline treatment of the tannin extract, causing hydrolysis of the gums to simple sugars, thus improving viscosity, strength, and water resistance of the unfortified tannin-formaldehyde adhesive [88].

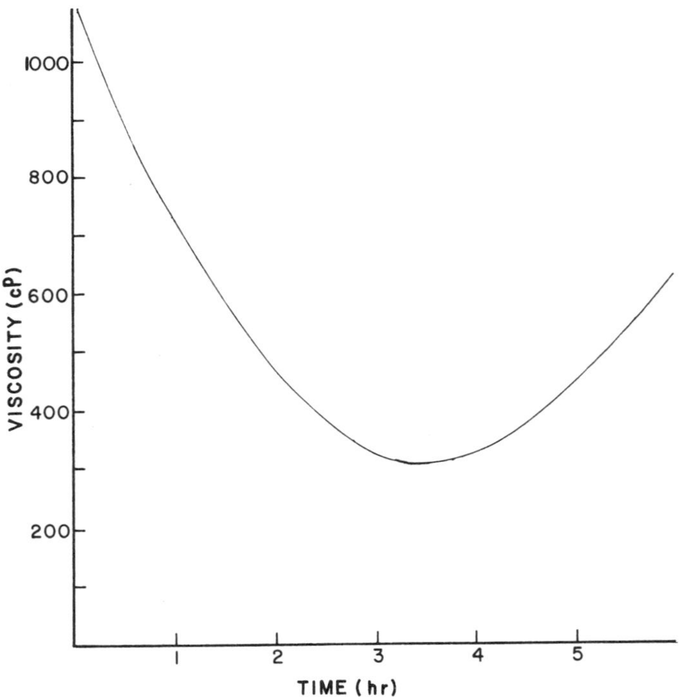

Fig. 4.8. Effect of time of hot caustic soda treatment on the viscosity of wattle extract at 50% solids; ph 7.0, 25°C.

Drastic hot caustic soda treatment of wattle extract causes tannin autocondensation and sometimes extract charring. Milder and prolonged hot caustic soda treatment causes, instead, a decrease in its viscosity. The viscosity decreases with increased length of the hot caustic soda treatment, reaching a minimum between 3 and 3½ hr and then increases again (Fig. 4.8). Adhesives prepared with extracts treated with caustic soda under heat for 3–3½ hr tend to give excellent cohesive-strength results caused by partial liberation of resorcinol by alkaline cleavage of the flavonoid etherocyclic ring [102].

Adhesives prepared with extracts in which the caustic soda treatment is overprolonged give poorer cured strength results [102]. Organic anhydrides were proved to be more effective than simple organic acids for the acid hydrolysis [88]. In addition to acid and alkaline treatments, the most common process used reduces further the viscosity of the tannin extract by addition of simple aromatic molecules as hydrogen bond breakers [88,89]. Unfortified tannin extracts which have undergone this process react with formaldehyde to give adhesives which are able to produce excellent exterior-grade particleboard. This type of process is effective when resorcinolic tannins are used. Some of the fortified tannin-formaldehyde adhesives for plywood already described can also be successfully used to produce excellent exterior-grade particleboard, although not as good as those obtained with the treated unfortified tannin-formaldehyde adhesive. The results obtainable with these unfortified tannin adhesives are comparable to those obtained with synthetic phenol-formaldehyde adhesives and are shown in Table 4.3.

Particular gluing and pressing techniques have been developed for tannin particleboard adhesives [30,103] to achieve pressing times much faster than those obtained with synthetic phenol-formaldehyde adhesives (2–3 min for tannin versus 5–5½ min for phenolic in boards of equal thickness). The success of these simple types of particleboard adhesives heavily relies on industrial application technology rather than on the chemistry and preparation technology of the adhesive itself [88,103].

In the case of spraying blenders, the following technique can be applied successfully. All the paraformaldehyde powder hardener and 30% of the wattle adhesive spray-dried powder is removed from the liquid glue mix. The paraformaldehyde and wattle adhesive powder are then added to the wood chips just before the adhesive blender where the liquid glue mix is sprayed onto the wood particles. Core material moisture contents of 10–12% and face material moisture contents of 20–22% can then easily be achieved in spite of the high viscosity of tannin solutions. The percentage of paraformaldehyde used should be 14% based on wattle extract solids. The adhesive pH should be, in optimal cases 6.5–6.7 for face material and 6.9–7.3 for core material. The percentage of resin solids on drywood should be 11% for core material and 14–18% for face material.

Table 4.3. Unfortified Tannin-Based Adhesives for Exterior-Grade Particleboard: Example of Industrial Boards Results

Panel density (g/cm^3)	Swelling after 2-hr boiling: Measured wet (%)	Swelling after 2-hr boiling: Measured dry (irreversible swelling) (%)	Original internal bond (IB) (tensile perpendicular) (kP/cm^2)	IB after 2-hr boiling[a] (kP/cm^2)	Cyclic test swelling after five cycles measured dry[a] (%)
0.700	11.0	0.0	13.0	9.0	3.0

[a]From German Standard Committee (Deutschen Normenausschuss), V100 and V313, specification for particleboard DIN 68761, Pt. 3, 1967.
Source: A. Pizzi, *J. Macromol. Sci., Rev. Macromol. Chem.* C 2(18):247–315 (1980).

In the case of spreading and spraying blenders, as in the case of Drais Turboplan blenders, the following technique can instead be successfully applied. The wattle-based adhesive is diluted with water to approximately 30% solids content and blended to the chips in the glue blender. All the adhesive is applied as a liquid glue mix. After resination, the wood particles have a very high moisture content, as high or higher than 30%. Before reaching the board-forming station, the wood particles are passed through a forced-air countercurrent predrier at a temperature variable between 70 and 90°C. The maximum period of permanence of the resinated wood particles in the predrier is approximately 4 sec. During the time of their permanence in the predrier, in which the wood particles describe a floating descending trajectory, their moisture content decreases to the low values required for fast pressing times, namely 10–14% for core particles and 16–22% for face particles. The air temperature in the predrier as well as the flow rate of the resinated wood particles through it must be regulated to obtain the required moisture content. The short time of permanence of the resinated particles in the predrier does not impair the performance of the adhesive. Also in this system the paraformaldehyde powder hardener must be added in powder form by means of a screw conveyor to the particles before they are resinated.

Whereas the second system described can be applied to all types of blenders, the first system cannot be applied to the second type of blender described. In the case of a Drais Turboplan machine the first system can produce good exterior particleboard in runs of no longer than 1 hr, as the machine is slowly choked by a solid ring composed of resin and the finer wood particles hardened by the paraformaldehyde present and the heat of particle friction. This chocking ring, of Bakelite-type consistency, forms in the first half of the gluing blender and on the inner wall of its casing. Its direction of growth is from the casings wall toward the central axis of the blender.

In the case of phloroglucinolic tannins, completely different adhesives and application techniques are used. A melamine-formaldehyde or a phenol-formaldehyde or a wattle-tannin-formaldehyde adhesive to which an excess of formaldehyde has been added (in the form of paraformaldehyde) and unmodified pine tannin extract are separately (simultaneously or subsequently) sprayed onto the wood chips [104]. The pine tannin can be added in liquid form, as a low-resin solids solution, or partly in powder form to avoid viscosity problems [104]. The separate application of the mixture of paraformaldehyde and fortifying adhesive (melamine, phenolic, or wattle tannin resin) and of the pine tannin is necessary to overcome any problem of short pot life and precuring of the tannin/formaldehyde network that can be experienced with such reactive tannins. The fortified tannin-formaldehyde adhesive for plywood (number 7, already described) can also be used, although with less encouraging results than the separate application system described above.

The best adhesive formulation for phloroglucinolic-type tannins such as pine tannin extracts (i.e., *P. radiata*, *P. patula*, and the loblolly pines such as *P. elliotii* and *P. taeda*) is instead a comparatively new adhesive formulation which is also capable of giving excellent results when using resorcinolic tannins such as a wattle tannin extract [105]. The adhesive glue mix consists only of a mix of an unmodified tannin extract solution to which has been added paraformaldehyde and 4,4'-diphenyldiisocyanate (commercial MDI) [105]. The proportion of tannin extract solids to MDI is 70:30, based on mass. If lower viscosities are needed, the tannin can undergo the modifications already described to decrease its viscosity. The extract solution, paraformaldehyde, and MDI can be mixed a short time before spraying onto the chips or sprayed onto the chips by separate applications which necessitates the use of separate nozzles for the tannin solution and the MDI. In both types of application systems, the paraformaldehyde is better added separately to the chips according the industrial techniques already described for resorcinolic-type tannin adhesives. In this manner, by fortification of the tannin with MDI, the poorer results generally obtained with phloroglucinolic tannins when the cross-linking mechanism used is based only on their reaction with formaldehyde are overcome. This is due to the fact that MDI will cross-link with the alcoholic and phenolic hydroxy groups of the tannins, sugars, and gums present in the extract, forming polyurethanes and increasing the cross-linking obtained by the reaction of the tannins phenolic A rings with the formaldehyde. As a consequence, certain factors limiting the successful use of phloroglucinolic tannins as wood adhesives, such as the usual low content of phenolic material (MDI also reacts with sugars and gums, hence these materials will participate in the cross-linking mechanism) and their uncontrollably short pot life due to the extremely high reactivity of their phloroglucinolic A rings with formaldehyde (controllable by separate application of paraformaldehyde) will be overcome by the use of a different reaction, as the phloroglucinolic nuclei of the flavonoids will not influence the rate of formation of the urethane links.

The properties of the particleboard manufactured with this system using phloroglucinolic tannin-based adhesives are the following:

	Swelling after 2-hr boil				
Density (g/cm^3)	Measured wet (%)	Measured dry (irreversible swelling) (%)	Original internal bond (IB) (kP/cm^2)	IB after 2-hr boiling (kP/cm^2)	IB retention after 2-hr boiling (%)
0.690	15.0	4.3	8.4	4.3	51

The results obtainable with this system are hence quite good and not too different from the results obtainable with some of the other tannin adhesives already described. Furthermore, such adhesive formulations can be used with any type of phloroglucinolic or resorcinolic (i.e., wattle) tannin and consequently constitute a really universal tannin adhesive for particleboard. In the case of a phloroglucinolic tannin extract being used, no pH adjustment of the solution is needed. In the case of the somewhat slower reacting resorcinolic tannin extracts, the pH is best adjusted to 6.5–7.0. The same formulation can be applied sucessfully to plywood manufacture.

One point that was given close consideration is the deactivating effect of water on the isocyanate group of MDI. It has been found that the amount of deactivation by water of this group when in a concentrated solution (50% or over) of a phenol is much lower than previously thought [106]. This is the reason aqueous tannin extract solutions and MDI can be reacted without substantial MDI deactivation by the water present.

F. Corrugated Cardboard Adhesives

The adhesives developed for the manufacture of damp-ply-resistant corrugated cardboard based on the addition of spray-dried wattle extract, urea-formaldehyde resin, and formaldehyde to a typical Stein-Hall starch formula of 18–22% starch content [107]. The wattle tannin-urea-formaldehyde copolymer is formed *in situ* and any free formaldehyde left in the glue line is absorbed by the wattle extract. The wattle extract powder should be added at a level of 4% of the total starch content of the mix (i.e., carrier plus slurry). Successful results can be achieved in the range 2–12% of the total starch content, but 4% is the recommended starting level. The final level is determined by the degree of water hardness and desired bond quality. While certain wattle-formaldehyde systems [108] have given considerable processing problems in the plant and have consequently been abandoned, the wattle extract-UF fortifier system is highly flexible and can be adopted to damp-proof a multitude of basic starch formulations. A typical starch corrugating mix and its strength results are listed in Table 4.4.

G. Generation of Resorcinol

All the flavonoid components present in wattle wood (not bark) tannins are only based on resorcinol (A rings), and alkali fusion of the wood extract gives resorcinol (also β-resorcilic acid if its decarboxylation is not complete) and protocatechuic acid [109]. The wood contains about 8% extractives, which afford 8% crystalline resorcinol on alkali fusion [109]. This alternative source of resorcinol from its synthesis based on benzene is not economical, notwithstanding rising commercial

Table 4.4. Typical Fortified Stein-Hall Starch Mix for Corrugated Cardboard

Primary tank	
Water (cold)	300 kg
Starch	75 kg
Heat to 40°C, then add:	
Caustic soda } premixed	17 kg
Water } premixed	15 kg
Heat to 70–72°C and hold for 10 min, then add:	
Water (cold)	300 kg (up to 650 kg is added in some factories to give desired viscosity characteristics)
Secondary tank	
Water (cold)	1450 kg (heat up to 30°C)
Starch	525 kg
Borax } premixed	13 kg
Water } premixed	15 kg
After dropping the content of the primary tank into the secondary tank, the components of the damp-proofing resin are added, in random order, to the secondary tank.	
Spray-dried wattle extract	25 kg (approx. 4% of starch mass)
Liquid UF resin (65% solids)	4 kg
Formaldehyde	8 kg
Mix for 10 min, then pump to holding tank or directly to glue trays.	
Typical average damp-ply bond strength = ±700 N/cm^2	

Source: A. Pizzi, *J. Macromol. Sci., Rev. Macromol. Chem. C 2*(18): 247–315 (1980).

resorcinol prices, as the resorcinol produced in this way is approximately 50% more expensive than that produced starting from benzene.

More interesting and economically more promising is the generation of resorcinol *in situ* as resorcinol already bound to the polymer [74, 87]. Examples of this approach are sulfitation [74] and/or acid hydrolysis of the flavonoid benzyl-ether links [90] already described, in which the A rings pass from the equivalent of *m*-methoxyphenol to the equivalent of a *m*-hydroxyphenol (resorcinol). Sulfited tannins are in effect more reactive toward formaldehyde than unsulfited flavonoids, indicating that the transformation of the A ring to linked resorcinol does truly have potential in adhesives application as reflected by an increase in reactivity of approximately 20–25%. The formulations presented in systems 1, 2, and 3 can be modified by using a sulfited tannin extract rather than a plain tannin extract. The process of manufacture of the adhesive, after sulfitation of the tannin extract, remains exactly the same, the only provision being that the amount of resorcinol chemical added is decreased in proportion to the amount of resorcinol generated from the tannin, hence in proportion to the level of sulfitation. On this basis the resorcinol chemical content of cold-setting adhesives described in previous paragraphs can be nearly halved by using a heavily sulfited (7.5–20%; optimum 10–12%) tannin rather than an unsulfited one, without any loss of performance of the adhesive.

H. Infrared Analysis of Resorcinol Content in Tannin-Based Adhesives [45]

The use of the ratio between the infrared (IR) absorption bands at 1150 cm^{-1}, characteristic of resorcinol, and at 1200 cm^{-1}, characteristic of the aryl-carbon-oxygen absorption of phenols, allows calculation of the resorcinol content in different wattle tannin-based resins: namely, simple mixtures of tannin and resorcinol, resorcinol-tannin grafts, and mixtures of phenol-resorcinol-formaldehyde resins with tannin. The empirical relationship between the 1150/1200 absorbance ratio and resorcinol content is different from that attained for synthetic phenol-resorcinol-formaldehyde resins [45,46]. This relationship is essentially linear if expressed as a percentage of resorcinol on the rest of the phenolic material present. This relationship through linear regression curve fitting can be represented by

$$y = 0.016x + 1.195 \quad \text{for } 0\% \geqslant x \geqslant 100\%$$

$$\text{correlation coefficient } r^2 = 0.988$$

where x is the percentage of resorcinol content and y is the 1500/1200 absorbance ratio.

Use of the ratio between the absorption bands at 1430 and 1450 cm^{-1} is also possible. The ratio gives a good linear relationship with the percentage of flavonoid A rings plus resorcinol on total resin solids. The relationship can be expressed by

$$y = 0.02375x - 0.00275 \quad \text{for } 25\% \geqslant x \geqslant 100\%$$

$$\text{correlation coefficient } r^2 = 0.950$$

where x is the percentage of A rings plus resorcinol on total resin solids and y is the absorbance ratio.

V. CONCLUSIONS

This chapter has focused on tannin structures and reactions related to their application in adhesives and resins and to their more successful industrial applications. Uses other than those described are possible or still under investigation: for example, substitution with tannins of phenolic resins used in paper impregnation for the production of industrial laminates. As the resins used in this and other applications are very similar to those described for plywood, it has not been considered necessary to describe them separately. This review, however, still offers a good example of how the fundamental knowledge of a chemical subject can bring interesting, unusual, and extensive industrial applications.

Although condensed tanning extracts have been used for the manufacture of leather since the turn of the century, extension of their industrial use to adhesives and resins reflects more recent developments. This use has contributed to the consumption of significant tonnages of tannin extracts, and emphasizes their importance as industrial raw materials in relation to alternatives to synthetic phenols based on petroleum. The future application of the resin systems discussed in this chapter should accelerate for economic reasons, and for the better and wider understanding of tannin extract composition, characteristic reactions, and macromolecular structure.

REFERENCES

1. E. Kulvik, *Adhes. Age 18*:3 (1975).
2. E. Kulvik, *Adhes. Age 19*:3 (1976).
3. S. E. Drewes and D. G. Roux, *Biochem. J. 87*:167 (1963).
4. H. G. C. King and T. White, *J. Soc. Leather Trades' Chem. 41*:368 (1957).
5. D. G. Roux and E. Paulus, *Biochem. J. 78*:785 (1961).
6. H. M. Saayman and D. G. Roux, *Biochem. J. 97*:734 (1965).
7. D. G. Roux, D. Ferreira, H. K. L. Hundt, and E. Malan, *Appl. Polym. Symp. No. 28*:335 (1975).
8. D. G. Roux, *Tydskr. Natuurwet.*, 133 (1970).
9. D. G. Roux, *Phytochemistry 11*:1219 (1972).
10. D. G. Roux, *Modern Applications of Mimosa Extract*, Leather Industries Research Institute Publication, Grahamstown, South Africa, 1965, pp. 34–41.
11. H. G. C. King, T. White, and R. B. Huges, *J. Chem. Soc.*, 3234 (1961).
12. J. W. Clark-Lewis and D. G. Roux, *J. Chem. Soc.*, 1402 (1959).
13. C. E. Boyette, M.Sc. thesis, Faculty of Forestry, Mississippi State University, 1978.
14. R. W. Hemingway and G. W. McGraw, *Appl. Polym. Symp.* 28 (1976).

15. D. du T. Rossouw, A. Pizzi, and G. McGillivray, *J. Polym. Sci. Polym. Chem. Ed. 18,* (1980); D. du T. Rossouw, M.Sc. thesis, University of South Africa, Pretoria, 1979; A. Pizzi and P. van der Spuy, *J. Polym. Sci., Polym. Chem. Ed. 18* (1980); A. Pizzi, *J. Polym. Sci., Polym. Chem. Ed.*, accepted for publication (1981).
16. L. J. Porter, *N.Z. J. Sci. 17*:213 (1974).
17. D. W. Engel, M. Hattingh, H. K. L. Hundt, and D. G. Roux, *Chem. Commun.*, 695 (1978).
18. D. Nisi and L. Panizzi, *Gazzetta 96*:803 (1966).
19. V. Krishnamoorthy and T. R. Seshadri, *Tetrahedron 22*:2367 (1966).
20. F. Delle Monache, I. L. D'Albuquerque, F. Ferrari, and G. B. M. Bettóle, *Tetrahedron Lett.*, 4211 (1967); *Farmaco 25*:96 (1970).
21. S. E. Drewes, D. G. Roux, S. H. Eggers, and J. Feeney, *Chem. Commun.*, 368 (1966); *J. Chem. Soc. C*, 1217 (1967).
22. S. E. Drewes, D. G. Roux, H. M. Saayman, J. Feeney, and S. H. Eggers, *Chem. Commun.*, 370 (1963); *J. Chem. Soc. C*, 1302 (1967).
23. D. G. Roux, *Phytochemistry 11*:1219 (1972).
24. S. R. Evelyn, *J. Soc. Leather Trades' Chem. 38*:142, 309 (1954); S. R. Evelyn, D. R. Cooper, and P. C. Van Berge, *J. Polym. Sci. 33*:53 (1954).
25. K. Weinges, K. Göritz, and F. Nader, *Annalen 715*:164 (1968).
26. K. Weinges, W. Kaltenhäuser, H. D. Marx, E. Nader, J. Perner, and D. Seiler, *Annalen 711*:184 (1968).
27. I. C. du Preez, Ph.D. thesis, University of the Orange Free State, South Africa, 1971.
28. I.C. du Preez, A. C. Rowan, and D. G. Roux, *Chem. Commun.*, 315 (1971).
29. D. Ferreira, H. K. L. Hundt, and D. G. Roux, *Chem. Commun.*, 1257 (1971).
30. A. Pizzi, Ph.D. thesis, University of the Orange Free State, South Africa, 1977.
31. H. Scharfetter, A. Pizzi, and D. du T. Rossouw, IUFRO Conf. Wood Gluing, Merida, Venezuela, Oct. 1977.
32. D. G. Roux, private communication, Dec. 1978.
33. J. J. Botha, D. Ferreira, and D. G. Roux, *Chem. Commun.*, 700 (1978).
34. D. G. Roux and S. R. Evelyn, *Biochem. J. 76*:17 (1960).
35. D. G. Roux, *J. Soc. Leather Trades' Chem. 35*:322 (1951).
36. D. G. Roux, private communications, 1977–1978.
37. N. P. Slabbert, *LIRI Res. Bull. 632,* Leather Industries Research Institute, Grahamstown, South Africa, 1971.
38. A. Lemme, U.S. Patent No. 3,232,897 (1966).
39. C. G. Gordon-Gray, *J. Soc. Leather Trades' Chem. 61*:269 (1957).
40. D. G. Roux, *J. Soc. Leather Trades' Chem. 61*:275 (1957).

41. H. Buchwald and E. Richardson, *J. Inorg. Nucl. Chem. 15*:133 (1960).
42. D. H. Brown and J. D. McCallum, *J. Inorg. Nucl. Chem. 25*: 1483 (1963).
43. J. L. Goldstein and T. Swain, *Phytochemistry 2*:371 (1963).
44. J. Halmekoski, *Ann. Acad. Sci. Fenn. AII*:96 (1959).
45. A. Pizzi and G. F. Daling, *Holzforsch. Holzverwert. 32*:2 (1980).
46. S. Chow and P. R. Steiner, *Holzforschung 32*(H.4):120 (1978).
47. E. Stiasny and F. Orth, *Collegium 24*:50, 88 (1924).
48. A. Wissing, *Sven. Papperstidn. 20*:745 (1955).
49. E. Stiasny, *Gerber*, 187 (1905).
50. R. Lauffmann, *Collegium*, 332 (1917).
51. A. Pizzi and H. O. Scharfetter, *J. Appl. Polym. Sci. 22*:1745 (1978).
52. A. Pizzi and D. G. Roux, *J. Appl. Polym. Sci. 22*:1945 (1978).
53. A. Pizzi, *Adhes. Age 20*(12):27 (1977).
54. W. E. Hillis and G. Urbach, *J. Appl. Chem.*, 474 (Sept. 1959).
55. W. E. Hillis and G. Urbach, *J. Appl. Chem.*, 665 (Dec. 1959).
56. A. Pizzi, *J. Appl. polym. Sci. 22*:2397 (1978).
57. K. F. Plomley, *Aust. For. Prod. Technol. Pap. 46*, 1959, p. 16.
58. H. M. Saayman, *LIRI Res. Bull. 466*, Leather Industries Research Institute, Grahamstown, South Africa, 1967.
59. H. M. Saayman, private communication, 1972.
60. F. W. Herrick and L. H. Bock, *For. Prod. J. 8*(10):269 (1958).
61. H. McLean and J. A. F. Gardner, *Pulp Paper Mag. Can.*, Aug. 1952.
62. D. du T. Rossouw and A. Pizzi, 26th Conv. S. Afr. Chem. Inst., Port Elizabeth, South Africa, Jan. 1979.
63. K. F. Plomley, *Div. Aust. For. Prod. Technol. Pap. 39*, 1966.
64. A. Pizzi and H. O. Scharfetter, *CSIR Spec. Rep. Hout 138*, Pretoria, South Africa, 1977.
65. R. Kroon, private communication, 1976.
66. N. J. L. Megson, *Phenolic Resins Chemistry*, Butterworth, Kent, England, 1958.
67. D. A. Fraser, R. W. Hall, and A. L. J. Raum, *J. Appl. Chem. 7*:676 (1957).
68. D. A. Fraser, R. W. Hall, P. A. Jenkins, and A. L. J. Raum, *J. Appl. Chem. 7*:689 (1957).
69. A. Pizzi, *J. Appl. Polym. Sci.*, Parts I and II, *24*:1247, 1257 (1979); A. Pizzi, *Polym. Lett. 17*:489 (1979).
70. J. P. Collman, Reactions of coordinated ligands and homogenous catalysis, *Adv. Chem. Ser. 37*(5):78 (1963).
71. A. E. Martell and M. Calvin, *Chemistry of Metal Chelate Compounds*, Prentice-Hall, Englewood Cliffs, N.J., 1952.
72. R. W. Kluiber, *J. Am. Chem. Soc. 82*:4839 (1960).

73. K. F. Plomley, *Aust. J. Appl. Sci. 11*:494 (1960).
74. A. Pizzi, *Colloid Polym. Sci. 257*:37 (1979).
75. N. P. Slabbert, Ph.D. thesis, Rhodes University, Grahamstown, South Africa, 1972.
76. A. J. Seavel, *J. Oil Colour Chem. Assoc. 61*:439 (1978).
77. A. Pizzi, unpublished work, 1978–1979.
78. R. Brown and W. Cummings, *J. Chem. Soc.*, 4302 (1958).
79. J. J. Botha, D. Ferreira, and D. G. Roux, *Report*, CSIR Unit for Flavonoid Chemistry, Bloemfontein, South Africa, 1976.
80. K. Freudenberg and J. M. A. de Lama, *Annalen 612*:78 (1958).
81. R. Brown, W. Cummings, and J. Newbould, *J. Chem. Soc.*, 3677 (1961).
82. D. G. Roux and M. C. Bill, *Nature 183*:42 (1959).
83. H. Richtzenhain and B. Alfredsson, *Chem. Ber.*, *89*:378 (1956).
84. D. G. Roux, *Wattle Tannin and Mimosa Extract*, Leather Industries Research Institute, Grahamstown, South Africa, 1965.
85. L. K. Dalton, *Aust. J. Appl. Sci. 1*:54 (1950).
86. L. K. Dalton, *Aust. J. Appl. Sci. 4*:54 (1953).
87. J. R. Parrish, *J. S. Afr. For. Assoc. 32*:26 (1958).
88. A. Pizzi, *For. Prod. J. 28*:12 (1978).
89. B. Taljaard and H. M. Saayman, Viscosities of wattle extract solutions, Wattle Bark Industry of South Africa, Adhesive Research Committee, *Report V13*, 1977.
90. A. Pizzi and D. G. Roux, *J. Appl. Polym. Sci. 22*:2717 (1978).
91. G. C. Pimentel and A. L. McLellan, *The Hydrogen Bond*, W. H. Freeman, San Francisco, 1960, p. 202.
92. P. R. Steiner and S. Chow, Factors influencing western hemlock bark extracts used as adhesives, *Inf. Rep. VP-X-153*, Western Forest Products Laboratory, Vancouver, Oct. 1975.
93. A. Pizzi, *J. Appl. Polym. Sci.*, Parts I and II, *24*:1247, 1257 (1979); A. Pizzi, *Polym. Lett. 17*:489 (1979).
94. A. Pizzi, *J. Appl. Polym. Sci. 23*:2999 (1979).
95. F. W. Herrick and L. H. Bock, *For. Prod. J. 8*(10):269 (1958).
96. Specification for synthetic adhesive resins for wood, British Standard BS 1204, Parts 1 and 2, 1965.
97. A. Pizzi, D. du T. Rossouw, W. Knuffel, and M. Singmin, *Holzforsch. Holzverwert. 32*(6):140 (1980); F. A. Cameron and A. Pizzi, *CSIR Spec. Rep. Hout 207, 211, 212, and 215,* (1981); *Hout 280 and 291* (1983), Pretoria, South Africa.
98. F. W. Herrick and R. J. Conca, *For. Prod. J. 10*(7):361 (1960).
99. P. K. van der Westhuizen, A. Pizzi, and H. Scharfetter, *Wood South. Afr. 3*(7):13 (1978).
100. R. E. Kreibich, *Adhes. Age 17*:26 (1974).
101. A. Pizzi, *J. Appl. Polym. Sci. 22*:3603 (1978).
102. A. Pizzi, *Int. J. Adhes. Adhes. 1*(2):107 (1980); *2*(1) (1981).
103. A. Pizzi, *Adhes. Age 21*(9):32 (1978). *Holzforsch Holzverwert,* accepted for publication.
104. A. Pizzi and D. du T. Rossouw, unpublished data, 1978.

105. A. Pizzi, *J. Macromol. Sci., Chem.*, accepted for publication, 1980; *Holz Roh-Werkstoff*, August 1982.
106. A. Pizzi, *J. Appl. Polym. Sci. 25*:2123 (1980).
107. A. E. McKenzie and J. P. Yuritta, *Appita*, 26 (1974); P. A. J. L. Custers, R. Rushbrook, A. Pizzi, and C. J. Knauff, *Holzforsch. Holzverwert.*, submitted for publication.
108. H. M. Saayman and N. N. Brown, *For. Prod. J. 27*:3 (1977).
109. D. G. Roux, D. Ferreira, and D. C. F. Garbutt, *Appl. Polym. Symp. No. 28*:1365 (1976).

5

Lignin-Based Wood Adhesives

HORST H. NIMZ* / *Polymer Institute, University of Karlsruhe, Federal Republic of Germany*

*Present affiliation: Institute of Wood Chemistry and Chemical Technology of Wood, Federal Research Centre for Forestry and Forest Products, Hamburg 80, Federal Republic of Germany

I. INTRODUCTION

The occurrence of lignin as a waste product in pulp mills has made it an attractive raw material for adhesives ever since the beginning of the sulfite pulping of wood. The first patents dealing with the application of spent sulfite liquor (SSL) as an adhesive for paper, wood, and other lignocellulosic materials date back to the end of the last century [1], and since then have shown an ever-growing number. According to statistics of the Food and Agriculture Organization of the United Nations, the worldwide pulp production in 1976 amounted to some 90 million tons, of which about 15 million tons were obtained by the sulfite process. By this process, only half of the wood raw material is obtained as pulp, the other half forming SSL. Currently, some 550,000 tons of SSL is produced annually in West Germany, 460,000 tons in Austria, 1.65 million tons in Sweden, 1.1 million tons in Finland, 2.9 tons in Canada, 2.2 tons in the United States, and 2.8 million tons in the USSR.

On the other hand, the technical utilization of lignin on a large scale is still an unsolved problem. Presently, most of the spent liquors in pulp mills are burnt. Only about 20% are used for various purposes, such as dispersants, oil-well-drilling muds, pelletizing materials, molding stabilizers and concrete grinding additives. In light of an ever-growing shortage of organic raw materials, this is of course an unsatisfactory situation.

The main drawbacks in the application of SSL as an adhesive for paper are its dark color and its high hygroscopicity. For this reason it could not really compete with other natural or synthetic adhesives. However, the brownish color of SSL is no disadvantage in the manufacture of particleboard, plywood, or fiberboard. Furthermore, the rapid increase in particleboard production during the past three decades demands an adhesive that is cheap, available in large quantities, and independent of crude oil. Besides meeting these requirements, lignin does not release formaldehyde, which has become a serious problem with urea-formaldehyde (UF) adhesives. This chapter describes several developments in the use of lignin derivatives as wood adhesives; none of these developments are at present used industrially in full-scale production. However, several of the systems presented have potential industrial interest.

As a major wood component, native lignin is neither hygroscopic nor soluble in water. However, during technical sulfite pulping, lignin becomes soluble in water, due to partial degradation and introduction of sulfonic acid groups ($-SO_3H$). In applying SSL as an adhesive, it has to be converted to an insoluble state during the curing period. Chemically, the curing of lignin is a cross-linking process, which leads to new carbon-carbon and ether bonds between different lignin molecules or within one macromolecule. Inter-as well as intramolecular cross-linking reactions decrease the solubility and swelling ability of lignin. Cross-links in lignin can be achieved either by condensation or by radical coupling reactions. A great number of patents have become known during the past three decades (see Tables 5.1, 5.3, and 5.6) dealing with the application of SSL as a wood adhesive, in which the lignin is cross-linked by condensation reactions. However, either high temperatures and long heating times or mineral acids are required for these condensation reactions, which cause structural changes or charring in the wood particles. Recently, cross-linking of the lignosulfonate molecules by radical combinations, which avoids mineral acids and high temperatures, has been developed. The chemistry of these two types of cross-linking reactions, condensations and radical coupling, is discussed briefly in the following section.

II. CHEMICAL BACKGROUNDS OF THE CURING REACTION OF LIGNIN

Lignin is composed of phenylpropane (C_9) units that are linked together by carbon-to-carbon as well as carbon-to-oxygen (ether) bonds. Our present knowledge of lignin structure is based on the assumption that it is formed from *p*-hydroxycinnamyl alcohols by oxidative coupling [2]. The main precursor of spruce lignin is coniferyl alcohol (I), which is oxidized by hydrogen peroxide and peroxidase to a phenoxy radical (R). The unrepaired electron in R is delocalized and reacts at three different sites of the radical, indicated by the four resonance structures R_a to R_d, leading to the dilignols A–F (Fig. 5.1).

Dilignols A–F, due to their free phenolic groups, can be oxidized further to dimeric phenoxy radicals which, by radical coupling, polymerize to a high molecular weight cross-linked lignin molecule. Figure 5.2 shows a structural scheme of spruce lignin composed of 18 C_9 units. Due to its infinite molecular weight and cross-linked structure, lignin in wood is insoluble in any solvent.

Treatment of wood with HSO_3^- ions under pulping conditions leads to partial degradation by the splitting of some ether bonds and introduction of sulfonic acid groups ($-SO_3H$) mainly at the benzylic carbon atoms, rendering the lignin soluble in water. Thus lignosulfonates differ from lignin (Fig. 5.2) by having a somewhat lower molecular weight and about 0.5 sulfonic acid group per C_9 unit. Technical spent sulfite liquor (SSL) contains 50–60% lignosulfonates, about 30% carbohydrates, and up to 20% inorganic salts, depending on the pulping conditions.

A. Cross-Linking of Lignosulfonate by Condensation Reactions

When lignosulfonate is treated with strong mineral acids at elevated temperatures or heated at temperatures above 180°C, condensation reactions leading to diphenyl methanes and sulfones take place (Figure 5.3). The reactivity of lignosulfonates depends to some extent

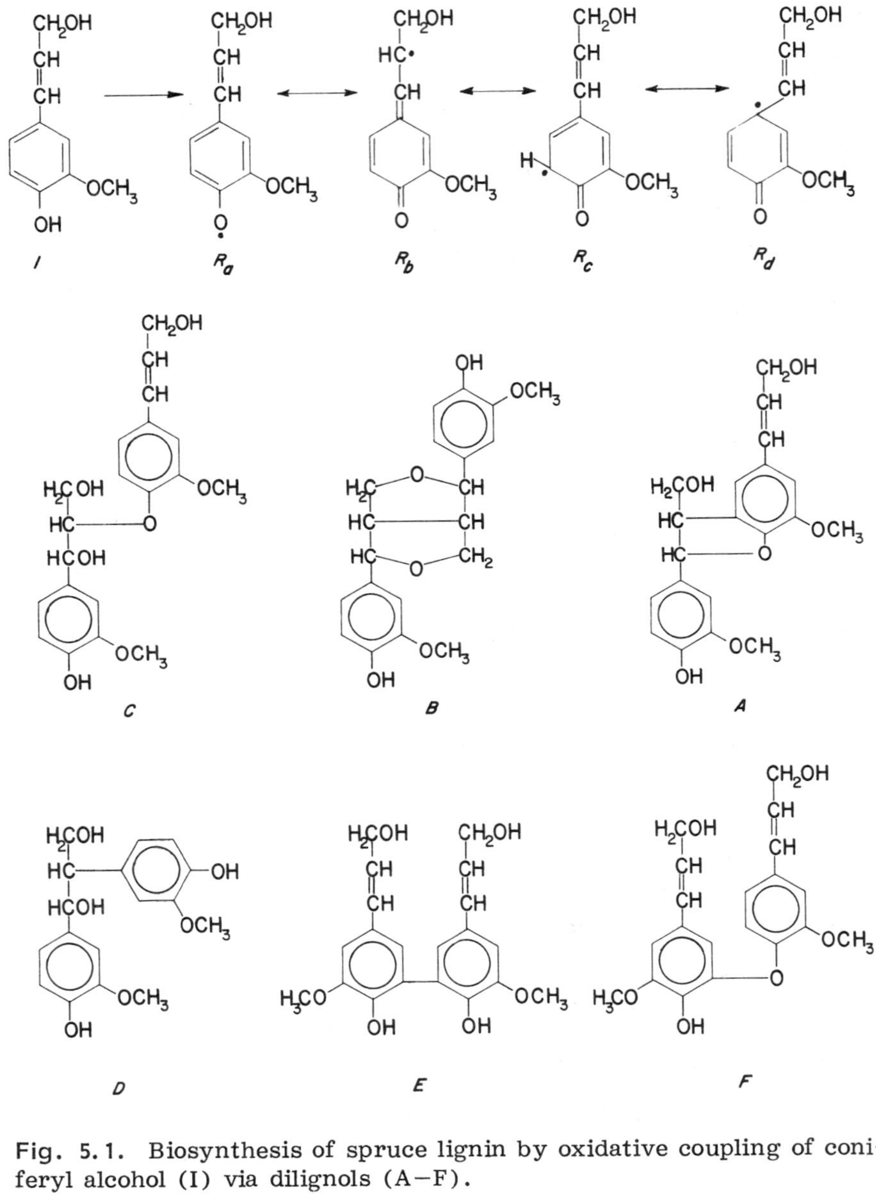

Fig. 5.1. Biosynthesis of spruce lignin by oxidative coupling of coniferyl alcohol (I) via dilignols (A–F).

on the cation. Of the four technically obtained lignosulfonates, the calcium-based exhibit the lowest and the ammonium-based the highest reactivity, while the sodium and magnesium lignosulfonates show a medium reactivity.

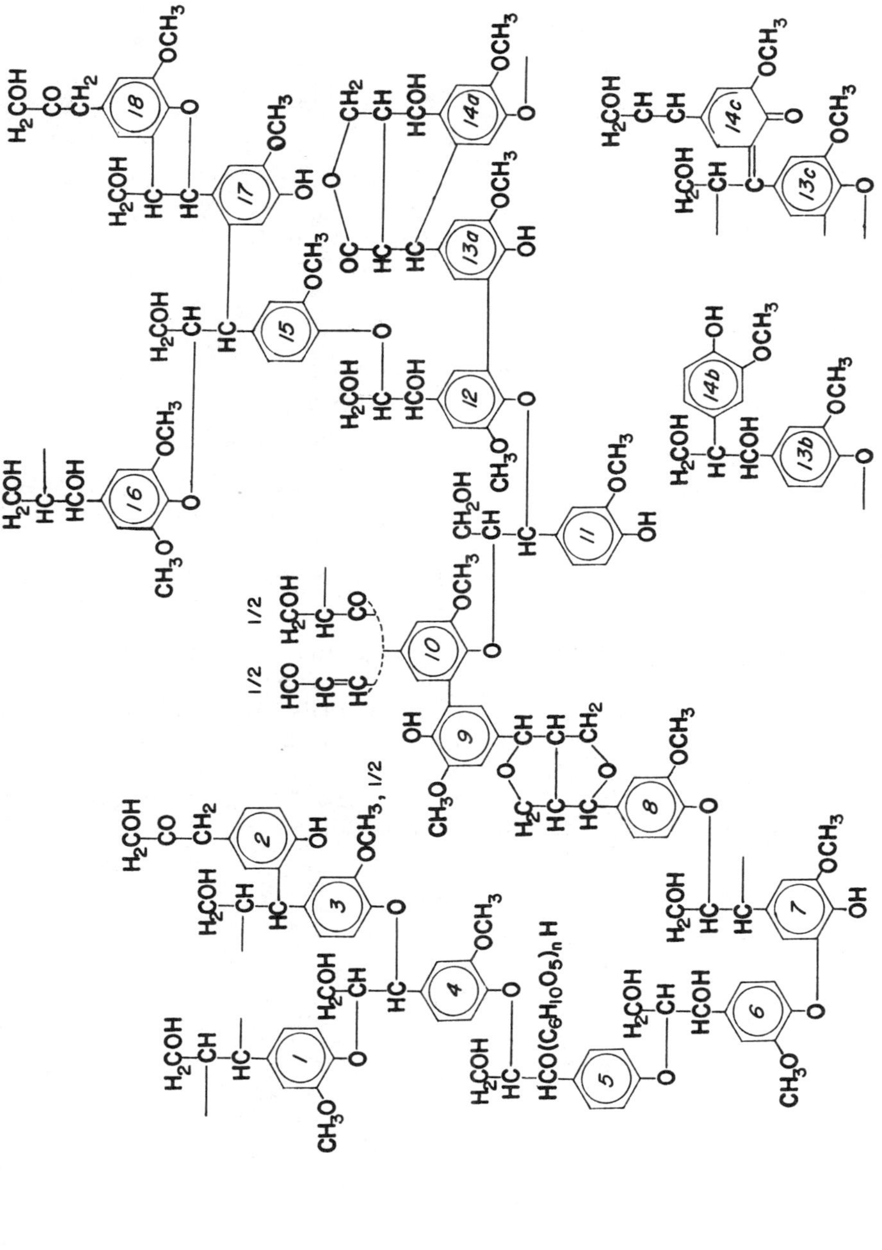

Fig. 5.2. Structural scheme of spruce lignin according to Freudenberg [2].

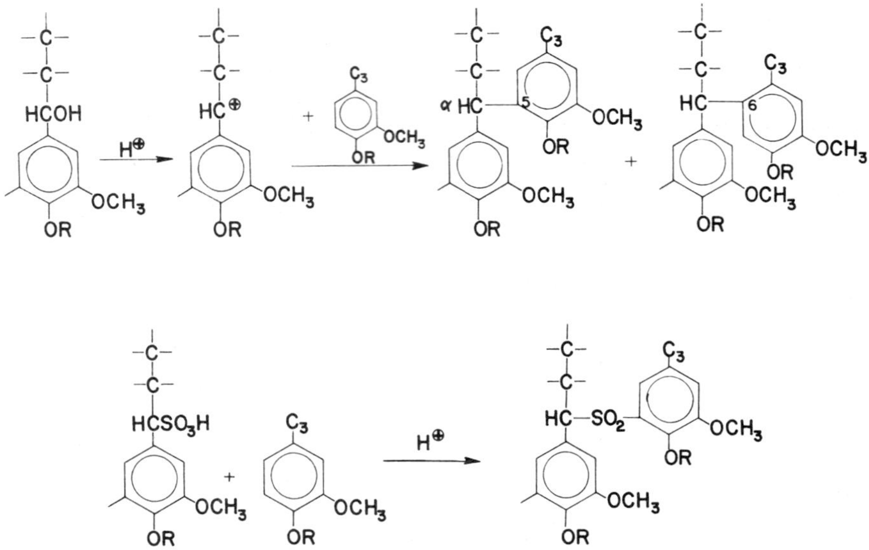

Fig. 5.3. Condensation reactions of lignosulfonates.

The first reaction in Figure 5.3, leading to diphenylmethanes, is of the same type as the formation of phenolic resins from phenol and formaldehyde (Fig. 5.4). Figure 5.4 also reveals that there are similarities between the two structures of a synthetic polyphenol such as phenol-formaldehyde (PF), and lignin (Fig. 5.2), which is a natural polyphenol. Thus lignin also reacts with formaldehyde and can be cross-linked with it, as indicated in Figure 5.5.

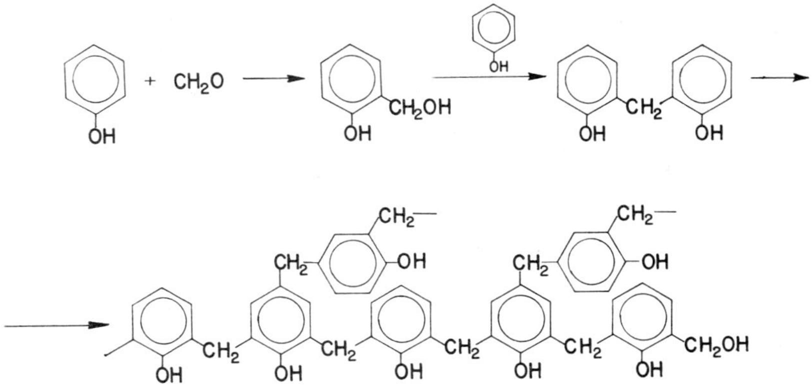

Fig. 5.4. Formation of novolak from phenol-formaldehyde.

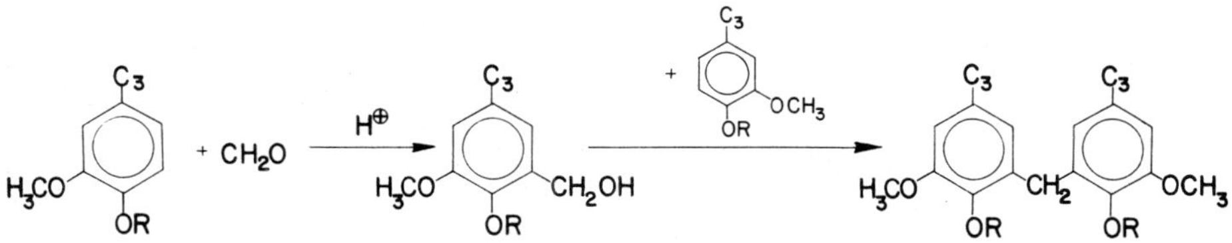

Fig. 5.5. Cross-linking of lignin by formaldehyde.

It has been found recently by Shen and Calvé [3] that the condensation of ammonium-based lignosulfonate is more effective with higher than with lower carbohydrate concentrations. This may be explained by the formation of furfural and hydroxymethylfurfural, which act like formaldehyde as linking bridges between two aromatic nuclei in lignin (Fig. 5.6).

B. Cross-Linking of Lignosulfonate by Oxidative Coupling

Lignosulfonic acid in technical SSL contains about 0.4 free phenolic hydroxy group per C_9 unit. Therefore, like the formation of lignin in plants (Fig. 5.1), cross-linking of lignosulfonate is possible by oxidative coupling (Fig. 5.7). Oxidants such as hydrogen peroxide, and catalysts such as sulfur dioxide or potassium ferricyanide, are most effective. Treatment of a 50% technical SSL with this redox system leads to a very vigorous exothermic reaction under evaporation of water. The resulting resin can be used as a cation exchange resin [4]. The yield of the resin under certain conditions exceeds 70%, indicating that some carbohydrates must have also been enclosed in the resin.

The crucial advantage of this type of cross-linking compared with condensation reactions is that it needs neither mineral acids nor high temperatures, due to the recombination of radicals, for which the activation energy is very low. The strongly exothermic reaction causes a uniform temperature profile during pressing of particleboard without external heat.

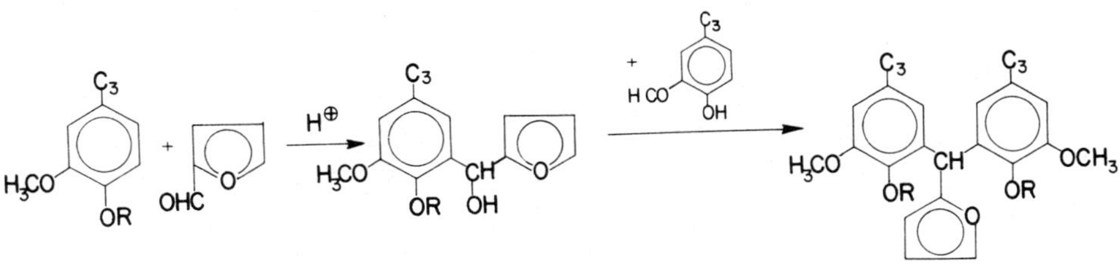

Fig. 5.6. Cross-linking of lignin by furfural.

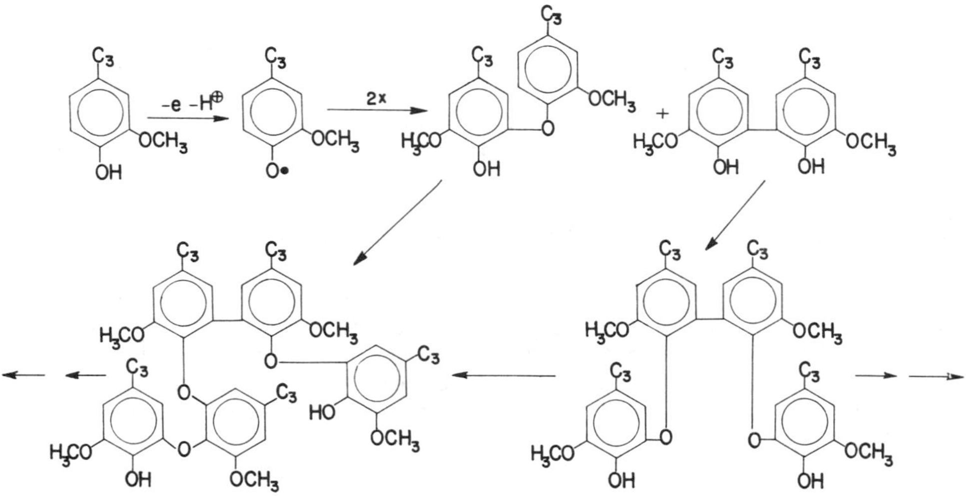

Fig. 5.7. Cross-linking of lignin by oxidative coupling.

III. APPLICATION OF LIGNIN AS AN ADHESIVE FOR PARTICLEBOARD, PLYWOOD, AND FIBERBOARD

According to its structure as a polyphenol (Fig. 5.2), lignin as an adhesive should be similar to phenol-formaldehyde resins. This is true for native lignin in wood, while technical lignins (lignosulfonate and black liquor) have to be additionally cross-linked in order to transfer them into insoluble resins. However, condensation reactions in lignin by heat or mineral acids cannot be as effective as in synthetic phenol-formaldehyde (PF) resins, due to the lower amount of free positions in the aromatic nuclei of lignin and their considerably lower reactivity than in PF resins. First, there is only 0.5 free 5-position (*ortho* to the phenolic groups) per C_9 unit, while the 6- and 2-positions are less reactive. Second, there is less than one benzyl alcohol or ether group per C_9 unit in lignin, while in synthetic PF resins up to three methylol groups can be introduced into one phenolic ring (Fig. 5.4). Third, the aromatic nuclei in lignin (mainly guaiacyl residues) are in a precondensed, rigid state, which hinders their reaction with benzyl alcohol groups in adjacent C_9 units. Finally, the aromatic nuclei in lignin are considerably less reactive toward hydroxybenzyl alcohol groups than phenol, due to the presence of methoxy or methoxy-equivalent groups rather than hydroxy groups on the lignin aromatic rings. For these reasons, lignin in technical spent liquors cannot be as effectively cross-linked as synthetic PF resins. At least higher press temperatures at longer heating times or higher acid concentrations are necessary.

Quite a number of patents have been pending during the past three decades dealing with lignin as adhesive for particleboard, plywood, and fiberboard in the absence of conventional PF or urea-formaldehyde (UF) adhesives (Table 5.1). Besides lignin, in most cases additional cross-linking agents for lignin are necessary, such as epoxides, polyisocyanates, polyols, polyacrylamides, polyethyleneimine, aldehydes, maleic anhydride, amines, proteins, melamine, hydrazine, and so on. So far, these procedures, for different reasons, have not led to any major practical application. Very few procedures use lignosulfonates or SSL without integrated cross-linking chemicals, like that of Pelikan et al. (1954), Pedersen and Jul-Rasmussen (1963), Shen (1973), and Nimz et al. (1972) (see Table 5.1). The patent of Pelikan et al. describes a procedure of using lignin as an adhesive for floor layers by cross-linking it oxidatively with chromium trioxide. The mechanism of cross-linking is the same as with hydrogen peroxide (Nimz et al., 1972), but it is much less effective. Its applicability to particleboards has been tried, but the boards exhibited low tensile strengths and disintegrated in water at 20°C in less than 2 hr. The other three methods, mentioned above, will be discussed in more detail.

A. Curing Lignosulfonate Boards by Long Pressing Time and Postheating Treatment

According to Pedersen and Jul-Rasmussen, DAS 1,303,355 (1963) (Table 5.1), wood chips are mixed with 20–25% of their weight with a 50% technical SSL and pressed at 185°C for 30 min giving a 12-mm-thick board which has to be postheated at 195°C for 80 min in an autoclave. The pH value of the SSL had been adjusted to 3 by citric acid. The particleboard obtained has a bending strength of 230 kP/cm^2, a tensile strength perpendicular to the grain of 5.3 kP/cm^2, and a density of 0.7 g/cm^3. Press temperatures may vary between 170 and 235°C and temperatures of the autoclave between 170 and 210°C.

High press and autoclave temperatures as well as long heating times are necessary for an effective cross-linking by condensation reactions in lignin, as pointed out above. The color of the boards is dark, due to decomposition reactions and charring, caused by the high temperatures, and the density of the boards usually reaches values at around 0.8 g/cm^3 if the required tensile strength is to be obtained. The temperature in the core layer during pressing reaches 140°C. This may also cause condensation reactions between wood and SSL as well as chemical and physical changes in the wood particles [5].

The relatively high dimensional stability of the particleboard toward water may be caused by these changes. Pedersen and Jul-Rasmussen (1963) found a thickness expansion after a 2-hr soaking in water at 20°C of only 1.5% and 13.8% water absorbance. Open-air tests, extending over 5 years, carried out by the wood panel products laboratory

Table 5.1. Patents on Lignin-Based Wood Adhesives Without Combination with Conventional UF or PF Binders (see Tables 5.3 and 5.6)

Year	Author or patentee	Adhesive formulation	Application	Literature
1951	Grolit Chem. Techn. Erzeug Ges, Vienna	SSL, aldehydes, sulfuric acid	Adhesive	OP 189,395 20.12.51/25.3.57
1954	T. Holzer	LS, protein, fire-retardent materials	Fiberboard	DAS 1,008,907 12.1.54/23.5.57
	Int. Paper Co.	SSL, polyacrylamide	Adhesive	US 2,923,693 2.8.54/2.2.60
	F. Pelikan et al.	SSL, CrO_3, gypsum	Floor layers	DBP 975,049 12.5.54/1.12.55 and DBP 977,363 (3.2.66)
1956	Crown Zellerbach Corp.	LS, proteins, aldehydes	Adhesive	US 2,846,409 23.1.56/5.8.58
1959	Owens Illinois Inc.	Lignin, polyurethanes	Adhesive	US 3,577,358 16.10.59/10.2.67
1960	Union Carbide Corp.	Lignin, polyethylene oxide	Adhesive	US 3,355,400 26.9.60/28.11.67
1963	Ambros AG	LS, protein	Fiberboard	DBP 1,445,435 17.1.63/21.11.68
	Congoleum-Nairn Inc.	SSL, clay, phosphates	Adhesives	US 3,352,700 1.11.63/14.11.67
	Dansk Spaanplade Ko, A. H. F. Pedersen, and Jul-Rasmussen	SSL, postheating	Particleboard	DAS 1,303,355 5.6.63/8.8.74

	Georgia Pacific Corp.	LS, hydroxylamine, proteins	Packaging materials	US 3,314,807 17.10.63/18.4.67
1964	Chemirad Corp.	Lignin, polyethylene imine	Adhesive	BR 1.069,223 24.6.64/17.5.67
	Nippon Kokutu Tetsudo	LS or thiolignin, epoxides, amine, acid	Particleboard	JA 19088/64 6.4.64/18.7.69
1965	Casella Farbwerke Mainkur AG	LS, melamin	Particleboard	DAS 1,234,016
	Pacific Corp.	LS, styrene oxide	Adhesive	US 3,347,842 1.3.65/17.10.67
1966	Weyerhauser Co.	Lignin, phosphorus derivatives	Adhesive	CA 865,736 10.11.66/9.3.71
1970	Darling and Co.	LS, protein, polyols	Wood adhesive	US 3,619,222 2.2.70/9.11.71
1971	Neste Oy, Helsinki	Lignin, hydrazine	Particleboard	DAS 2,141,404 18.8.71/14.4.77
1972	Helmitin-Werke, H. Nimz et al.	SSL, H_2O_2	Particleboard	DOS 2,221,353 2.5.72/3.1.74
	Wood Chem AS, Latv. SSR	Ammonium-extracted lignin	Particleboard	SU 398,413 21.4.72/31.1.74
1973	D'Alelio GF	Lignin esterified with epoxy groups	Adhesive	US 3,905,926 14.12.73/16.9.75
	Canadian patents and dev, K. C. Shen	SSL, sulfuric acid	Particleboard	DT 2,410,746 6.3.73/12.9.74

Table 5.1. (Continued)

Year	Author or patentee	Adhesive formulation	Application	Literature
	A. von Haebler	LS, polyisocyanate, starch	Chipboard	DT 2,336,381 17.7.73/20.2.75
	Leningrad Forestry Acad.	LS, textured paper	Wood adhesive	SU 501,886 18.12.73/8.4.76
1974	Form-Altstoff-Hande	LS, polyol, surfactant, polyisocyanate	Adhesive	DT 2,403,656 25.1.74/30.10.75
	G. A. Serlachius Oy, Mantha, Finland	LS, urea, sugar	Particleboard	DAS 2,439,660 19.8.74/15.3.79
1975	Alstoff Handels, GE	LS, polyol, surfactant, polyisocyanate, dextrin	Adhesive	JA 057,480 16.5.75/2.12.76
	D'Alelio GF	Lignin esterified with epoxycarboxylic acid	Wood panels, etc.	US 3,984,363 12.5.75/5.10.76
	S. Chow	Lignin, bark	Barkboards	*For. Prod. J.* *25*(11):32–37 (1975)
1977	Georgia Pacific Corp.	LS, melamine, aldehyde	Particleboard	US 4,130,515 30.6.77/19.12.78

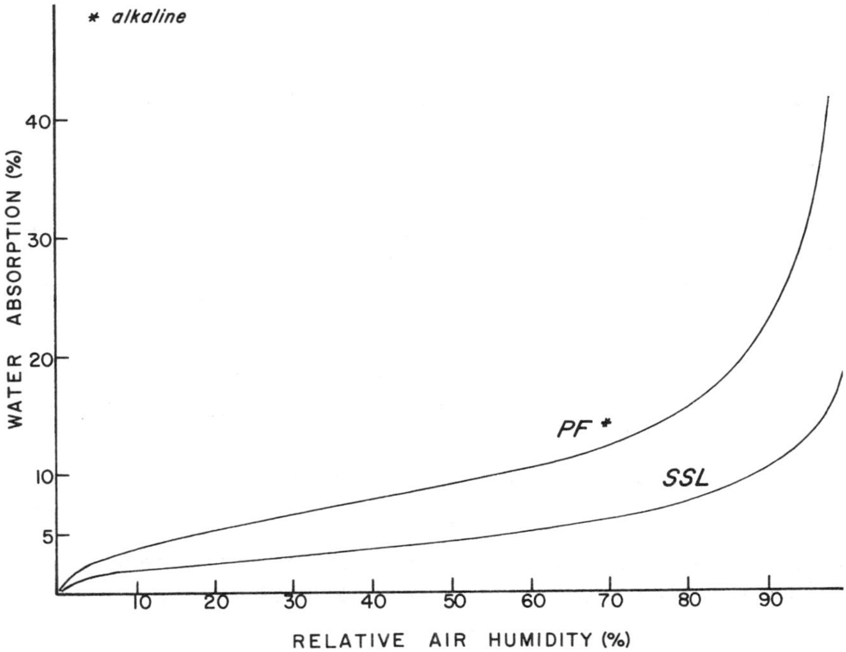

Fig. 5.8. Moisture content of SSL particleboard (according to Pedersen) in relation to relative air humidity compared with conventional PF particleboard. (From Ref. 7.)

of the Technical Research Centre of Finland [6] revealed that SSL boards, obtained according to the Pedersen procedure, were superior in strength and in surface properties to UF as well as to PF particleboard. Recently Roffael [7] has shown that water absorption of Pedersen SSL particleboard at different air humidities is only about half as high as with conventional PF particleboard (Fig. 5.8). Weathering for 1 year gave a nearly constant humidity at around 6% for SSL particleboard, while PF boards gave humidities between 12 and 15% (Fig. 5.9). Also after soaking the boards in water at 20°C for 24 hr, the LS boards lost only 25% of their initial tensile strength, while that of conventional PF particleboard decreased by 70% (Fig. 5.10). In contrast, the mechanical strength properties of SSL particleboard were inferior to those of PF board [8].

The Pedersen procedure has been applied in mill-scale tests in Denmark, Switzerland, and Finland, but has been discontinued in all cases. One reason for this failure may be the high energy cost caused by the two-stage heating treatment. The autoclave must consist of refined steel, due to the evolution of corrosive gases such as sulfur

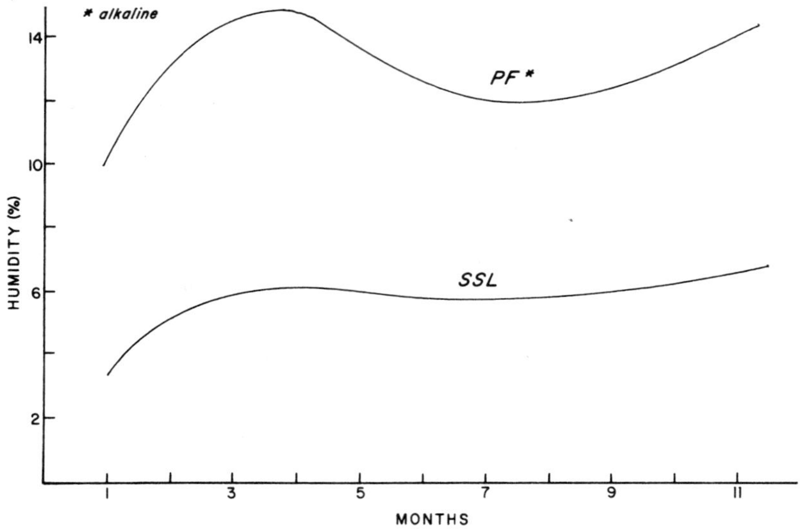

Fig. 5.9. Average humidities of SSL (according to Pedersen) and PF particleboard during weathering over a 1-year period. (From Ref. 7.)

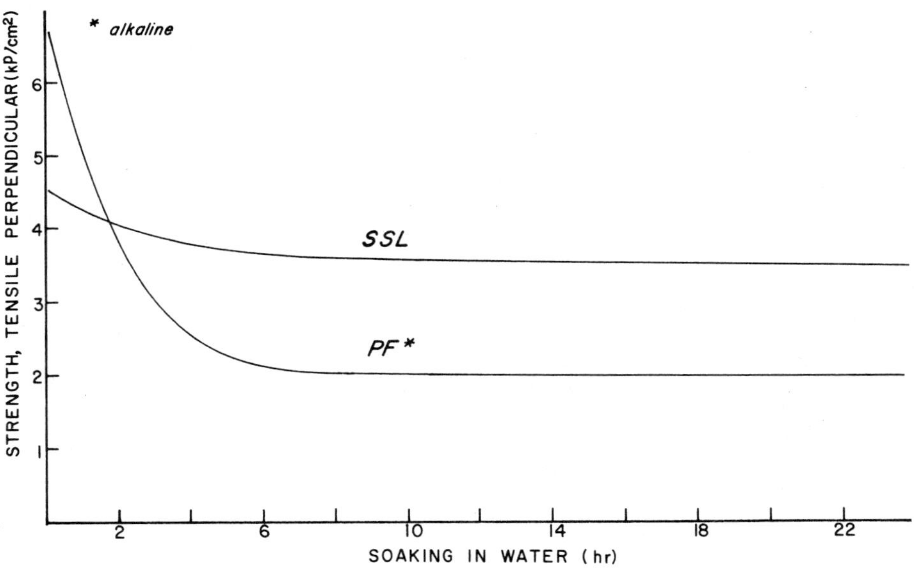

Fig. 5.10. Tensile strength of SSL (according to Pedersen) and conventional PF particleboard, plotted versus time of soaking in water at 20°C. (Roffael 1969)

dioxide, causing additional high costs. Another reason was the long pressing and fast curing time needed for manufacture. However, one of the main reasons for the discontinuation of the procedure was, according to private information, the frequent fires induced by the high pressing and posttreatment temperatures.

As mentioned above, the condensation rate of lignosulfonates depends on the cation, with ammonium ions exhibiting the highest reactivity. Recently, Shen and Calvé [3] used fractionated ammonium-based SSL as binder for particleboard and found the highest reactivity, leading to the best mechanical board properties, with a low molecular weight fraction. Unexpectedly, the tensile strength of dry particleboard obtained with a low molecular weight ammonium-based SSL fraction increased with the sugar content of the SSL. Best board properties were obtained with 6% of a low molecular weight (0–5000) ammonium-based SSL fraction having 50–60% sugar. In this case, a pressing time of 8 min at 210°C was sufficient for manufacture of 11-mm-thick waferboard to meet the Canadian standard requirements of exterior grade particleboard.

Obviously, the sugars take part in the condensation reactions of lignosulfonate according to Figure 5.6. While the bending strength of dry boards steadily increased, with the sugar content of the SSL going up to 80%, the bending strength, after a 2-hr boiling of the boards reached a maximum at about 50–60% carbohydrates, indicating that the condensation between lignin and carbohydrates leads to better water resistance than that between carbohydrates only.

B. Curing Lignosulfonate Particleboard with Sulfuric Acid

In 1973, K. C. Shen of the Eastern Forest Products Laboratory in Ottawa, Canada, proposed sulfuric acid as a curing agent for SSL waferboard (Table 5.1). The pressing conditions were the same as that of conventional PF particleboard, when poplar wafers were first sprayed with 1% of a 15–20% sulfuric acid and then with 4–5% SSL powder, which adheres at the surface of the wet wood wafers. Later [9] concentrated sulfuric acid (9%) was added to the SSL before spray-drying, and the powder adhered to the wax-coated wafers. High pressing temperatures of about 205°C were also necessary, the catalytical effect of the sulfuric acid merely reducing the pressing time to that of industrial conditions for PF particleboard.

The strength properties of the boards, having an average specific gravity of 0.67 g/cm^3, were measured by the torsion shear at the center plane of 1 × 1 in. specimens, from which the internal bond strength (tensile strength perpendicular to the surface) can be obtained by multiplication with the factor 0.7 [10]. Values obtained for internal bond strength and modulus of rupture (MOR) for dry samples as well as after 2 hrs of boiling met the Canadian standards for particleboard [11].

The torsion shear strength and MOR of dry boards was independent of the pressing time, while the wet strength increased proportionally with the pressing time. This means that for exterior-grade requirements, distinct pressing times are necessary.

Best board properties were obtained with 1% concentrated sulfuric acid, based on dry wood particles. At higher acid concentrations the strength of dry boards decreased, while that of wet boards showed a further increase. However, charring of the wood particles takes place at acid concentrations higher than 0.9%. The thickness expansion of SSL boards obtained with 1% sulfuric acid lay between 26 and 46%, after soaking in water at 20°C for 1 week, and between 51 and 66% after 2 hr of boiling. These values are considerably worse than those of exterior-grade PF particleboard.

The acidity of the particleboard was found to be pH 3, after disintegration of the boards in 10 times their weight of water [12]. It has been reported that the acidity had no long-term influence on the mechanical board properties, checked by conventional accelerated aging treatments [12]. In this case, 11-mm boards had been pressed for 6 min and a part of them had been postheated at 149°C for 2 hr. Of the accelerated aging treatments only 20 days of heating at 149°C showed faster aging of SSL than of PF boards, which is due to the higher acidity of the SSL boards.

On the whole, the SSL particleboard obtained by Shen can not be compared technically with exterior-grade PF particleboard. Its great advantage, however, might be on the economic side, saving between 22 and 28% of total production costs relative to PF boards. This calculation was made by Shen in 1974 and might even have changed in favor of the SSL particleboard, due to a further increase in phenol prices in the meantime. The Shen procedure, however, has found no practical application yet, as the results obtained are still far from those obtainable with PF particleboard. In 1977 Shen mentioned that a short production trial run had been carried out at a waferboard plant; although the preliminary results were promising, additional work is still required to modify the binder formulation and production parameters to meet the requirements of plant operation, and to obtain results comparable to those of PF particleboard.

C. Curing SSL Boards with Hydrogen Peroxide

The drawback inherent in the Pedersen and Shen procedures—high pressing temperatures and long pressing times or strong mineral acids—can be avoided if the cross-linking of the lignin molecules is achieved by radical coupling instead of condensation reactions (Nimz et al. 1972, Table 5.1). In this case, the formation of new carbon-carbon as well as carbon-oxygen bonds between two radicals is a very fast reaction with a low activation energy, which needs no external heating or mineral acids as catalyst. This means that the reaction is very specific, and side reactions such as decompositions and charring can be avoided, while linkages between wood and SSL may also occur.

The essential radicals are formed from phenolic groups in the lignosulfonate molecules by oxidation with hydrogen peroxide in the presence of a catalyst. Out of a number of catalysts, sulfur dioxide (SO_2) has been proven to be the most effective [13]. A 50% calcium-based SSL containing about 1% SO_2 at pH 2 reacts vehemently with a 35% hydrogen peroxide solution in a strongly exothermic reaction, forming an insoluble gel. The reaction time is less than 1 min but depends on the source and composition of SSL. At higher pH values, the reaction takes some minutes or needs heating up to about 70°C, but after reaching 70°C, the reaction is also very fast.

It has been reported [13] that the SSL, containing the catalyst, and the hydrogen peroxide solution have to be sprayed separately on the wood chips. Under certain conditions, the hydrogen peroxide can be mixed with the SSL and sprayed together on the wood chips, in a single operation [13]. Another possibility consists of adding half of the SSL as spray-dried powder, lowering the humidity of the blended chips to about 13% [14], which is the upper limit according to German standards (DIN 68761). The powder may either be mixed together with the liquid SSL and the hydrogen peroxide or added separately after the wood chips have been sprayed with the mixture of SSL solution and hydrogen peroxide. The humidity of the wood chips can thus be adjusted to predetermined values. The pot life of the blended wood chips, which is the assembly time between spraying and pressing, would then be extended.

At the present state of the procedure, medium-density interior-grade particleboard can be obtained from wood chips with 20% SSL, based as dry material on dry wood ships, at pressing temperatures between 100 and 120°C under otherwise conventional manufacturing conditions for UF particleboard. To illustrate the influence of different parameters on board properties, the other parameters are kept constant at 20% SSL, 9% of a 35% hydrogen peroxide solution, 4% ammonium chloride, and 0.6% SO_2, all based on oven-dried wood chips, in manufacturing particleboard (70 × 70 × 15 mm) with a pressing time of 5 min at 120°C.

1. Amounts and Composition of SSL

Figure 5.11 demonstrates the dependence of tensile strength perpendicular to the grain and swelling degrees in water at 20°C on the amounts of SSL. The SSL was a fermented calcium-based 54% solution, obtained from the pulp mill in Karlsruhe-Maxau (Holzman AG), West

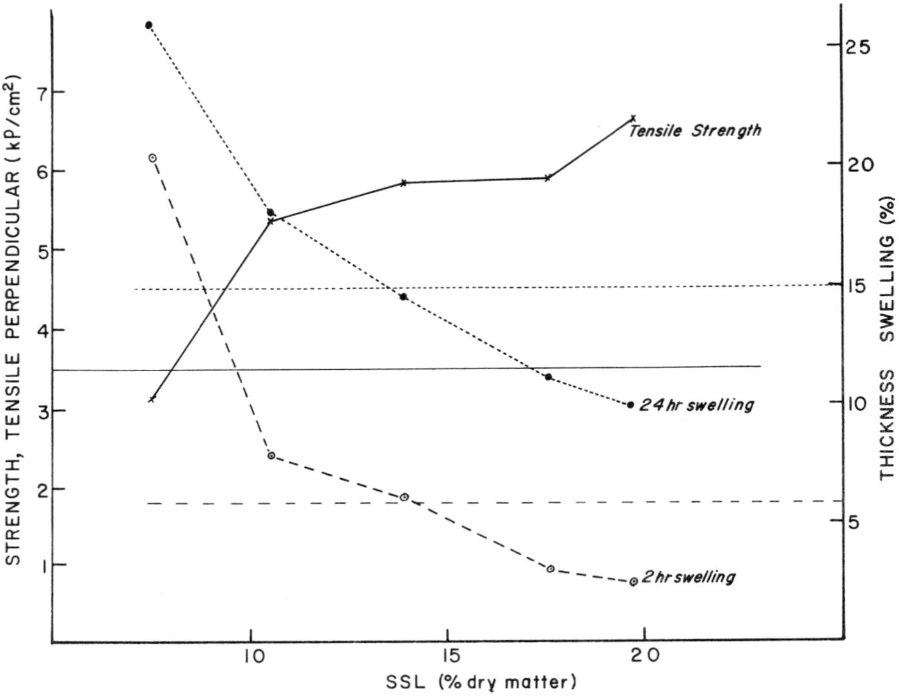

Fig. 5.11. Dependence of tensile strength (solid line) and thickness swelling (s hr, dashed line; 24 hr dotted line) on the amount of SSL, based on dry wood chips, in 15-mm-thick one-layer particleboard. Parallel lines to the abscissa indicate German standards for particleboard according to DIN 761.

Germany. In Figure 5.11 the German standards for tensile strength (3.5 kP/cm^2, solid line) and for thickness expansion after soaking in water at 20°C for 2 hr (6%, dashed line) and after 24 hr (15%, dotted line) are indicated by lines parallel to the abscissa. From Figure 5.11 it can be seen that the tensile strength meets the required standard value at about 10% SSL, and the dimensional stability requirements are met with at least 15% SSL. An amount of 20% SSL further improves the board properties, so this concentration was chosen in all other experiments in which variations of the other parameters were effected.

It has been observed that the board properties sometimes changed, with the pH value of the fermented SSL varying between 4.3 and 5.4. This was found to be due to a higher ash content, which resulted from addition of calcium oxide to the SSL before fermentation in order to improve the yield of alcohol in the pulp mill. Generally, ash contents below 20% are tolerable. The ash content may also be the reason for

Table 5.2. Influence of SSL Composition on Chipboard Properties

SSL	Calcium (%)	Sodium (%)	Sugar (%)	Total sulfur (%)	Ash (%)	pH	Molecular weight (%)		(kP/cm^2)	Swelling (%)	
							>5000	>40,000		2 hr	24 hr
Waffolin	5.2	—	25	4.5	12.3	3.1	—	—	5.5	3.5	9.3
Ultra B002	—	3.7	2.9	6.1	12.1	5.4	79	24	6.2	4.2	10.8
EP003	—	5.4	11.1	5.9	16.6	8.2	65	23	5.7	4.7	11.5
SLN2	—	9.4	12.0	6.6	25.4	7.4	44	10	—	—	—

the astonishing fact that nonfermented calcium-based SSL leads to improved particleboard properties, as indicated in Table 5.2. In Table 5.2, Waffolin is a nonfermented calcium-based spray-dried SSL obtained from Holmen KF, Düsseldorf, West Germany; the samples Ultra B002, EP003, and SLN2 were sodium-based spray-dried SSL obtained from Rauma-Repola Oy, Finland. Their analysis together with the board properties obtained are listed in Table 5.2.

According to Table 5.2, the sugar content, going up to 25%, has no significant influence on the board properties. Rather, the differences in board properties can be explained by different ash contents of the SSL powders. This is especially seen with SLN2, where an ash content of 25.4% is obviously responsible for the inferior board properties. Of the three sodium-based SSLs, Ultra B002 has the highest and SLN2 the lowest molecular weight. If there is any influence of the molecular weight of SSL on the board properties, according to Table 5.2, a higher molecular weight should improve board properties, in accordance with Forss (see below), while Shen (see above) found the opposite effect with ammonium-based SSL.

Of the four samples listed in Table 5.2, Waffolin is the cheapest, because it is obtained from nonpurified, untreated calcium-based SSL by spray-drying. When a 54% solution of calcium-based SSL, from which Waffolin was obtained by just spray-drying, was used, similar board properties were obtained as with Waffolin (Table 5.2).

2. *Temperature in the Particleboard Core Layer During Pressing*

It has been reported [13] that pressing temperature and time had no major influence on the mechanical board properties. Similar board properties were obtained at pressing temperatures between 80 and 200°C and times between 2 and 6 min. This is obviously due to the strongly exothermic reaction. At pressing temperatures of 100 or 120°C, the temperature in the core layer of the board rises very fast, reaching a maximum at about 130°C after about 1 min, and then decreases slowly to about 105°C after 5 min (Fig. 5.12). In the example in Figure 5.12, the humidity of the blended wood chips, before pressing, was about 22%, but with a humidity of only 13%, the temperature in the core layer during pressing rose to 150°C. In any case, the temperature of the press is needed only to start the reaction, which is then autocatalyzed by its own reaction heat.

An important factor in the procedure is that the temperature in the board during pressing is homogeneous over the whole board thickness from the very beginning of the reaction, leading to a uniform appearance of a board cross section. In contrast, during pressing of particleboards with conventional glues, the temperature at the surface layers is much higher than that of the core layer. The procedure also allows the manufacture of particle board thicker than 20 mm at short pressing

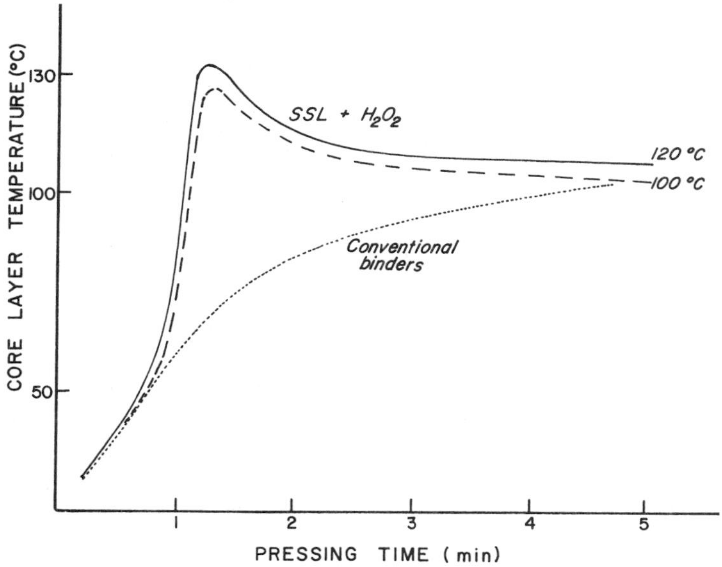

Fig. 5.12. Temperature in the core of one-layer SSL-H_2O_2 particleboard ($46 \times 60 \times 1.5\ cm^3$) plotted versus pressing time at pressing temperatures of 120°C (solid line) and 100°C (dashed line). Dotted line indicates approximate temperature in the core of particleboard during pressing with conventional glues.

times. Furthermore, complicated moldings can be pressed from wood chips, by starting the reaction at just one narrow heating point.

The influence of higher, homogeneous temperatures in the core layer of particleboard during pressing has been demonstrated by Roffael [15] with PF particleboard, where the temperature by conventional contact heating in the core layer reaches only 110°C, while the surface layers may reach that of the press platens, which is above 170°C. The board properties improved significantly when high-frequency heating (HFH) was applied instead of contact heating.

3. *Amounts of Hydrogen Peroxide and Sulfur Dioxide*

It has been shown [13] that optimum board properties are obtained with 9% of a 35% hydrogen peroxide solution. The values outlined in Figure 5.13 were obtained with 20% by weight on dry wood chips of a 54% SSL solution at pH 2; however, similar values are obtained with 4% ammonium chloride at pH 4.5. If the hydrogen peroxide solution is sprayed separately on the wood chips, for safety reasons, the concentration should not exceed 35%. However, where the hydrogen peroxide

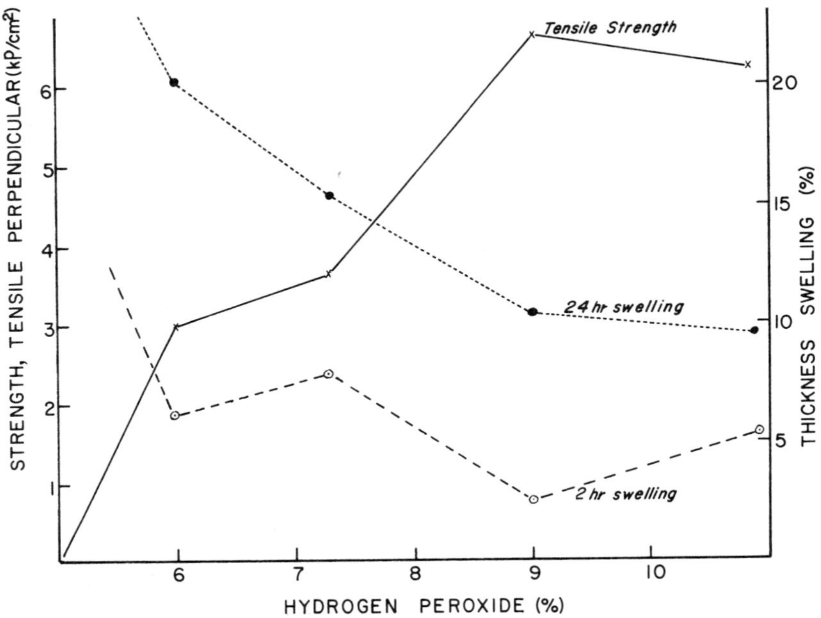

Fig. 5.13. Dependence of particleboard properties on the amount of 35% hydrogen peroxide, based on weight of dry wood particles.

can be mixed with the SSL before spraying, the concentration of the hydrogen peroxide may be higher. This would lower the moisture content and allow a higher ratio of 54% aqueous SSL to spray-dried SSL, diminishing the costs of the adhesive formulation. Furthermore, the observation [13] that the distribution of the hydrogen peroxide on the SSL-blended wood chips is very crucial if the hydrogen peroxide is separately sprayed on the wood chips is no longer valid in one-mix formulations.

The amount of SO_2 is less crucial to the board properties. Best results were obtained with 0.5% SO_2, based on dry wood chips. At lower amounts, the mechanical board properties decreased, but with higher amounts, up to 1.5%, no major influence was observed. The SO_2 may be added to the SSL either as gaseous sulfur dioxide or as sodium hydrogensulfite ($NaHSO_3$). In the first case, the pH value has to be adjusted with ammonia.

4. *pH Value and Amounts of Ammonium Chloride*

If 35% hydrogen peroxide is added to a 54% SSL containing 0.5% SO_2, the reaction time is strongly dependent on the pH value of the SSL [13]. At pH 2, the reaction is nearly complete within 1 min, while at pH 4.5, the reaction starts only after external heating to about 70°C. Similarly, when wood chips are first sprayed with SSL and the catalyst and thereafter with hydrogen peroxide, at pH values below 3.5 the wood chips heat up, especially when stored in large quantities, within 10 to 20 min. This is not observed at pH value 4.5; in this case ammonium chloride as a hardener must be added. In the more recent procedure, mixing the SSL with the hydrogen peroxide before spraying, the blended wood chips may be stored for some hours before pressing without heating up. The mixture of SSL with hydrogen peroxide can also be stored for at least 12 hr without losing its activity, even if the pH value is lower than 4.5. At pH values between 3 and 4, none or less than 4% ammonium chloride can be applied without impairing the board properties.

5. *Moisture Content and Shape of Wood Particles*

Until recently, the main drawback in this procedure has been the high moisture content of the particleboard obtained, which exceeds 20% when a 54% SSL and 35% hydrogen peroxide solutions are used. This situation, however, has been drastically changed by mixing the SSL together with the hydrogen peroxide. When using half of the SSL as a 54% solution and the other half as spray-dried powder, the moisture content of the blended wood chips will be about 13%, meeting the German standards for particleboard. The moisture content of the blended wood chips can be further diminished by increasing the ratio of spray-dried powder to 54% solution of SSL. In this case, the powder must be added separately after addition of the SSL-hydrogen peroxide solution, or at least part of it, due to the high viscosity of the adhesive formulation. This method, however, is somewhat sensitive to the shape of the wood particles. No problems occur with thin wood chips having a thickness of about 0.3 mm, but with more compact wood particles, such as flakes, strands, slivers, or granules, having a smaller surface, the distribution of the spray-dried powder adhering to the wet particle surface is not good enough. In this case, most of the spray-dried powder has to be mixed with the adhesive solution and sprayed on the wood particles.

6. *Conclusions*

This procedure [13] may be more expensive than those of either Shen or Pedersen, due to the hydrogen peroxide consumption. On the other hand, less heating energy is required and mineral acids are avoided. Also, no corrosion problems occur. Due to the exothermic reaction, the homogeneity of the board is expected to be better; charring is avoided and thicker boards can be pressed at shorter pressing times. When compared with conventional UF boards, no formaldehyde is released and the price of the board should not be higher. The cost will depend primarily on the hydrogen peroxide, and the latter depends on the amounts consumed. In the procedure described, high amounts of adhesive (SSL) are used. However, taking into account the low price of SSL and the corresponding price of saved wood chips at a given specific gravity of the boards, this should not mean a major increase in total costs. It was also found that low-density boards, with a specific gravity of about 0.53, obtained with the system described still meet the required mechanical strength properties. This should save costs and wood raw material in manufacturing low-density particleboards.

Due to the chemical structure of lignin, it may be expected that the H_2O_2-SSL boards may be further improved to meet the exterior-grade standards. Research focused on this aim is in progress.

IV. LIGNIN IN COMBINATION WITH PHENOL-FORMALDEHYDE ADHESIVES

The number of patents pending on lignin as a substitute or extender for phenolic wood adhesives during the past three decades is quite impressive (Table 5.3). Under certain conditions, up to 40% of phenol-formaldehyde (PF) adhesive can be replaced by lignosulfonate or black liquor without significantly extending the curing time or worsening board properties. Lignin-PF formulations have been used in manufacturing particleboard, fiberboard and plywood. The reason for their application has to be seen in the lowering of costs, resulting from the difference in cost between phenol-formaldehyde and lignin. However, in most cases the lignin has to be pretreated by deionization, ultrafiltration, or cation change. Two recent procedures that have become better known are discussed next in more detail.

A. Lignin-PF Formulations

In 1971, Roffael and Rauch [16] claimed that the curing time of SSL particleboard could be reduced and, according to Pedersen, the post-heating treatment in an autoclave avoided when phenolic resins of the novolak type were added to the SSL [16]. Due to coagulations between calcium-based SSL and phenol-formaldehyde, the calcium-based SSL has to be transferred into sodium-based SSL. The board properties are strongly dependent on the pH value of the glue, as can been seen from Figure 5.14, in which 10% SSL, 4% novolak, and 2.1% hexamethylenetetramine were applied to dry pine wood chips to prepare 9-mm-thick boards at a pressing time of 12 min.

While the highest bending and tensile strengths were obtained between pH 5 and 7, the percentage swelling in water at 20°C, after 24 hr, had a minimum at pH 3.5. For this reason, a pH value of 4.7 has been suggested by the authors as a compromise [16]. Both the mechanical strength and the dimensional stability of the particleboard can be improved by higher ratios of novolak in the glue formulation or increasing pressing temperatures up to 250°C. Besides conventional

Table 5.3. Patents on Lignin-Based Wood Adhesives in Combination with PF Resins

Year	Author or patentee	Adhesive formulation	Application	Literature
1950	Inst. Int. Financier	SSL, cresol, formaldehyde	Adhesive	FP 1,023,308 1.6.50/17.3.53
1951	Inst. Int. Financier	Alkalilignin, PF	Plywood	DP I 5,676/39b 22.3.51/4.2.54
1953	Agrashell Inc.	Lignocelluloses, PF	Plywood	AP 2,781,286 6.5.53/12.2.57
	Ontario Research	SSL, PF	Wood Adhesive	US 2,794,790 2.11.53/4.6.57
1955	Agrashell Inc.	Lignin, PF, aminoaldehyde, epoxides	Adhesive	DAS 1,012,412 21.4.55/18.7.57
	American-Marietta Comp.	Lignin, lignocellulose PF	Adhesive	EP 777,502 2.2.55/26.6.57
1956	F.H. Snyder and Associates Inc.	Lignin, PF	Fiberboard	US 3,019,155 21.12.56/30.1.62
1958	Minnesota and Ontario Paper Co.	LS, resorcinol, formaldehyde	Fiberboard	US 2,926,722 20.6.58/1.3.60
1960	Agrashell Inc.	Lignin, PF, aminoaldehyde, epoxides	Fiberboard Plywood	US 3,093,604-7 30.6.60/11.6.63
	Agrashell Inc.	Alkalilignin, PF or aminoaldehyde	Fiberboard	US 3,099,633 30.6.60/30.7.63
	The Borden Company	Lignin, PF	Plywood, etc.	US 3,227,667 21.9.60/4.1.66

Table 5.3. (Continued)

Year	Author or patentee	Adhesive formulation	Application	Literature
1961	C. F. Schulerud et al.	Alkalilignin, PF	Extender for PF	*Tappi 44*(11): 823–30 (1961)
	West Virginia Pulp and Paper Co.	Oxidized lignin, PF	Plywood	US 3,071/570 6.6.61/1.1.63
	Wood Conversion Corp.	LS, PF	Fiberboard	US 3,308,218 24.5.61/7.3.67
1962	Diamond Alkali Company	SSL, PF, UF	Wood adhesive	US 3,296,159 19.10.62/3.1.67
	D. Narayanamurti et al.	Lignin, PF	Plywood	*Kunststoffe 52*, 341–43 (1962)
1963	Dow Chemical Co.	Lignin, PF, starch	Plywood	GB 1,012,067 24.7.63/8.12.65
	H. Ryozo	Lignin, PF	Wood powder	*Adhes. Adhes. 7*(11):865–866 (1963)
1964	D. Narayanamurti and R. C. Gupta	Lignin, PF	Wood adhesive	*Kunststoffe 54*(3): 166–169 (1964)
1965	Natl. Polychemicals Inc.	LS, PF	Plywood	RU 223, 691 6.1.65/Feb. 69
	Quaker Oats Co.	Lignocelluloses, PF	Plywood	US 3,441,528 16.12.65/29.4.69

1968	Canadian Int. Paper Co.	Desulfonated SSL, PF	Hardboard	US 3.551.405
	Centre Int. Econ. Technol. I.C.E.T.	Lignin, PF, melamine, resin layer	Sawdust boards	BE 729,278 11.3.68/18.8.69
1969	Archangelsk Inst., Lesa, Lesokim, SSR	Lignin, PF	Wood adhesive	*Chem. Abstr. 71*: 40452 (1969)
	Georgia Pacific Corp.	LS, PF	Adhesive	US 3,597,375 30.7.69/3.8.71
	Georgia Pacific Corp.	LS, PF	Plywood	US 3,658,638 30.7.69/2.12.70
1970	I. Abe	Black liquor, PF	Particleboard, plywood	*Hokkaido For. Prod. Res. Inst. Rep. 55*:118–131 (1970)
1971	Georgia Pacific Corp.	LS, PF, alcohol	Plywood	US 3,677,884 3.2.71/18.7.72
	Israel Ministry, development	Brominated LS, PF or UF	Flame-retardant chipboard	DT 2,247,036 28.9.71/5.4.73
	Mitsui Tostsu Chem. Inc., Tokyo	LS, PF	Adhesive	DOS 2,152,285 20.10/71/4.5.72
1972	Champion Int. Corp.	LS, PF	Particleboard, plywood	US 3,931,072 24.8.72/6.1.76
1973	AB Casco, Stockholm	LS, PF	Chipboard	DT 2,408,441 21.2.73/29.8.74

Table 5.3. (Continued)

Year	Author or patentee	Adhesive formulation	Application	Literature
1973	Cellulose Paper Ind. Res.	LS, PF	Wood adhesive	SU 496,290 28.12.73/22.3.76
	Cellulose Paper Ind. Res.	Lignin, PF, Sulfuric acid	Fiberboard	SU 480,577 29.12.73/22.10.75
	T. Enkvist	Black liquor, PF	Plywood	DOS 2,334,540 6.7.73/24.1.74
	Keskuslab Central-lab, K. Forss et al.	LS (mol wt > 5000), PF	Chipboard	DOS 2,346,009 12.9.73/21.3.74
1974	AB Casco, Stockholm	LS, PF	Resin	DBP 2,408,441 21.2.74/29.8.74
	EIDAI KK	Lignin, PF	Adhesive	JA 086,638 30.7.74/5.2.76
	Fraunhofer Ges., E. Roffael et al.	LS (free of earth metal ions), PF (resol)	Chipboards	DOS 2,406,887 13.2.74/14.8.75
	Georgia Pacific Corp.	LS, PF	Plywood	US 3,957,703 25.2.74/18.5.76
				US 3,956,207 25.2.74/11.5.76
		LS, PF, epichloro-(bromo)hydrin		US 3,886,101 8.7.74/27.5.75

	Keskuslab Central-lab, K. Forss et al.	Lignin (mol wt > 5000), PF	Plywood, fiber-particleboard	DOS 2,447,590 5.10.74/24.4.75
	Sumitamo Durez KK	Lignin, PF, molasses, aldehyde	Binder	JA 141,476 11.12.74/14.6.76
	Voran Tech. Inst.	Hydrolysis, lignin, tricresol, sulfuric acid	Binder	SU 842,483 27.5.74/28.12.75
1975	EIDAI KK	Black liquor, PF	Particleboard, plywood	J5 1,092,892 14.2.75/14.8.76
		Alkalilignin, PF	Wood adhesive	J5 1,148,731 17.6.75/21.12.76
		Black liquor, PF	Particleboard	J5 2,000,848 24.6.75/6.1.77
		LS, or black liquor, PF	Particleboard, plywood	J5 2,008,042 10.7.75/21.1.77
		Black liquor, PF	Plywood	J5 2,021,305 8.8.75/17.2.77
	Georgia Pacific Corp.	LS, PF, trialkylphosphate surfactant	Plywood	US 3,931,070 15.1.75/16.1.76
	Keskuslab Central-lab, K. Forss et al.	Alkalilignin (mol wt > 5000), PF	Chipboard, plywood	DT 2,601,600 28.1.75/29.7.76
	Neuszwyss Anstalt	LS, PF, UF, or melamine	Chipboard	BE 832,584 20.8.75/16.12.75
	Svenska Träforsk-ninginstitutet	Lignin, PF	Plywood	DT 2,624,673 13.6.75/30.12.76

Table 5.3. (Continued)

Year	Author or patentee	Adhesive formulation	Application	Literature
1976	I. Johansson	Lignin, PF	Wood adhesive	DOS 2,624,675 2.6.76/30.12.76
1976	Keskuslab Central-lab, K. Forss et al.	Lignin (mol wt > 5000), PF	Plywood, fiber-particleboard	DOS 2,601,600 17.1.76/29.7.76
	Svenska Träforsk-ninginstitutet	Lignin, PF	Wood adhesive	DOS 2,624,673 2.6.76/30.12.76
1977	Serlachius GA Oy, Mantha, Finland	LS, PF	Plywood	DAS 2,701,071 12.1.77/17.1.80
1979	MacMillan Bloedel Ltd., Vancouver	Methylolated kraft lignin, PF	Fiberboard, particleboard	DOS 2,949,182 6.12.79/3.7.80

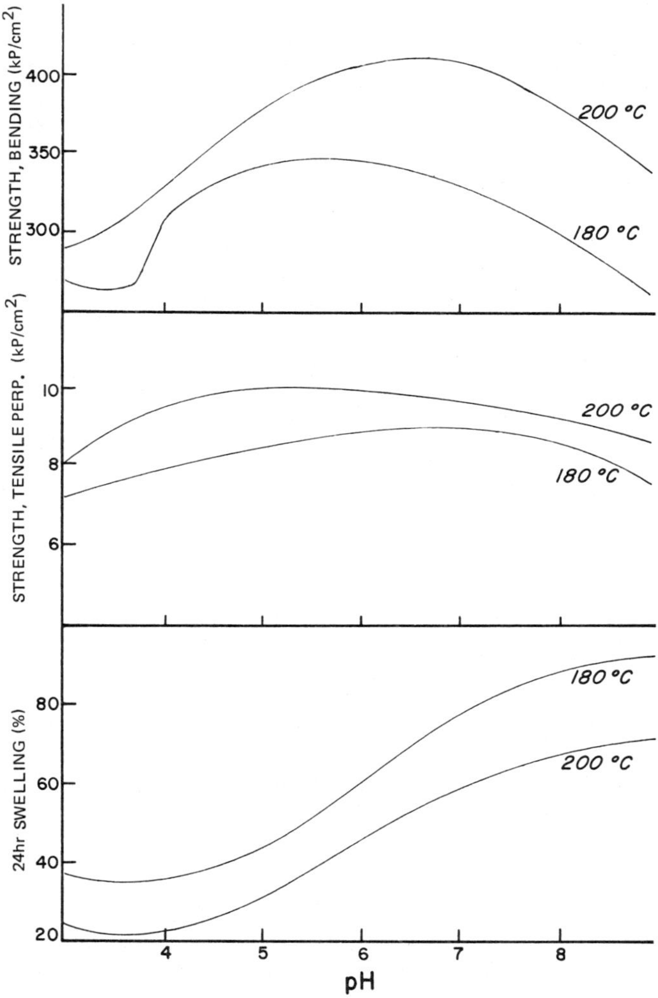

Fig. 5.14. Dependence of bending strength, tensile strength, and thickness swelling of 9-mm-thick one-layer pine wood chipboard, obtained with 10% sodium-based SSL and 4% novolak binder on the pH value. (From Ref. 16.)

contact heating in a flat press, high-frequency heating was applied, raising the temperatures during pressing in the core layer to 220°C for 1 min, which diminished the pressing time. In contrast to their publication in 1971 [16], Roffael and Rauch in 1973 [17] found that phenolic resins of the resol type (Kauresin 250, BASF) also improve the binding properties of SSL in particleboard, and the percentage swelling, shown in Figure 5.14, can be improved to meet the German standard specification [18] (6% after 2 hr in water at 20°C) by applying a postheating treatment at 200°C for 1 hr.

The postheating treatment could be avoided when higher amounts of resol-type resin were used. In conventional PF particleboard the PF resin amounts to about 8%, based on dry wood particles. It has been found [17] that up to 33% of the resol-type adhesive in conventional PF particleboard can be substituted in the surface layers of a three-layer 22-mm board by sodium-based lignosulfonate without any major deterioration in the mechanical board properties. In 20-mm one-layer particleboards at pH 9, up to 25% of the phenol-formaldehyde resin could be replaced by sodium-lignosulfonate under convertional pressing conditions, leading to particleboard meeting the German standards specification [7,18] (Table 5.4). Furthermore, 10% of the PF resin in beech/plywood could be substituted for by sodium-based lignosulfonate at a pressing temperature of 165°C, and up to 30% at 190°C [8]. Highest shear strength of the plywood was obtained with an adhesive formulation of pH 12–13.

B. Karatex Adhesive

According to Forss and Fuhrmann (1972) (Table 5.3) [6,19], the amount of lignin in lignin-PF adhesives for particleboard, plywood, and fiberboard can be increased to 40–70% if a high molecular weight fraction (molecular weight > 5000) of either lignosulfonate or black liquor, obtained from alkaline pulping of wood, is applied. Fractionation of SSL or black liquor can be achieved by ultrafiltration [6,19]. A typical

Table 5.4. Properties of 20-mm-Thick, One-Layer Pinewood SSL-PF Particleboard[a]

Adhesive (%)		Tensile strength (N/mm^2)	After 2-hr boiling (N/mm^2)	Bending strength (N/mm^2)	Swelling (%)	
PF	SSL				2-hr	24 hr
9	–	1.04	0.36	29.7	12.8	14.9
8.1	0.9	1.04	0.36	29.4	12.9	14.8
7.65	1.35	1.05	0.32	29.0	13.2	15.3
6.75	2.25	1.04	0.26	29.0	14.1	16.5
5.85	3.15	0.88	0.1	28.5	16.7	19.9

[a]pH 9, pressing time 6 min at 190°C, specific gravity of boards 0.7 g/cm^3.
Source: Ref. 7.

Table 5.5. Composition of Karatex Compared with a Typical PF Adhesive

	PF (%)	Kraft lignin or LS, mol wt > 5000 (%)	Sodium hydroxide (%)	Additives (%)	Water (%)
PF adhesive	33	–	2	11	54
Karatex	19	12	1	13	55

Source: Ref. 6.

Karatex formulation for plywood is given in Table 5.5, compared with a commercial PF-glue formulation.

According to the authors, the higher effectivity of high molecular weight lignin molecules is due to their higher level of cross-linking, which requires less phenol-formaldehyde for the formation of an insoluble copolymer than do low molecular weight lignin molecules. However, bearing in mind the findings of Shen [3] that low molecular weight ammonium-based SSL is more effective, Forss and Fuhrmann appear not to have checked the influence of inorganic salts in SSL or black liquor that are separated off during ultrafiltration. Forss and Fuhrmann assume that condensates between smaller lignin molecules and phenol-formaldehyde "are unable to contribute to the three-dimensional network" [6], which is unlikely because low molecular weight lignin molecules are more reactive than are high molecular weight molecules.

In the manufacture of particleboard either high-frequency (HF) heating or combined contact HF heating has been applied. In the latter case, the press platen temperature has been 180°C. German standard requirements for weather-resistant particleboards were met at pressing times between 10 and 12 sec/mm and 8–12% adhesive, based on dry wood particles.

One advantage inherent in the fractionation by ultrafiltration is that the lignin becomes more uniform and less dependent on variations in pulping conditions and wood source, which sometimes cause serious problems in the application of technical lignins. Full-scale plywood mill tests, some of them running continuously for several weeks, appear to have been performed in two Finish plywood mills.

V. LIGNIN IN COMBINATION WITH UREA-FORMALDEHYDE RESINS

In 1965, W. Arnoldt of BASF (Table 5.6) found that the pressing time of SSL particleboard obtained with sulfuric acid as catalyst (see the Shen procedure), can be reduced by 50% and the specific gravity of

Table 5.6. Lignin-Based Wood Adhesives in Combination with UF Binders

Year	Author or patentee	Adhesive formulation	Application	Literature
1951	Permanente Cement Co.	SSL, UF, ammonia	Fiberboard	US 2,846,431 1.10.51/5.8.58
1959	Diamond Alkali Co.	LS, UF or PF	Particleboard	US 3,076,772 19.2.59/5.2.63
1963	Inca Inks Inc.	Black liquor, UF	Packaging materials, etc.	US 3,395,033 14.1.63/11.4.66
1964		LS or black liquor, UF	Particleboard, plywood	DBP,B 75851/22i 11.3.64
1965	BASF	LS or black liquor, UF, mineral acid	Plywood, particleboard	BE 660,938 11.3.65/13.9.65
	BASF, W. Arnoldt	SSL, UF	Particleboard	DAS 1,223,534 23.6.65/25.8.66
1970	Ukrainian Mech. Treat. Wood Res. Inst.	LS, UF	Chipboard	SU 332,064 11.5.70/20.4.72
1972	Sanko Plywood KK	LS, UF	Formaldehyde-free plywood	J4 9,071,118 18.11.72/10.7.74

1973	BASF, Ch. Schmidt-Hellerau	SSL, UF	Cold adhesion of UF	DOS 2,354,928 2.11.73/15.5.75
	EIDAI KK	LS, UF, polyethylene-glycol	Formaldehyde-free adhesive	J5 0,034,332 30.7.73/2.4.75
	Koyo Sangoy Co. Ltd., Asahi KK	Lignin, UF, thiourea	Formaldehyde-free plywood	JA 036,328 2.8.73/3.4.75
	L. K. Putseiko et al.	LS, UF	Timber adhesive	SU 313,343 2.12.73
1974	Leningrad Forestry Acad.	LS, UF, phosphoric acid	Fiberboard	SU 518,363 24.7.74/14.7.76
1975	Tokai Pulp Co.	SSL, UF, aldehyde	Adhesive	JA 107,248 3.9.75/8.3.77
1976	Champion Int. Corp.	LS, UF	Particleboard	US 3,994,850 5.1.76/30.11.76
1977	BASF	SSL (1–6%), UF	Cold adhesion of UF	DOS 2,745,809 12.10.77/26.4.79
1978	Svensk Selenindustri, AB, F. J. Edler	SSL (33%), UF	Wood adhesive	DOS 2,835,752 16.8.78/31.10.79

Table 5.7. Particleboard Properties Obtained with LS-UF Adhesives

UF (%):	100	90	80	75	70
LS (%):	–	10	20	25	30
Gelling time (min)	81	102	143	176	219
Specific gravity (g/cm^3)	0.686	0.680	0.678	0.678	0.681
Bending strength (N/mm^2)	17.5	18.6	17.9	15.8	15.1
Tensile strength (N/mm^2)	0.57	0.58	0.44	0.32	0.23
Swelling after 2 hr	26.2	26.4	26.8	36.8	42.5
Swelling after 24 hr	34.0	34.2	40.5	51.7	57.10

Source: Ref. 20.

the boards by 7–10%, if 10–30% of the SSL blended wood chips are replaced by UF blended wood chips. However, at pressing times of still 0.6–1 min per millimeter of board thickness, the pressed particleboard has to be posttreated at high temperatures to meet German standards for mechanical strength properties. Again, the necessary posttreatment of the boards have hindered the practical application of this finding.

On the other hand, small amounts (up to 10% of the UF resin) of SSL improve the cold adherence of blended UF particleboard (Schmidt-Hellerau, BASF, 1973, 1977, Table 5.6). This has found practical application as is in current industrial use in some western European particleboard mills.

Roffael [7] has shown that 20% of the UF in the surface layers of UF particleboard can be replaced by ammonium-based lignosulfonate without significantly worsening the mechanical board properties. The release of formaldehyde decreased only slightly, which was attributed to the reduction of UF resin rather than to a reaction of formaldehyde with the lignin. According to Figure 5.5, the reaction of lignin with formaldehyde takes place only in strongly alkaline or strongly acid conditions or at elevated temperatures, which are not the conditions under which conventional UF particleboard is prepared.

The binding of formaldehyde by lignin in UF particleboard is claimed in three Japanese patents, listed in Table 5.6, together with other patents, dealing with lignin UF formulations as wood adhesives. Other improvements achieved by lignosulfonate in UF resins are decrease of adhesive viscosity, wettability of wood particles, and water resistance of finished boards (Table 5.6).

According to a recent report [20], substitution of up to 15% of the UF adhesive in particleboards by SSL does not cause any major impairment in particleboard properties. This can be seen from Table 5.7, in which the properties of 17.7-mm-thick one-layer particle board obtained with 8% UF binder (F/U = 1.27), replaced partially by 10–30% magnesium-based SSL, are listed. The adhesive contained 0.5% paraffin emulsion and 3% ammonium chloride. Pressing time was 10 sec/mm at 200°C.

From Table 5.7 it is obvious that substitution of 20% or more of the UF binder by magnesium-based SSL worsens both the strength and water resistance of the boards, while the gelling time (pot life) of the adhesive is increased. When three-layer 20-mm particleboard was manufactured with 15% of calcium-based (A), sodium-based (B), or ammonium-based (C and D) lignosulfonate and 85% UF binder, with pressing times of 9 sec/mm at 200°C and 10.5% adhesive in the surface layer and 8.5% in the core layer, the board properties shown in Figure 5.15 were obtained.

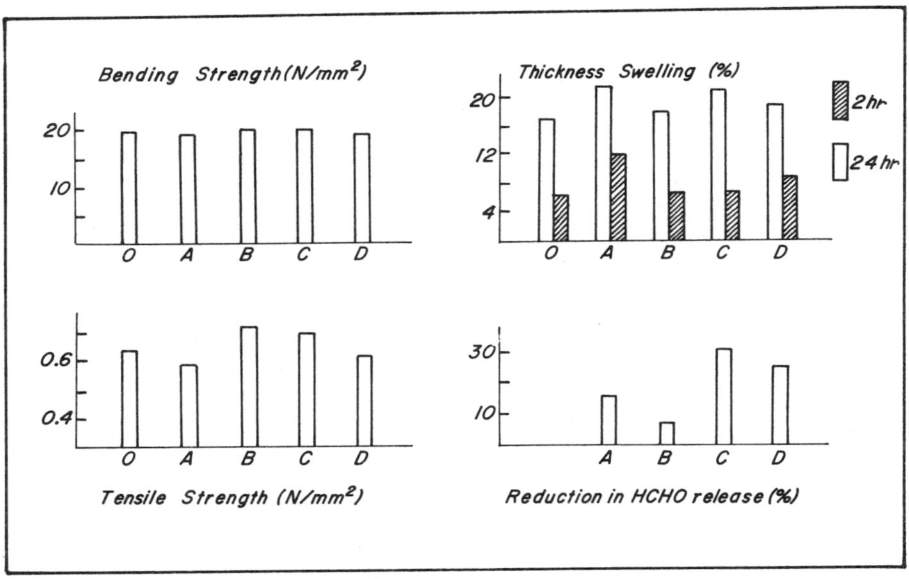

Fig. 5.15. Bending strength, tensile strength, thickness swelling, and reduction in formaldehyde release of 20-mm-thick three-layer particleboard obtained with either pure urea-formaldehyde (UF) binder (0) or with UF containing 15% calcium-based (A), sodium-based (B), or ammonium-based SSL (C and D). (From Ref. 20.)

Table 5.8. Properties of Particleboard Containing 33% LS and 67% UF (Resin Types A and B) Compared with Those of Pure UF Binder (Resin Type C)[a]

Resin type	LS (%)	Board thickness (mm)	Specific gravity (g/cm^2)	BM (kPa)	EM (kPa)	IB (kPa)	Water absorption, 2 hr (%)	Swelling, 2 hr (%)
C	–	10	0.76	19,237	2875×10^3	986	29.7	8.4
A	33	13	0.77	16,858	2703×10^3	972	18.9	6.1
B	33	10	0.79	18,548	3034×10^3	800	24.6	7.9
B	33	19	0.75	17,031	3082×10^3	738	18.9	4.0

[a]Board properties checked according to ASTM C 1037-72. BM, bursting modulus; EM, elasticity modulus; IB, internal bond strength.

Source: Edler, 1978

At board densities of about 0.7 g/cm^3, the bending and tensile strengths of the UF-LS boards are not decreased compared to boards prepared with 100% UF binder, while the percentage of swelling is increased. The formaldehyde release is considerably decreased by ammonium- (C and D) rather than by calcium- and sodium-based SSLs, indicating that the ammonium ions but not the lignin react with formaldehyde under the conditions existing in the boards. In the case of calcium- and sodium-based SSLs, the reduction in formaldehyde release lies between 10 and 18%, which corresponds to the amount of SSL in the UF-SSL formulation.

In a recent patent by Edler (1978, Table 5.6) it has been claimed that about 33% of UF binder in particleboards can be replaced by ammonium-based SSL if certain conditions are maintained. First, the UF resin should have a relatively high number of methylol (CH_2OH) groups, characterized by a Witte number of 1—1.8, preferably 1.6, which leads to better compatibility between UF and SSL. Second, the concentration of ammonium ions has to be adjusted to 0.2—4%. The ammonium ions react with free formaldehyde, forming less reactive urotropin (hexamethylenetetramine), which leads to excessive sulfonic acid groups in lignin. The free sulfonic acid groups catalyze condensation reactions in lignin. If the ammonium-ion concentration is higher than 4%, based on dry lignosulfonate, the acidity becomes too high, resulting in very fast curing. The latter causes soft board surfaces and diminished strength properties. On the other hand, if the ammonium ion concentration is below 0.2%, the curing time becomes too long. In Table 5.8 the properties of four types of particleboard, obtained with three different types of adhesive—A having a Witte number of 1.58, B of 1.50, and C of 1.02— are listed. The latter was prepared using only UF resin, while types A and B contained 33% lignosulfonate and 67% UF. Pressing times were about 15 sec/mm at 160°C. Wood chips consisted of 67% pine and 33% Douglas fir.

From Table 5.8 it can be seen that the mechanical strength properties of boards obtained with resins of types A and B (33% lignosulfonate) show no major impairment when compaired with those of conventional UF boards (resin type C). Significant improvements in the water resistance are gained, due to the polyphenolic structure of lignin.

VI. OUTLOOK

Despite its availability in large quantities as a waste product of pulp mills, lignin hitherto has found only scarce utilization as a chemical raw material. Although many efforts have been undertaken in the application of lignin as an adhesive for wood materials (Tables 5.1, 5.3, and 5.6), no full-scale industrial application in either particleboard, plywood, or fiberboard production is known presently. Forss (1976) has claimed that several mill tests have been carried out in two Finnish plywood mills, which have acquired the right to manufacture and use Karatex, but it is not known to the author if and to what extent Karatex is used for the production of commercial boards.

The prevailing advantage in the application of SSL instead of PF resins for the manufacture of exterior-grade boards is a considerable saving in adhesive costs. According to Shen [3], about $11 could be saved per 100 m^2 of 9.5-mm-thick medium-density particleboard, which amplifies to over $1 million per year for a particleboard plant of an annual capacity of 10 million square meters of 9.5-mm-thick particleboard. These values are valid only for the Shen procedure (see above). In case of the Pedersen procedure, additional energy costs are caused by long curing times and costs of refined steel autoclaves. Also, the density of the boards is higher. The crucial drawbacks of both these procedures, which so far have hindered their industrial application, seem to be corrosion and high-temperature problems, and long processing times, the latter problem considerably increasing overhead costs.

Less savings are to be expected from interior-grade particleboards, which currently use UF adhesives. The main advantage in this case may be seen in formaldehyde-free particleboard. However, this aim cannot be reached by only partial substitution of UF by SSL, as has been already explained, although a great number of patents have dealt particularly with partial substitution of UF by SSL in particleboard (Table 5.6).

The application of hydrogen peroxide as a curing agent for SSL as a wood adhesive [13] causes additional costs, but overcomes the drawbacks of the aforementioned procedures. According to this procedure, the particleboard is free from formaldehyde. They do not need any additional UF or PF binders. The pressing temperatures are low and curing is fast, due to the evolution of heat from the curing reaction itself. A uniform temperature profile over the board thickness renders the boards more homogeneous than conventional boards and avoids charring on the surfaces, which is often observed with other typical SSL boards. Also, corrosion problems so far do not exist and are not to be expected.

No environmental pollution can occur because SSL itself is a waste product of pulp mills and its application as a wood binder helps to solve environmental problems. Also, hydrogen peroxide does not cause pollution like many other chemicals, due to its decomposition to water and oxygen.

The only disadvantage in this procedure [13], the high moisture content of the boards, has, in principle, been overcome in the meantime. The procedure now is at an industrial trial stage. It is also intended to further improve the water resistance of the particleboard in order to meet exterior-grade standards requirements.

ACKNOWLEDGMENTS

The author wishes to express his cordial thanks to Dr. W. Clad, BASF, Ludwigshafen, West Germany, for many fruitful discussions and for supplying relevant literature. I would also like to thank Miss G. Hitze for her skillful performance of some experiments that have been published in this chapter for the first time.

REFERENCES

1. F. Melms and K. Schwenzon, *Verwertungsgebiete der Sulfitablauge*, VEB Deutscher Verlag Für Grunstoffindustrie, Leipzig, 1967.
2. K. Freudenberg, Lignin: its constitution and formation from p-hydroxycinnamyl alcohols, *Science 148*:595–600 (1965); K. Freudenberg, in K. Freudenberg and A. C. Neish (eds.), *Constitution and Biosynthesis of Lignin*, Springer-Verlag, Berlin, 1968.
3. K. C. Shen and L. Calvé, Ammonium-based spent sulphite liquor for waferboard binder, *Adhes. Age*, 25–29 (Aug. 1980).
4. H. H. Nimz, I. Gurang, and I. Mogharab, Untersuchungen zur Vernetzung technischer Sulfitablauge, *Liebigs Ann. Chem.*, 1421–1434 (1976).
5. R. J. Mahoney, 1980, Physical change in wood particles induced by the particleboard hot-pressing operation, *Proc. 14th Int. Particleboard Symp.*, Washington State University, Pullman, Wash., 1980.
6. K. G. Forss and A. Fuhrmann, Karatex—the lignin-based adhesive for plywood, particleboard and fiberboard, *Paperi ja Puu (Finland) 11*:817–824 (1976).
7. E. Roffael, Fortschritte in der Verwendung von Sulfitablaugen als Binde- und Zusatzmittel bei der Herstellung von Holzspanplatten, *Adhaesion 11*:334–336; *12*:368–370 (1979).
8. E. Roffael and W. Rauch, 1974, Phenolharz- und sulfitablaugengebundene Spanplatten. Vergleich einiger Eigenschaften, *Holz-Zentralbl.* No. 43/44, Apr. 10, 1974.
9. D. P. C. Fung, K. C. Shen, and L. Calvé, 1977, Spent sulphite liquor-sulphuric acid binder: its preparation and some chemical properties, *Report OPX 180 E*, Eastern Forest Products Laboratory, Ottawa, 1977.
10. K. C. Shen and M. N. Carroll, A new method for evaluation of internal strength of particleboard, *For. Prod. J. 19*(8):17–22 (1969); K. C. Shen, Modified powdered spent sulphite liquor as binder for exterior waferboard, *For. Prod. J. 24*(2):38–44 (1974).
11. Canadian Standard for Particleboard CSA 0188/68.

12. K. C. Shen, Spent sulphite liquor binder for exterior waferboard, *For. Prod. J. 27*(5):32–38 (1977).
13. H. H. Nimz and G. Hitze, The application of spent sulphite liquor as an adhesive for particleboards, *Cellul. Chem. Technol. 14*:371–382 (1980).
14. H. H. Nimz and A. Pizzi, private communications, 1980.
15. E. Roffael and W. Rauch, Über die Herstellung von Holzspanplatten auf Basis Sulfitablauge: III. Möglichkeiten zur Verkürzung der thermischen Nachbehandlung von Sulfitablauge-gebundenen 9 mm dicken Spanplatten, *Holzforschung 26*:197–202 (1972).
16. E. Roffael and W. Rauch, Über die Herstellung von Spanplatten auf Basis von Sulfitablauge: II. Über ein neues und schnelles Verfahren zur Herstellung Sulfitablauge-gebundener Spanplatten, *Holzforschung 25*:149–155 (1971).
17. E. Roffael and W. Rauch, Über die Herstellung von Holzspanplatten auf Basis von Sulfitablauge: IV. Verwendung von Sulfitablauge in Kombination mit alkalischen Phenolharzen, *Holzforschung 27*:214–217 (1973).
18. German Standard Specification DIN.
19. K. G. Forss and A. Fuhrmann, Finnish plywood, particleboard and fiberboard made with a lignin-based adhesive, *For. Prod. J. 29*(7):39–43 (1979).
20. LCWH-Ligninsulfonate in Holzspanplatten, Studie 2 and 3, Lignin-Chemie Waldhof-Holmen BmbH, Dusseldorf, West Germany, 1979.

6
Diisocyanates as Wood Adhesives

K. C. FRISCH and L. P. RUMAO* / *Polymer Institute, University of Detroit, Detroit, Michigan*

A. PIZZI / *National Timber Research Institute, Council for Scientific and Industrial Research, Pretoria, South Africa*

I. INTRODUCTION

The use of diisocyanates as adhesives for wood has recently attracted considerable interest. Although diisocyanates have been used for at least 30 years for the manufacture of polyurethane resins for a variety of industrial products, their use as wood adhesives is relatively new. The pioneering work of Deppe and Ernst (1951) [1, 2] on the use of such materials for binding particleboard is well known. As a consequence of their work, the commercial manufacture of diisocyanate-bound particleboard started in Germany in 1975. Notwithstanding the fact that a few plants in North America and Europe are already

---

*Present affiliation: Ford Motor Co., Dearborn, Michigan.

using these adhesives, their commercial use must still be considered in its infancy.

Together with tannin- and lignin-derived resins, diisocyanate adhesives make part of a "club" of new wood adhesive resins which will attract more interest and assume more importance in the timber industry in the near and not so near future. While the other adhesives in this small group base their future prospects on their ability to outperform phenolic resins at considerably lower cost, and to their better availability due to their non-oil-derived status, the chances for the success of diisocyanates wood adhesives are based on the novelty of the wood gluing concept they represent as much as on improved technical performance. They are, fundamentally, oil-derived materials, but this, and their high cost, should not deter further expansion of the use for which they appear to be earmarked.

II. CHEMISTRY OF DIISOCYANATES

A. Reactions of Isocyanates with Active Hydrogen-Containing Compounds

The chemistry of the reaction of isocyanates and diisocyanates with wood is inextricably bound with the chemistry of preparation of polyurethanes. The two most important reactions in the preparation of polyurethanes are the reactions between isocyanates and hydroxyl-containing compounds and the reaction between isocyanate and water. The former is the basic reaction for the formation of urethane groups and can be considered as the chain propagation reaction

$$R{-}N{=}C{=}O + R'{-}OH \rightarrow R{-}NH{-}\overset{\overset{\displaystyle O}{\|}}{C}{-}O{-}R'$$

(R' = alkyl or aryl)

The second reaction takes place with liberation of carbon dioxide and with simultaneous formation of substituted urea groups. The first step in this reaction is the formation of unstable carbamic acid, which decomposes to form an amine and carbon dioxide.

$$R{-}N{=}C{=}O + H_2O \rightarrow [R{-}NH{-}\overset{\overset{\displaystyle O}{\|}}{C}{-}OH] \rightarrow R{-}NH_2 + CO_2$$

The amine immediately reacts with additional isocyanate to form a substituted urea as follows:

$$R{-}NH_2 + R{-}N{=}C{=}O \xrightarrow{\text{fast}} \underset{\text{urea}}{R{-}NH{-}\overset{\overset{\displaystyle O}{\|}}{C}{-}NH{-}R}$$

The overall reaction can be represented as

$$2R{-}N{=}C{=}O + H_2O \rightarrow R{-}NH{-}\overset{\displaystyle O}{\overset{\|}{C}}{-}NH{-}R + CO_2$$

An alternative mechanism provides for formation of a carbamic acid anhydride, followed by decomposition to a substituted urea and carbon dioxide.

$$R{-}N{=}C{=}O + H_2O \rightarrow [R{-}NH{-}\overset{\displaystyle O}{\overset{\|}{C}}{-}OH] + \xrightarrow{R{-}N{=}C{=}O}$$

$$(R{-}NH{-}C(=O))_2O \rightarrow R{-}NH{-}\overset{\displaystyle O}{\overset{\|}{C}}{-}NH{-}R + CO_2$$

Still another reaction that may occur to some extent is the formation of the amine salt of carbamic acid, which decomposes under the influence of heat to yield water and urea:

$$[R{-}NH{-}\overset{\displaystyle O}{\overset{\|}{C}}{-}OH] + R{-}NH_2 \rightarrow [R{-}NH{-}\overset{\displaystyle O}{\overset{\|}{C}}{-}O]\,[NH_2R] \xrightarrow{T^\circ}$$

$$R{-}NH{-}\overset{\displaystyle O}{\overset{\|}{C}}{-}NH{-}R + H_2O$$

Other reactions that lead to branching and cross-linking and can be used for wood adhesives are the reaction of the isocyanate group with amines and the formation of allophanate and biuret linkages by reaction with urethanes.

$$R{-}N{=}C{=}O + R'{-}NH_2 \rightarrow R{-}NH{-}CO{-}NH{-}R'$$

$$R{-}N{=}C{=}O + R{-}NH{-}CO{-}OR' \rightarrow R{-}NH{-}CO{-}\underset{\underset{\displaystyle R}{|}}{N}{-}CO{-}O{-}R'$$

allophanate

$$R{-}N{=}C{=}O + R{-}NH{-}CO{-}NH{-}R \rightarrow R{-}NH{-}CO{-}\underset{\underset{\displaystyle R}{|}}{N}{-}CO{-}NH{-}R$$

biuret

Another cross-linking reaction is trimerization, leading to the formation of isocyanurate rings which are hydrolytically and thermally very stable.

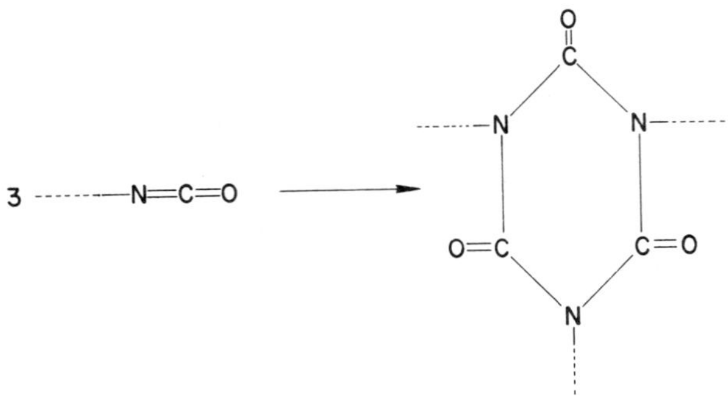

Thus several reactions of the isocyanate group are useful to bond wood with diisocyanates. Molecules presenting two —N=C=O groups (thus the name "diisocyanates") are necessary to form bridges of various lengths with the alkyl and phenolic hydroxy groups of wood cellulose and lignin, respectively. The reactions taking place between diisocyanates and wood are thus as follows:

1. *Reaction with Cellulose and Lignin Hydroxy Groups*

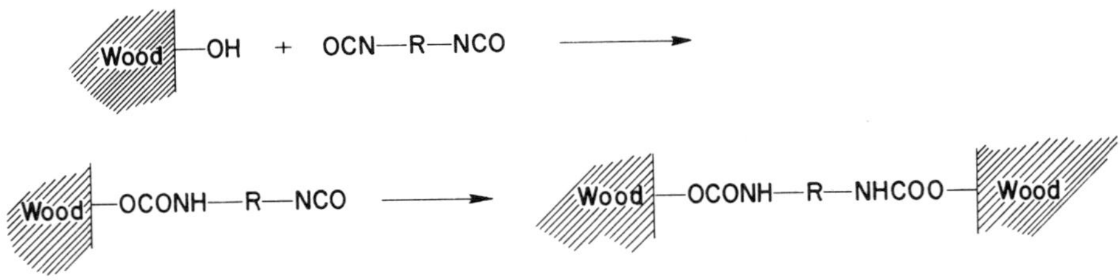

2. *Polyurea Formation*

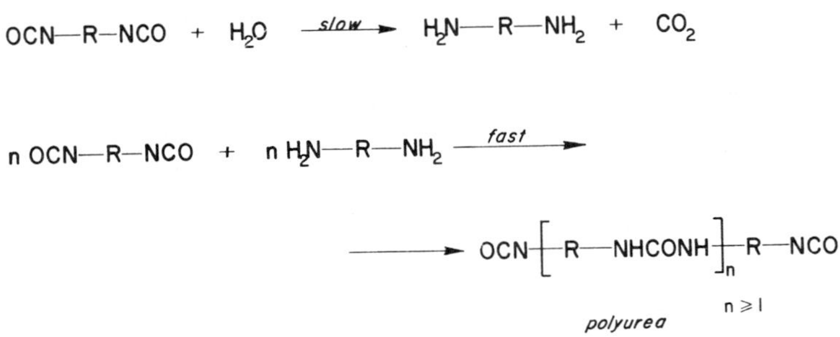

The polyurea formed when terminal —NCO groups are still present will then react with wood in the same manner as a diisocyanate. Thus, through the formation of polyureas, a certain variation in the diisocyanate polymer length is introduced which contributes to successful wood bonding. If too much water is present, however, too high an amount of diisocyanates is deactivated, too much polyurea with no —NCO terminal groups is formed, and the system looses considerably in adhesive efficiency, as not enough urethane linkages between the adhesive and the wood can be formed.

B. Catalysis

Since many of the reactions of the diisocyanates which have been described may be too slow, catalysts can be employed whose functions is not only to bring about faster reactions rates but also to establish a proper balance between the wood-to-adhesive cross-linking reaction (the hydroxyl-isocyanate reaction) and the polyurea formation (isocyanate-wood water content reaction) which controls the molecular distribution of the final polyurethane network. Another important function of catalysts is to bring about completion of the reactions, resulting in an adequate "cure" of the glue line. Completion of cure results in maximum strength properties and maximum chemical and weathering resistance.

Up to now it has not been found necessary to use catalysts for diisocyanate adhesives for particleboard due to the high temperatures used in this particular process. Furthermore, the diisocyanate used for the manufacture of particleboard is commercial "polymeric" MDI (MDI = 4,4'-diphenylmethane diisocyanate).

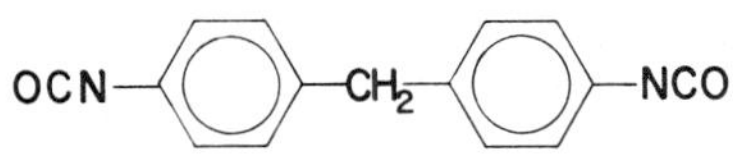

MDI
4,4' - diphenylmethane - diisocyanate

I

Such a commercial product is known as "polymeric" MDI, as it is a mixture of the isomers of dimer, trimer, and higher polymeric isocyanates derived from side reactions. Thus in the case of the use of "polymeric" MDI for bonding particleboard diisocyanate, catalysis has been of little or no importance, except for the basic catalytic effect brought about by some of the strongly alkaline wax emulsions used in the glue mixes. However, diisocyanate catalysis is bound to assume more importance in the future, especially in the application of cold-setting diisocyanate adhesives for wood laminating. Cold catalysis might also assure some importance in the particleboard field if the need for faster pressing times or milder pressing temperatures will arise in the future.

The catalysts most commonly employed are tertiary amines and metal catalysts, especially tin catalysts. Tertiary amines are catalysts for both the isocyanate-hydroxyl and the isocyanate-water reactions. The efficiency of tertiary amine catalysts depends on their chemical structure. It generally increases as the basicity of the amine increases and the steric shielding of the amino nitrogen decreases [3–5]. Table 6.1 shows the importance of steric factors in the catalytic activity of tertiary amines. Some of the most commonly used tertiary amine catalysts are triethylenediamine, N-alkyl morpholines, N,N,N',N'-tetramethylethylenediamine, N,N.N',N'-tetramethyl-1,3-butanediamine, N,N'-substituted piperazines, and dialkylalkanolamines.

Organotin compounds are extremely effective catalysts for the isocyanate-hydroxyl reaction. They include stannous octoate, stannous

Table 6.1. Relative Catalytic Activity of Amine Catalysts

Chemical name of Amine Catalyst	Chemical Formula	pK_a	Relative Catalytic Activity
1,4-Diazabicyclo (2,2,2)-octane	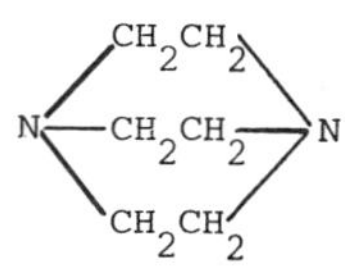	5.4	23.9
N,N-Diethylcyclohexylamine	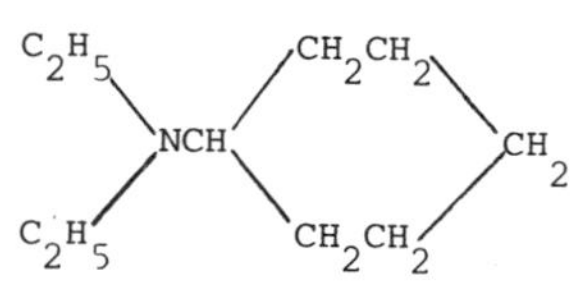	10.0	0.7
N-Methylmorpholine	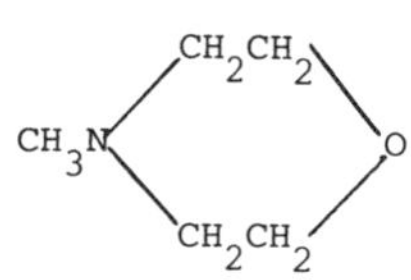	7.4	1.0
N,N,N',N'-Tetramethyl-methanediamine	$(CH_3)_2NCH_2N(CH_3)_2$	10.6	0.085

Source: Refs. 3 and 6.

Table 6.2. Relative Reactivity of Catalysts in Isocyanate-Hydroxyl Reaction

Catalyst[a]	Concentration (%)	Order of Activity
Uncatalyzed	–	1
TMBDA	0.1	56
DABCO	0.1	130
TMBDA	0.5	160
DBTDL	0.1	210
DABCO	0.2	260
DABCO	0.3	330
SnOct	0.1	540
DBTDL	0.5	670
DBTDL + TMBDA	0.1 + 0.2	700
SnOct + TMBDA	0.1 + 0.2	1000
DBTDL + DABCO	0.1 + 0.2	1000
SnOct + TMBDA	0.1 + 0.5	1410
SnOct + DABCO	0.1 + 0.5	1510
SnOct	0.3	3500
SnOct + DABC)	0.3 + 0.3	4250

[a]TMBDA, tetramethylbutanediamine; DABCO, 1,4-diaza(2.2.2)bicyclo-octane; DBTDL, dibutyltin dilaurate; SnOct, stannous octoate.
Source: Refs. 3 and 7.

oleate, dibutyltin dilaurate, and dibutyltin di-2-ethylhexoate. In diisocyanate catalysis it is common practice to use combinations of tertiary amines and tin catalysts when requiring a "hot" catalyst system.

Table 6.2 shows the relative order of activity of various catalysts and catalyst combinations in a model isocyanate-hydroxyl reaction. It also points out clearly the synergistic effect between tin catalysts and tertiary amines. It is of interest to note that a synergistic effect has also been observed with two tertiary amine catalysts, tetramethylbutanediamine and triethylamine. This is probably due to a combination of two different factors leading to relatively strong basicity. The

Table 6.3. Relative Reactivity of Catalysts in Isocyanate-Water Reaction[a]

Catalyst	Order of Activity
Stannous octoate	1.0
N-Ethyl morpholine	1.1
Dibutyltin dilaurate	1.3
Triethylamine	1.5
N.N.N',N'-Tetramethyl-1,3-butane diamine	1.6
1,4-Diaza(2.2.2)bicyclooctane	2.7

[a]Concentration: 0.1 part catalyst to 100 parts resin.
Source: Refs. 3 and 8.

relative order of activity of some catalysts for the water-isocyanate reaction is shown in Table 6.3.

C. Mechanism and Kinetics of Catalyzed Reactions of Isocyanates with Hydroxyl Compounds

The reaction between phenyl isocyanate and methanol in the presence of tertiary amines follows second-order kinetics: first order with respect to the isocyanate, and first order with respect to the alcohol [3,9]. The second-order rate constant k_2 can be expressed by the relation

$$k_2 = k_0 + k_c[\text{amine}]$$

where k_0 is the rate constant for the uncatalyzed reaction and k_c is the catalytic constant characteristic of the activity of the catalyst. They assume the following reaction steps:

$$\text{ArNCO} + \text{NR}_3 \underset{k_2}{\overset{k_1}{\rightleftarrows}} \underset{\displaystyle \text{NR}_3}{\underset{|}{\text{ArNCO}}}$$

$$\underset{\displaystyle \text{NR}_3}{\underset{|}{\text{ArNCO}}} + \text{R'OH} \xrightarrow{k_3} \text{ArNHCOOR'} + \text{R}_3\text{N}$$

Using stationary-state conditions, the concentration of the complex is given by the expression

$$[\text{complex}] = \frac{k_1[\text{ArNCO}][\text{NR}_3]}{k_2 + k_3[\text{R'OH}]}$$

and hence the rate of the catalyzed reaction can be expressed by

$$\frac{-d[\text{ArNCO}]}{dt} = \frac{k_1 k_3[\text{ArNCO}][\text{NR}_3][\text{R'OH}]}{k_2 + k_3[\text{R'OH}]}$$

The overall rate constant is then

$$k = k_o + \frac{k_1 k_3[\text{NR}_3]}{k_2 + k_3[\text{R'OH}]} = k_o + k_c[\text{NR}_3]$$

Hostettler and Cox [3,10] studied the effect of catalyst concentration on the phenyl isocyanate 1-butanol reaction and found the data to correspond to the relationship

$$k_2 = k_o + k_c[\text{catalyst}]^n$$

where n = 0.89 for di-n-butyltin diacetate, and 1.04 for triethylamine.

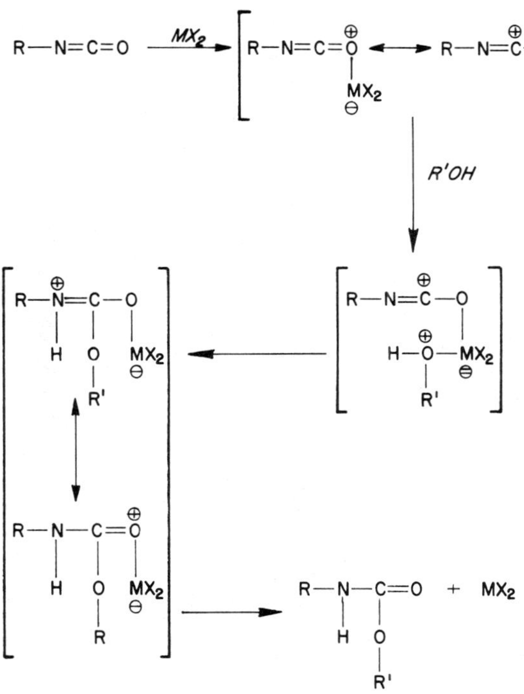

Scheme I

The metal-catalyzed isocyanate-hydroxyl reaction involves the formation of a ternary complex between the metal compound, the isocyanate, and the hydroxyl compounds (as shown in Scheme I) [3,11]. The complex formation could occur in two steps, with either the isocyanate or the hydroxyl compound reacting first. In the ternary complex, the isocyanate and hydroxyl compound are positioned on the same side of the metal compound in close proximity to each other.

The following expression was proposed to show a simple relationship between the specific velocity and catalyst level with reference to hydroxyl concentration [3,12]:

$$\log v = Q + n \log R$$

where v is the overall catalytic velocity constant, n the slope of the curve, R the molar ratio (hydroxyl)/(catalyst), and Q a constant representing the extrapolated value of log v at equal concentration of hydroxyl and catalyst (see Fig. 6.1).

The stepwise mechanism of activation by organotin compounds of alcoholic hydroxy groups of polyols and isocyanate is represented in Scheme II. A number of amino ethers have also been investigated for their catalytic and synergistic effects [3,12]: Compounds such as $O[CH_2CH_2CH_2N(CH_3)_2]_2$; $O[CH_2CH_2OCH_2CH_2CH_2N(CH_3)_2]_2$;
$O[CHCH_2N(CH_3)_2]_2$; $S[CH_2CH_2N(CH_3)_2]_2$; and $N[CH_2CH_2N(CH_3)_2]_2$.
  | |
CH_3 CH_3

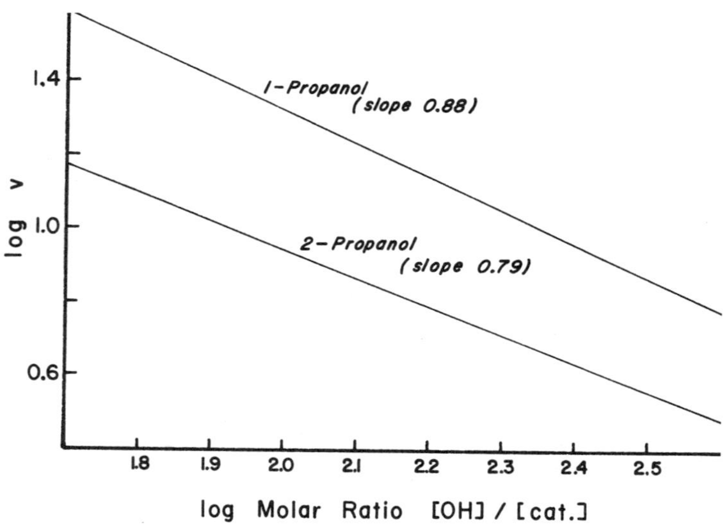

Fig. 6.1. Effect of stannous octoate on the rate of the phenyl isocyanate-hydroxyl reaction. (From Ref. 3.)

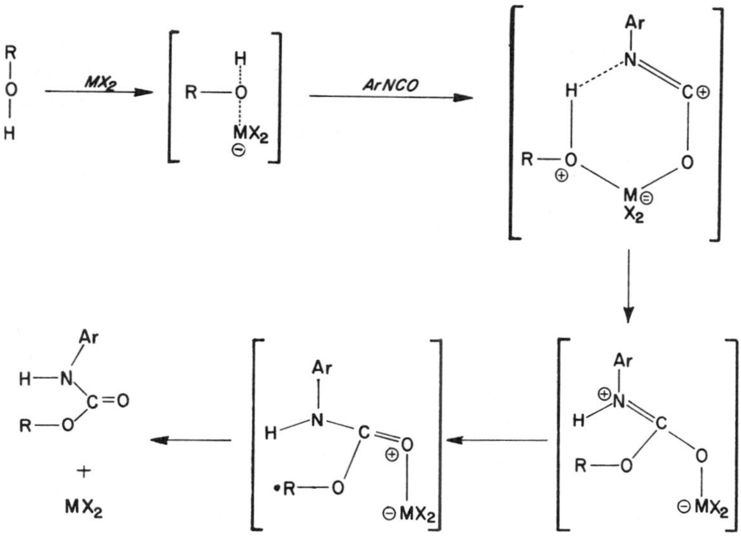

Scheme II

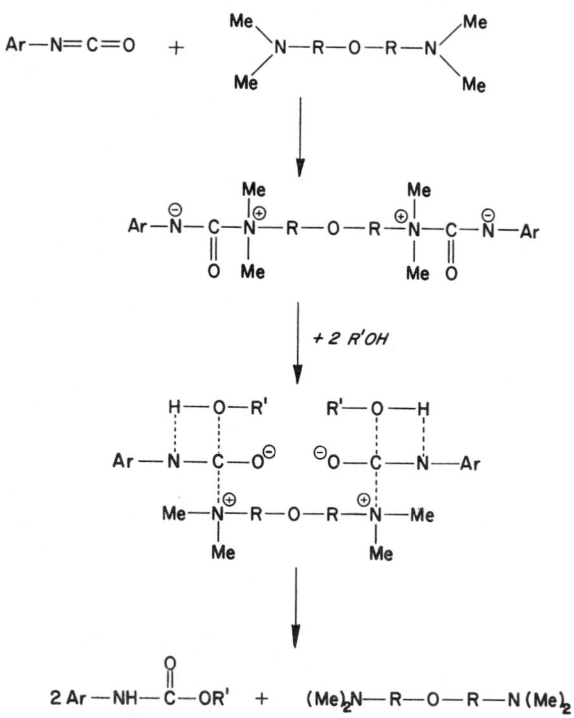

Scheme III

The mechanism proposed for hydroxyl/isocyanate reactions by bis(dimethylaminoalkyl) ethers is shown in Scheme III.

The synergistic nature of the amine ethers in conjunction with stannous octoate was explained by an activated complex intermediate stabilized through attachment of the basic amine to the acidic carbonyl carbon of the isocyanate group:

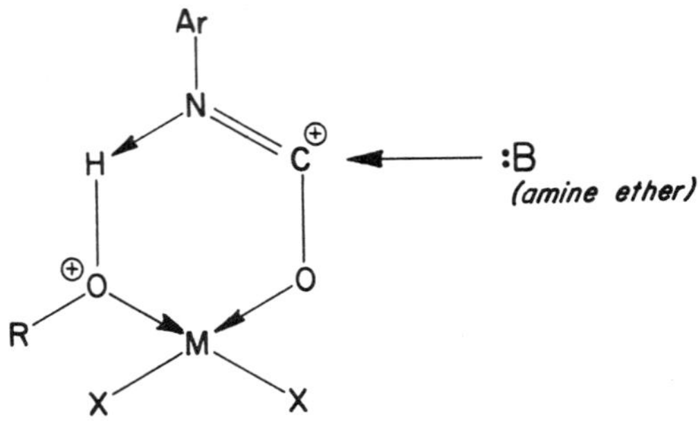

A mechanism for urethane bridge formation in the reaction between phenyl isocyanate and the hydroxy group of methyl alcohol catalyzed by metal acetylacetonates has also been proposed. An octahedral complex having the following structure is formed:

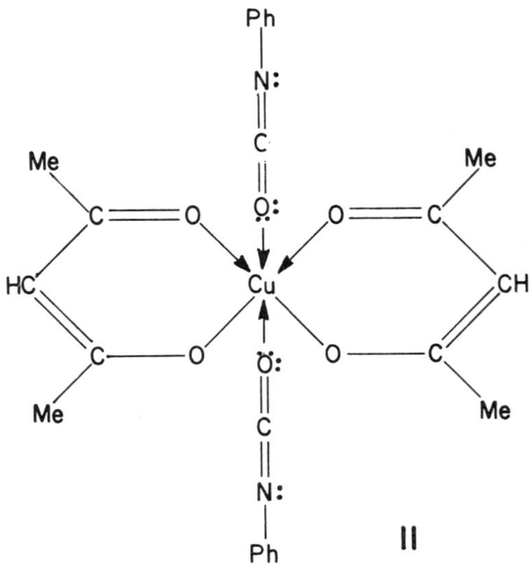

II

In this complex, the electrophilic nature of the carbonyl carbon of the phenyl isocyanate is increased, which enhances the nucleophilic attack of the oxygen atom of the methyl alcohol to form the phenyl urethane. The latter is apparently coordinated with the copper ion Cu^{2+} at first and is then separated as the reaction product after decomposition of complex IV:

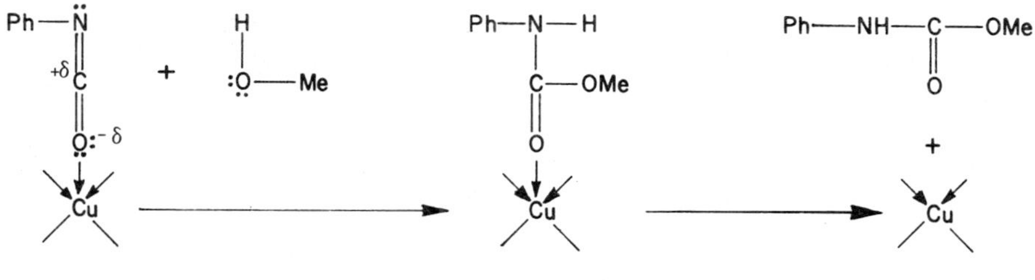

III IV V

With time, more complicated complexes are formed. One such product, a dianilide of copper acetylacetonate VI, was actually isolated when the reaction was carried out in chloroform. Entelis, Nesterov, and Tiger

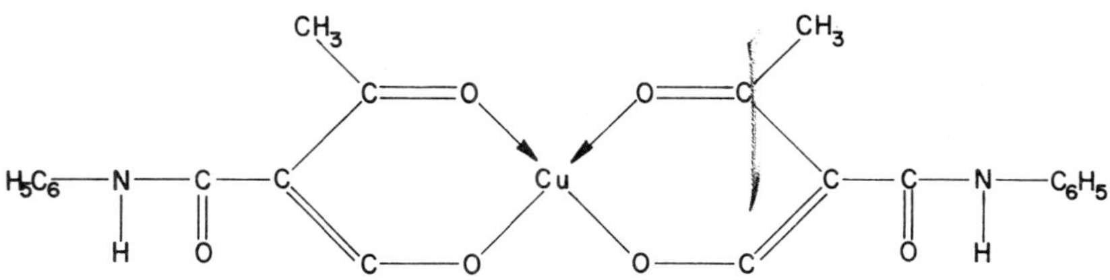

VI

[3,13] have used the large positive values of the parameter ρ in the Hammett equation to show that in the absence of catalysts the reaction between NCO— and OH— groups begins with the nucleophilic attack of the carbonyl carbon atom in the isocyanate, which has a significant deficit in electron density, by the alcohol:

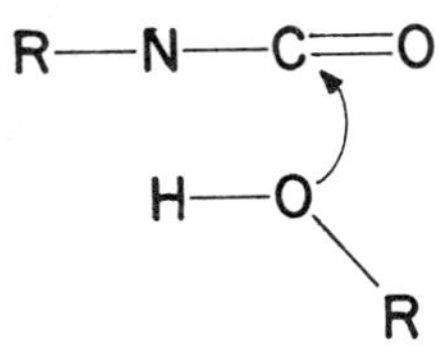

However, upon introduction of an active organotin catalyst into the system, two phenomena were observed: first, there was a significant decrease in the parameter ρ, and second, the effect of the catalyst appeared to be the more significant the less active the isocyanate in the noncatalytic reaction, as shown in Table 6.4.

It has been shown that the reaction of the isocyanate with an alcohol in the presence of compounds of tetravalent tin takes place through an intermediate complex of the catalyst with the alcohol. Hence, the accelerating action of the catalyst on the reaction is, apparently, through the activation of the OH— group due to formation of a donor-acceptor

Table 6.4. Reaction Rate of Substituted Phenyl Isocyanates with Methyl Alcohol in Heptane in the Presence of Dibutyltin Dilaurate at 25°C[a]

σ	R in RPhNCO	K (liters/mol sec) Uncatalyzed	Catalyzed	$K_{cat}/K_{uncat} \times 10^4$
-0.07	m-CH_3	2.9×10^{-4}	6.6×10^2	228.0
0.00	H	5.4×10^{-4}	7.4×10^2	137.0
0.23	p-Cl	3.6×10^{-3}	1.2×10^3	33.3
0.37	m-Cl	7.6×10^{-3}	1.5×10^3	19.8
0.71	m-NO_2	9.2×10^{-2}	2.1×10^3	2.3
		$\rho = 3.3$	$\rho = 0.9$	

[a]Methylalcohol concentration = 1.6 mol/liter.
Source: Ref. 3.

type of bond between the oxygen of the alcohol and the tin atom in the complex. The hydroxyl hydrogen of the alcohol becomes quite mobile and active with respect to the NCO— group, due to a decrease in the electron density on the hydroxyl oxygen. In this case, in contrast to the noncatalytic reaction, the first step in the reaction of the NCO— and OH— groups is the electrophilic attack of the nitrogen of the isocyanate by the acidic hydrogen of the alcohol-catalyst complex with the preliminary formation of a hydrogen bond which changes into the covalent N—H bond. This leads to a sharp decrease in the electron density and to the appearance of a positive charge on the carbonyl carbon of the isocyanate. In the second step, a breakdown in the donor-acceptor bond between oxygen and tin complex occurs due to the competing action of the carbon of the isocyanate for the complexed oxygen of the alcohol. This explains the marked decrease in the parameter ρ for a catalyzed reaction where the NCO— group enters the reaction, first as a nucleophilic (nitrogen atom) reagent. This is also in line with the observation that the effect of the catalyst appears to be more significant the less active the isocyanate in the noncatalytic reaction. This is undoubtedly due to the dual nature of the reaction center of the isocyanate, which is capable of adding an electron donor atom to the carbonyl carbon as well as an electron acceptor atom to the nitrogen of the NCO— group.

The catalytic effect of urethane groups in the reaction between isocyanates and alcohols has been observed [3,14,15]. The uncatalyzed reaction of phenyl isocyanate with urethanes did not proceed at a measurable rate at 40°C. The increased reaction rate is due to the increasing concentration of urethane groups formed during the alcohol-isocyanate reaction and the catalytic effect of these urethane groups on the further alcohol-isocyanate reaction. However, the point in the reaction where this catalysis becomes noticeable, as indicated by deviation of the kinetic plot from linearity, appears to vary with the catalyst used. In the case of stannous 2-ethylhexoate, deviation occurs at 37–40% reaction, with cobaltous stearate at 45–50% reaction, whereas with triethylenediamine, the deviation occurs only after more than 65% reaction has taken place.

The type of hydroxyl group, primary or secondary, or the presence

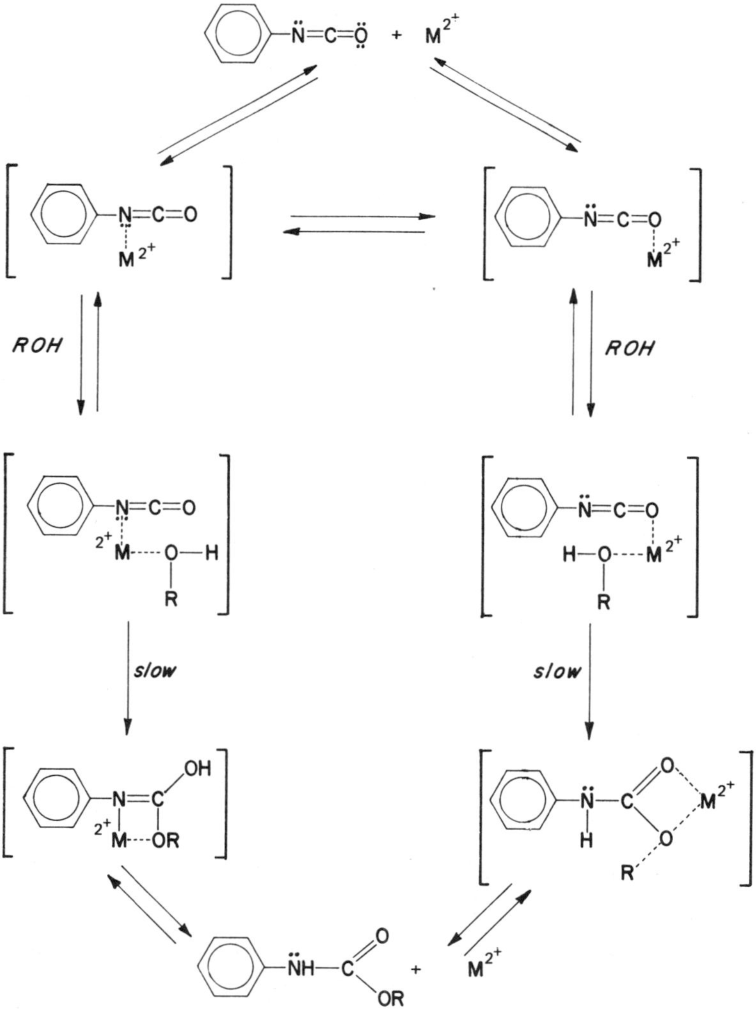

Scheme IV

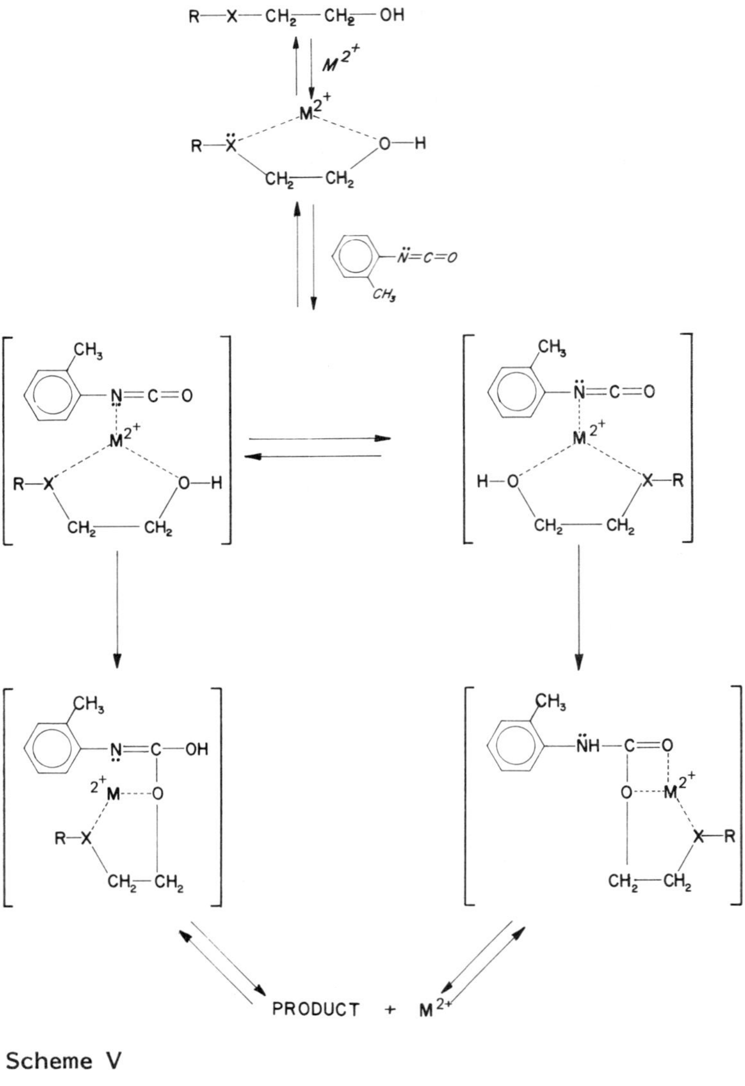

Scheme V

of other functional groups in the molécule, such as ethers, does not appear to influence the point where the deviation starts [3,16]. The sole effect of these groups is reflected in the respective rate. Substituting a diisocyanate such as tolylene diisocyanate for phenyl isocyanate, the tertiary amine-catalyzed reaction is similar to that of the reaction with the monoisocyanate [3,16].

Robins [2,17] screened a large number of metal ions, including manganese, iron, cobalt, nickel, lead, and tin, for their catalytic activities in isocyanate-alcohol reactions. He noted that the metal ion catalysis of isocyanate-hydroxyl reactions was considerably enhanced by the presence of certain "activating" groups α or β (even γ) to the carbon carrying the reactive hydroxyl, and also that certain metal ions (e.g., Co^{2+}, Zn^{2+}, Cd^{2+}) were very sensitive to changes in steric factors at the isocyanate group. He proposed two mechanisms: one for a monofunctional alcohol (Scheme IV) and one for an "activated" alcohol (Scheme V). According to these mechanisms, the metal ion seems to fulfill a twofold function: (1) it increases the electrophilic character of the carbon in the isocyanate group by coordinating either at the oxygen or nitrogen, and (2) it brings both the isocyanate and alcohol in closer proximity by forming a ternary complex.

The results that were obtained for complex formation between metal catalysts, such as tin and lead compounds, and 1-methoxy-2-propanol and phenyl isocyanate, respectively, were strong experimental confirmation of ternary complex formation, first proposed by Britain and Gemeinhardt [3,11]. Although the relative speed of complex formation is usually very fast, the rate-determining step, the reaction between the ligands after complex formation has taken place, depends on the coordination number of the metal ion, the configuration of the complex, and the ionic radius of the ion.

The catalytic properties of a polymeric tertiary amine, poly-4-vinylpyridine, and its copolymers with styrene and methyl methacrylate, were compared with that of pyridine in reactions of mono- and diisocyanates with alcohols and glycols [3,18]. It was found that the reaction rate was proportional to the concentration of the polymeric catalyst up to that concentration at which the reaction system becomes very viscous, which indicates the accessibility of catalytic sites in the polymer for the reactants.

The polymer-catalyzed reactions exhibited higher rates in comparison with those catalyzed by pyridine at low reactant concentrations. However, when the concentration was increased, the rates of the polymer-catalyzed reactions reached limit values and ultimately became less than in the case of pyridine. It was also found that the catalytic activity of the polymeric catalyst increased as the styrene content in the copolymer increased.

The increase in the rate of reactions catalyzed by a polymer in comparison with monomeric catalyst may be due either to higher catalytic activity of active centers or to the increase of the effective concentration of the reactants in the reaction zone. The coincidence of total activation energies of the phenyl isocyanate-methyl alcohol concentration (6 kcal/mol) in the presence of pyridine and of a copolymer (35.5% polyvinyl pyridine/64.5% styrene) as measured in the range 20–60°C suggests the latter explanation to be the more likely one. These phenomena can be explained by the solution behavior of the polymers, which form coils or aggregates that are easily permeated by the reacting molecules. These coils or aggregates have a certain ability to dissolve the reactants. It may be assumed that the distribution of the aromatic isocyanate in the solvent (chloroform) and in the polymeric catalyst, containing benzene rings, is shifted in favor of the latter as a result of hydrophobic interaction of the hydrocarbon aromatic residues. Therefore, the isocyanate equilibrium concentration in this type of catalyst would be higher than its average concentration in solution and would increase with the styrene content in the copolymer. The limiting value of the catalytic activities of the polymeric catalyst can be explained on the basis of saturation of the macromolecular coils or aggregates with reactants.

III. TECHNOLOGY OF DIISOCYANATE ADHESIVES FOR WOOD

A. Particleboard Adhesives

1. *Characteristics of "Polymeric" MDI as a Wood Adhesive*

4,4'-Diphenylmethane diisocyanate (MDI) is the preferred type of diisocyanate for particleboard manufacture. It has a low vapor pressure ($\pm 10^{-5}$ mm Hg) [19] at ambient temperature, which contributes to its safety of use; it has a low viscosity, which renders it particularly suitable for the particleboard industry; and its cost is lower than that of other diisocyanates. At ambient temperature commercial "polymeric" MDI is a dark reddish brown liquid with a storage capability of approximately 12 months and a viscosity lower than 400 cP. 4,4'-MDI is more reactive than other forms (i.e., 2,4'-MDI, etc.) and the isocyanate manufacturing industry has now geared itself to produce commercial products in which 4,4'-MDI is predominant (±90% of total product). In addition to 4,4'-diphenylmethane diisocyanates, a commercial MDI wood adhesive will also contain up to 60% of higher molecular weight isocyanates as isomeric forms of a degree of polymerization of up to 6.

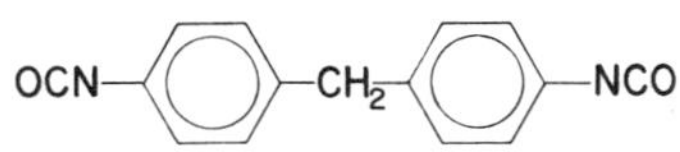

4,4' MDI

VII

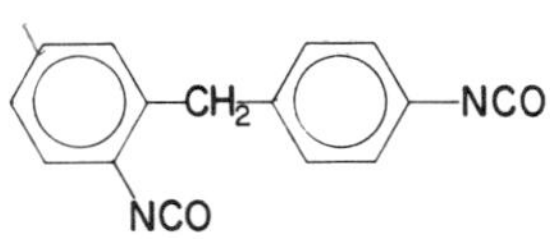

2,4' MDI

VIII

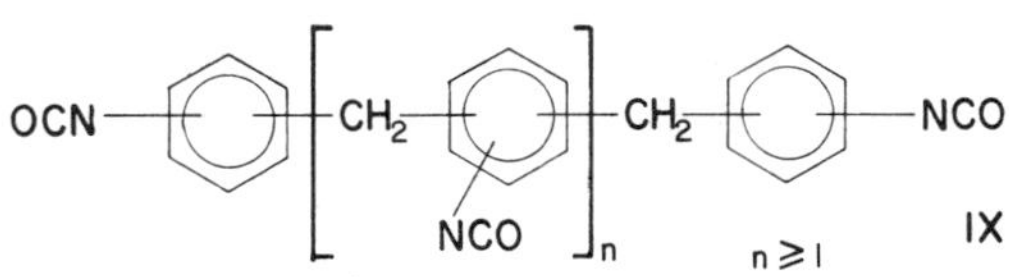

"polymeric" MDI

IX

Its reaction with the hydroxyl groups of wood constitutents to form covalent urethane bonds between wood and glue line and its reaction with the water in wood to form polyureas are the main reactions contributing to its performance as a wood adhesive. Wittman has calculated that more than 50% of the MDI introduced for bonding leads to the formation of polyureas. The formation of polyureas during MDI bonding of wood is beneficial for several reasons. First, by increasing the molecular weight distribution of the adhesive, better gap-filling properties are obtained. Second, the formation of amidic groups and of urethane groups, obtained by the reaction of the amidic groups with further isocyanate, will accelerate the curing reaction and the wood-MDI combination will behave in effect as a self-catalyzing system. Third, in "hybrid" adhesives in which MDI is coupled with PF, UF, or MUF adhesives, the amidic groups will readily coreact with the methylol groups of the formaldehyde-based resins with a considerable improvement in both strength and bonding efficiency, as described later in the section on combined resin systems.

Advantages of MDI bonding of wood products over more traditional adhesive systems are its tolerance to faster pressing times, the water and weather resistance of the glue lines obtained, and the need for lower amounts of adhesive [1,2,19]. Disadvantages are the tendency of MDI-bonded particleboard to stick to the hot caul sheets of the press, the need to provide a metering and pumping system devoid of water contamination, and doubts as regards MDI level of toxicity and its high cost [1,2,19]. Some of these disadvantages have now been completely overcome. Excellent release agents have now been developed by several MDI manufacturers [2,19]. The manufacturers recommend cleaning of the press caul sheets followed by wiping with a saturated solution of the release agent. In industrial scale application about 10–20 g/m^3 of the release agent is sufficient to prevent sticking of the panel to the press [2]. Initial fears that the use of such release agents could affect paintability of the finished particleboard have been proved to be unfounded when the release agent is used in moderation. As regards the need to work with a completely anhydrous metering and pumping system, this disadvantage has been overcome by the recent introduction of fairly stable "emulsifiable" isocyanates: namely, emulsions of MDI in water [19]. These materials, which have improved handling characteristics over MDIs that are not in emulsion form, appear to have completely solved this problem. Toxicity problems also appears to have been overcome [19,20].

2. *Processing Characteristics*

"Polymeric" MDI is very hygroscopic and the adhesive must be stored in closed containers. When nonemulsified MDIs are used, the gluing station, gluing blender, and pipings must be cleaned with solvents when changing from a waterborne resin to MDI. When using 4% MDI

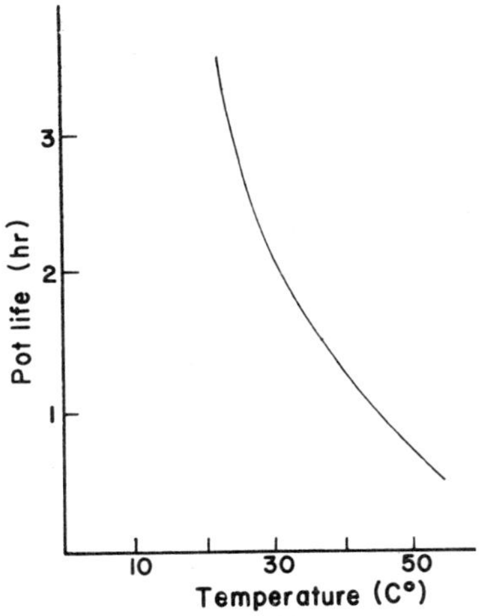

Fig. 6.2. Emulsifiable MDI: pot life as a function of temperature.

on core chips, wet internal bond strengths of the order of ±0.5 MPa are obtained [2]. Lowering the viscosities and thus improving resin distribution will give improved internal bonds. The viscosity can be lowered by preheating the resin to 70–80°C or by adding a solvent [2]. In the latter case resin contents of 2.5–3% are sometime sufficient to obtain satisfactory results [2]. However, water repellents will have to be added, as the percentage swelling of the board in water would

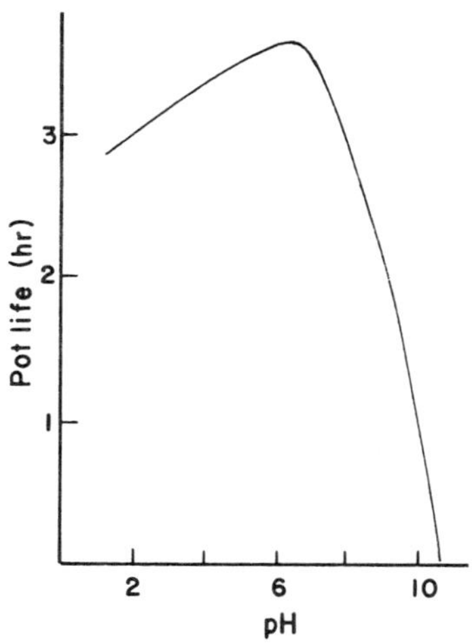

Fig. 6.3. Emulsifiable MDI: pot life as a function of pH.

Table 6.5. Percentage on Dry Wood of Glue Mix Components

	Core	Surface
Emulsifiable MDI	5.0–6.0	6.0–7.0
Total water added (wax emulsion included)	6.0	10.0–12.0
Wax emulsion (solids only)	1.0	1.0
Total percent water content (included wood at 2% moisture content)	8.0	12.0–14.0

Source: Ref. 3.

otherwise increase noticeably [2]. However, the use of solvents has the disadvantage to enhance the flammability hazard and the liberation of toxic vapors. The bond quality is not affected over a wide moisture content range of the glued particles (0–25%). The effects of temperature and of the pH of the glue mix over its pot life are shown in Figures 6.2 and 6.3.

The storage life of glued particles before pressing is of ±5 hr at 20°C. At 40°C the storage life is reduced to approximately 2 hr [2, 21]. Press times are as much as 20% lower than those obtainable for PF-bonded boards in which no metallic accelerators have been added [2,22]. At a pressing temperature of 170°C about 13 sec/mm is necessary; this decreases to about 10 sec/mm at 200°C. Average glue mixes for three-layer particleboard bonded with polymeric MDI are shown in Table 6.5.

3. *Properties of MDI-Bonded Particleboard*

The properties of MDI-bonded particleboard produced using different levels of MDI in the core of the board and a fixed amount of 6% MDI on dry wood in the board surfaces are shown in Table 6.6.

From the results shown in Table 6.6 it appears evident that MDI-bonded particleboard of traditional chips size distribution is satisfactory. There is an indication in these results as well as in many other authors' results [2,19,24–26] that the loss of internal bond after the V-100 treatment is slightly worse than that obtainable with phenolic-type resins. This indicates that MDI-bonded boards might be slightly more sensitive to water attack than water-resistant boards bonded with the more traditional PF resins. However, this effect is quite small. It is also evident that the use of MDI as particleboard

Table 6.6. Characteristics of Traditional Particleboard Bonded with Different Levels of MDI

MDI (% on dry wood)	2	3	5	6
Press temperature (°C)	175	175	175	175
Pressing time (sec/min)	15	15	10	10
Board thickness (mm)	12	12	12	12
Density (g/cm^3)	0.680	0.680	0.680	0.680
Dry internal bond (IB) (kP/cm^2)	1.5	5.4	7.7	8.3
V100 wet IB (kP/cm^2)	—	0.35	1.7	2.2

Source: Refs. 3 and 23.

binder affords a saving of about 15–20% adhesives over more traditional binders when used for bonding particleboard with a quite traditional particle size and geometry distribution pattern. Greater savings in adhesive amount, claimed by certain authors, are for boards with certain different particle geometry distributions, such as waferboard.

4. *Combined Resins Systems*

Because diisocyanate-based particleboard adhesives are relatively expensive, attempts have been made to combine them with cheaper binding agents. More traditional, formaldehyde-hardened adhesives such as UF, MF, and tannin-based resins have been used for this purpose. Deppe and Ernst [1, 2] have sprayed the particleboard chips with commercial UF and MF resins subsequently followed by spraying with a MDI adhesive resin. The two successive operations were carried out in the same glue blender. Such a combination of MDI with aminoplasts yielded considerable increases in internal board even at low resin additions. Their strength results indicated that formulations containing 2% MDI and 5% UF on oven-dried wood in the board core are sufficient to stand the boil test. The development of MDI resins which are soluble in aminoplasts [2, 27] also offers new possibilities. Shelf life and pot life of these combinations are shorter than that of aminoplastic resins [2], and the wood particles need a certain minimum moisture content when these resin combinations are used [2].

When the two resin components, MDI and aminoplastic, are applied separately, good dry and wet internal bond strengths are obtained. However, when the two components are mixed before application, only 50–70% of the internal bond strength obtained by a separate application system is possible.

Also important is the combination of MDI with tannin-formaldehyde adhesives [28, 29]. This combination is important, as tannins are an inexpensive source of resorcinolic material which tend to perform better than phenolic resins as particleboard binders. They allow considerably faster press times than both PF and MDI adhesives (7 sec/mm at 190°C). Strength and water and weather-resistance are better than those of PF resins due to their partly resorcinolic nature, and furthermore they are non-oil-derived. In this respect two types of flavonoid tannins are eminently suited for combination with MDI: acacia-derived tannins (such as wattle or mimosa tannin) and pine-derived tannins (such as those derived from *Pinus radiata* bark). While acacia-derived tannins are already employed successfully, by themselves, for the industrial production of exterior-grade particleboard, this is not the case for pine-derived tannins, due to their difficulty in handling when used alone. The combination of pine tannin extracts with MDI is a particularly useful one, as the combination of these two materials achieves excellent performance at a very low cost. Here, too, there is a difference between the results obtained by premixing the resins before spraying onto the wood particles and separate application of the two resin systems, with the latter being the better of the two. In these combined systems several cross-linking reactions contributing to final strength of the particleboard take place: (1) all the reactions of MDI with wood hydroxyls and with water take place; (2) MDI also reacts with the hydroxy groups of the tannin; (3) the tannin is also cross-linked by formaldehyde; and (4) the formaldehyde cross-links polyureas formed by MDI reaction with water, with the crossing-links being between polyurea-polyurea and polyurea-tannin phenolic nuclei. This combination of possible bonds gives a very strong cured resin network. The glue mixes used and the results obtained with *P. radiata* tannin are shown in Tables 6.7 and 6.8, respectively.

The proportions of resin solids applied onto the chips, this being calculated as the sum of tannin extract solids + MDI, were of 10 and 18% for core and surface chips, respectively, in glue mixes 1–4. This means that in glue mix 1, 10% tannin extract was added to the core chips; in glue mix 2, 9% tannin extract solids and 1% MDI; in glue mix 3, 8% tannin extract solids and 2% MDI; in glue mix 4, 7% tannin extracts solids and 3% MDI. Glue mixes 5, 6, and 7 were control glue mixes in which 1%, 2%, and 3% MDI alone were added to the core chips to study the contribution of MDI to the results obtained with glue mixes 2, 3, and 4, respectively. In glue mixes 5, 6, and 7, 5.5% MDI was added to the surface chips. In glue mixes 5, 6, and 7 the water was sprayed onto the wood chips before the spraying of the MDI, to avoid any deactivation of the isocyanate groups which would impair the efficiency of the adhesive.

The pressing conditions used to press duplicate laboratory 12-mm particleboard were the following: moisture content (glue mixes 1, 2, 3, and 4) core = 18%, surface = 22% (glue mixes 5, 6, and 7) core and

Table 6.7. Glue Mixes in Parts by Mass

	Glue mix						
	1	2	3	4	5	6	7
Formulation type	100 tannin	90 tannin, 10 MDI	80 tannin, 20 MDI	70 tannin, 30 MDI	100 MDI	100 MDI	100 MDI
Tannin extract solutions 44%	250	225	200	175	–	–	–
Paraformaldehyde 96%	15	14	12,5	11	–	–	–
MDI	–	11	22	33	11	22	33
Wax emulsion	22	22	22	22	22	22	22
Water	60	80	110	135	100	100	100

Source: Refs 23 and 28.

Table 6.8. Particleboard Results

Glue mix no.	Adhesive type	Core glue solids (%)	Density (g/cm^3)	Swelling after 2-hr boil: Measured wet (%)	Swelling after 2-hr boil: Measured dry (irreversible swelling) (%)	Original internal bond (tensile perpendicular (kP/cm^2)	Internal bond after 2-hr boil (V100) (kP/cm^2)	Internal bond retention (%)
1	100 Tannin	10 Tannin	0.700	44.6	31.8	3.25	–	–
2	90 Tannin:10 MDI	9 Tannin:1 MDI	0.663	44.2	31.5	4.95	–	–
3	80 Tannin:20 MDI	8 Tannin:2 MDI	0.666	28.0	15.9	4.69	1.30	27.7
4	70 Tannin:30 MDI	7 Tannin:3 MDI	0.690	15.0	4.3	8.39	4.24	50.6
5	100 MDI	1 MDI	0.667	–	–	0.45	–	–
6	100 MDI	2 MDI	0.670	–	–	1.54	–	–
7	100 MDI	3 MDI	0.700	72.2	38.5	5.39	0.35	6.5
Commercial phenol formaldehyde control		–	0.654	13.8	3.0	5.92	3.37	57.0
Commercial modified tannin/formaldehyde control		10 modified tannin	0.702	12.8	2.2	13.1	6.9	52.7
German standard DIN 68761, Pt. 2							minimum 1.5	

Source: Refs. 23 and 28.

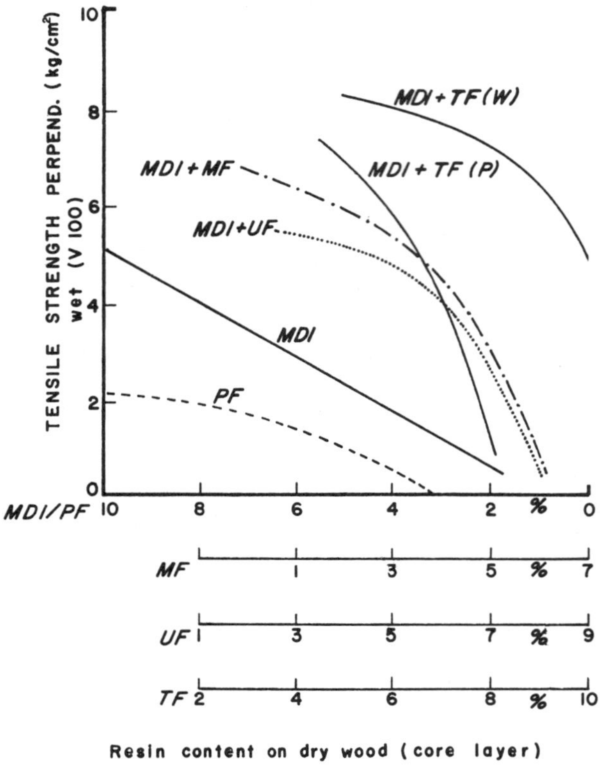

Fig. 6.4. Internal bond of particleboards bonded with MDI-aminoplast, or with MDI-tannin-formaldehyde (W, wattle; P, pine) resin conditions.

surface = 10%, maximum pressure = 25 kg/cm^2, pressing time = $7\frac{1}{2}$ min, pressing temperature = 170°C (much faster pressing times can be obtained in industrial conditions) (Fig. 6.4).

The results obtained for laboratory boards prepared with mixtures of a flavonoid-type tannin extract, a diisocyanate, and formaldehyde show that this type of formulation, in the case of a 70:30 mass/mass ratio of tannin extract/MDI, can be used for the production of good exterior-grade particleboard. The results are very similar for all the tannin extracts used (the results in the table are obtained with a *P. radiata* extract). These results were obtained by the use of unmodified tannin extracts. The advantages of the 70:30 tannin extract/MDI adhesive formulation are:

1. The ease of preparation. The tannin extract does not need to be chemically modified before use.

2. The adhesive does not need to be prepared by an adhesive manufacturer but can be assembled directly by the end user by purchasing materials easily available commercially.
3. Any tannin extract of the flavonoid type can be used with good results in this formulation, allowing the utilization of tannins more difficult to handle than those currently available and the use of materials which are currently considered as waste.
4. The tannin adhesive formulation described can also be used industrially by spraying separately onto the wood chips the tannin solution and the diisocyanate in (1) the same glue blender or (2) each into a separate glue blender, with the two blenders in line as described for other wood adhesives by other authors [30].

The use of these techniques allows indefinite pot life during industrial application of such an adhesive system. The paraformaldehyde needed for the tannins can be added into the tannin glue mix or separately in the glue blender [31]. Similar experiments were also successfully carried out with mixtures of diisocyanates and PF resins [32].

B. Plywood

As the characteristics of isocyanate-bonded boards are their excellent moisture resistance, freedom from formaldehyde vapors, and stiffness, as well as other physical properties, they have also been considered as alternatives for plywood adhesives. However, their use as plywood adhesives has been unsuccessful up to now [19]. Perhaps, as new diisocyanate adhesives directed to this specific application are prepared, the technical problems involved might be overcome [19].

Up to now the only successful applications for plywood of "polymeric" MDI has been obtained with a MDI-tannin-formaldehyde combined resin system (wattle tannin was used) [29,33] similar to that described for particleboard application. Good results were obtained on both beech and pine veneer plywood with mixtures of MDI-wattle extract (solids) in the range 17:83–25:75 by mass. In some of the formulations up to 12% UF resin were also incorporated.

The results obtained were of the order of 400 psi and 90% wood failure for dry and cold-water soak tests in the case of beech plywood and of 200–250 psi and 80–90% wood failure for dry, cold soak and then 72-hr boil test in the case of pine plywood. Pressing times were comparable with those of phenolic resins for this application. An added advantage of these combined formulations is the considerable improvement in ease of spread of the adhesive afforded by the addition of "polymeric" MDI to tannin-based adhesive solutions, notwithstanding the significant increase in viscosity caused by the MDI-tannin and MDI-H_2O reactions. Deactivation of isocyanate groups by the water present did not appear to interfere significantly with the results [33].

The combination of an expensive material as MDI with an inexpensive material such as a wattle tannin adhesive renders this type of formulation cost competitive with more traditional PF adhesives. Such formulations did not appear to give good results when the MDI was combined with pine tannin adhesives, due to their high reactivity with formaldehyde and their difficulty in handling.

REFERENCES

1. H. J. Deppe and K. Ernst, *Holz Roh.-Werkst. 29*(2):45 (1971).
2. H. J. Deppe, *Proc. 11th Int. Particleboard Symp.*, Washington State University, Pullman Wash., 1977.
3. K. C. Frisch and L. P. Rumao, *J. Macromol. Sci. C: Rev. Macromol. Chem.* 5(1):103 (1970).
4. B. G. Alzner and K. C. Frisch, *Ind. Eng. Chem. 51*:715 (1959).
5. A. Farkas, G. A. Mills, W. E. Erner, and J. B. Maerker, *Ind. Eng. Chem. 51*:1299 (1959).
6. F. Mort, *J. Oil Colour Chem. Assoc. 45*:95 (1962).
7. H. W. Wolf, Jr., Tin catalyst activity in urethane foam, *Elastomers Foam Bull.*, E. I. Du Pont de Nemours Co., Feb. 24, 1961.
8. G. G. Greth, R. G. Smith, and G. O. Rudkin, *J. Cellul. Plast. 1*:159 (1965).
9. J. W. Baker and J. B. Holdsworth, *J. Chem. Soc.*, 731 (1947).
10. F. Hostettler and E. F. Cox, *Ind. Eng. Chem. 52*:609 (1960).
11. J. W. Britain and P. G. Gemeinhardt, *J. Appl. Polym. Sci. 4*: 207 (1960).
12. F. G. Willeboordse, F. E. Critchfield, and R. L. Meeker, *J. Cellul. Plast. 1*:79 (1965).
13. S. G. Entelis, O. V. Nesterov, and R. P. Tiger, *Dokl. Akad. Nauk SSSR 178*:661 (1968).
14. H. Okada and Y. Iwakura, *Makromol. Chem. 64*:91 (1963).
15. M. Sato, *J. Org. Chem. 27*:819 (1962).
16. H. A. Smith, *J. Polym. Sci. (A-1) 6*:1299 (1968).
17. J. Robins, *J. Appl. Polym. Sci. 9*:821 (1965).
18. V. S. Pshezetsky, I. Massov, and V. A. Kabanov, *J. Polym. Sci. C, 22*:309 (1968).
19. G. W. Ball and R. P. Redman, *Proc. FESYP Int. Particleboard Symp.*, Hamburg, 1978.
20. W. E. Johns, *Proc. 14th Int. Particleboard Symp.*, Washington State University, Pullman, Wash., 1980.
21. I. Sachs, *Holz Zentrabl. 103*(20):295, and *103*(25):384 (1977).
22. A. Pizzi, *J. Appl. Polym. Sci. 24*:1247, 1257 (1979).
23. A. Pizzi, unpublished data, 1978–1979.
24. A. Mclaughlin, *Proc. 14th Int. Particleboard Symp.*, Washington State University, Pullman, Wash., 1980.

25. A. D. Adams, *Proc. 14th Int. Particleboard Symp.*, Washington State University, Pullman, Wash., 1980.
26. J. B. Wilson, *Proc. 14th Int. Particleboard Symp.*, Washington State University, Pullman, Wash., 1980.
27. A. Wooler, German Patent No. Dt-OS 2,610,552; British Patent No. 10,264 (1975).
28. A. Pizzi, *J. Macromol. Sci., Chem. C*(16):7 (1981).
29. M. Gamo, S. Kocho, M. Kondo, and Y. Katsuzawa, *Nippou Setchaku Kyokaishi. 14*:420 (1978).
30. C. C. Campbell and A. J. Shohet, *Proc. 13th Particleboard Symp.*, Washington State University, Pullman, Wash., 1979.
31. A. Pizzi, *Holzforsch. Holzverwert. 31*:4 (1979).
32. C. Y. Hse, U.S. Patent No. 4,209,433 (1980).
33. A. Pizzi, *J. Appl. Polym. Sci. 25*:2123 (1980).

7

Polyvinyl Acetate Wood Adhesives

TERENCE MARTIN GOULDING / *General Chemical Corporation, Johannesburg, South Africa*

I. INTRODUCTION

Polyvinyl acetate is a thermoplastic polymer that has gained wide acceptance over the years as a raw material for the adhesives industry. Modified or unmodified, in solution or emulsion form, and as homopolymer or copolymer, it exhibits a versatility that makes it suitable for bonding a wide variety of substrates. In particular, it is capable of producing strong and durable bonds on wood and wood-derived products, and this has been a major contributing factor to the tremendous growth of polyvinyl acetate-based adhesives in recent years, from almost nothing in the early 1930s to an estimated worldwide production of all types of 2 million tonnes in 1977. It is familiar to people all over the world as the binder for interior and exterior emulsion paints, as the so-called "cold glue" that replaced the heated pot of animal glue for carpentry, and as the "white glue" used in millions of households as a general-purpose adhesive.

II. BACKGROUND

The first recorded mention of vinyl acetate occurs in German State Patent number 271,381, granted to Dr. F. Klatte of the Grisheim-Elektron Chemical Works in June 1912 [1]. Vinyl acetate is a colorless flammable liquid with a viscosity of 0.4 cP at 20°C, a solubility in water of about 2% at 25°C, a boiling point of 72.7°C, and a characteristic odor [2]. This is the starting point for the production of polyvinyl acetate (PVA), and, in conjunction with other vinyl monomers, acrylic esters, dialkyl maleates and fumarates, ethylene, and certain other monomers, of a range of speciality copolymers and terpolymers, while graft polymers can be produced with monomers such as styrene that will not copolymerize with vinyl acetate.

It is not certain when the first polymerization of vinyl acetate to polyvinyl acetate was performed. During the period between 1915 and 1925 the free radical initiation of polymerization of various vinyl monomers was widely studied and by 1930 polyvinyl acetate was commercially available. It was, however, only in the years following World War II that polymers of vinyl acetate began to be used in significant quantities, particularly in the paint and adhesive industries, and since then, PVA has shown the same rapid expansion as most other well-known thermoplastic materials. Today some 50 major manufacturers and countless small suppliers around the world make material available in solid form, dissolved in solvent, or dispersed in water, as homopolymer, copolymer, or terpolymer, for a variety of applications that include paint manufacture, production of general-purpose and specialized adhesives, textile coating, sizing and sealing of paper and related products, concrete additives, and production of sealants. Formulated products span the entire range of viscosities, from thin liquid to heavy paste, and may be fast or slow drying. Dried films may be clear or opaque, pigmented or unpigmented, flexible or brittle, hard or soft, as required. The product may be internally or externally plasticized or unplasticized, and may dry hard and tack-free or pressure sensitive. With this versatility it is hardly surprising that polyvinyl acetate is now available almost anywhere in the world, with factories in most developed countries. At this point, the United States and Germany are the major producers, with companies such as Air Products, Borden, Monsanto, and Union Carbide in the United States, and Hoechst in Germany. It is estimated that the 1977 production of all types was 300,000 tons in Germany, while production for the same period was 236,000 tons in the United States of America [3], 90,854 tons in the United Kingdom [4], and 89,800 tons in Japan [5].

III. CHEMISTRY OF POLYVINYL ACETATE

The steps involved in the manufacture of polyvinyl acetate are shown schematically in Figure 7.1.

A. Production of Vinyl Acetate Monomer

Originally, vinyl acetate monomer was produced by reacting acetylene and acetic acid together with suitable catalysts.

$$HC{\equiv}CH + CH_3COOH \longrightarrow H_2C{=}CH\,OOCCH_3$$

The earlier technique was to carry out this reaction in liquid phase [6], but this method was replaced in the early 1930s by more efficient gaseous-phase processes performed at high temperature [7]. The acetylene and acetic acid were both derived from coal, acetylene directly by the action of water on calcium carbide

$$CaC_2 + H_2O \longrightarrow HC{\equiv}CH + CaO$$

and acetic acid by reacting the acetylene so produced with water to yield acetaldehyde, and subsequently oxidizing this acetaldehyde to acetic acid.

$$HC{\equiv}CH + H_2O \longrightarrow CH_3CHO$$

$$CH_3CHO + [O] \longrightarrow CH_3COOH$$

This process remained viable for as long as calcium carbide was readily and cheaply available. With the increasing cost of energy, however,

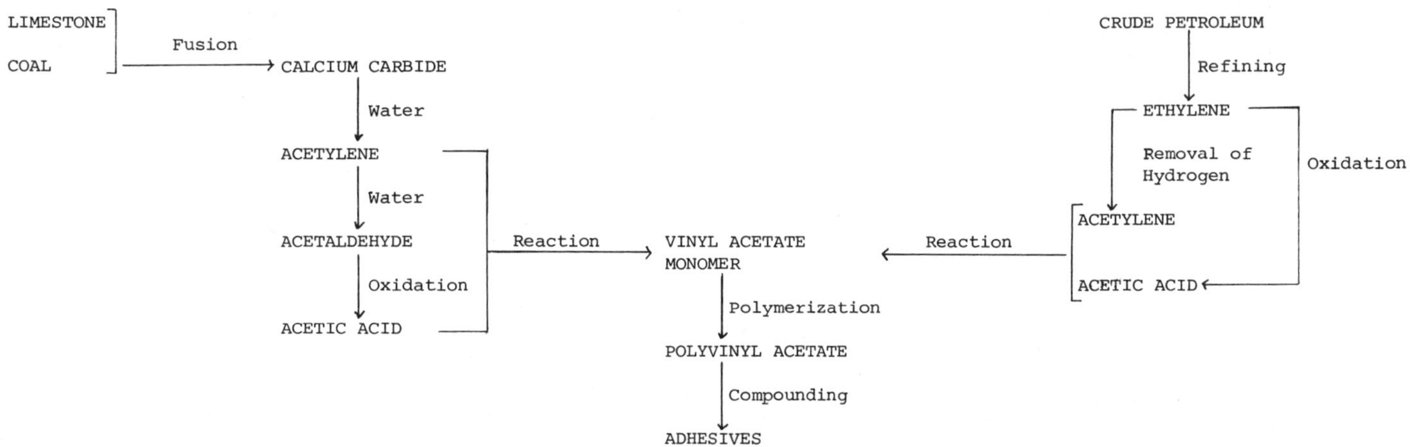

Fig. 7.1. Schematic representation of manufacture and compounding of polyvinyl acetate.

calcium carbide has become steadily more expensive and less readily available, while an ever-increasing range of chemicals has been made available by the rapid expansion of the petrochemical industry. These two factors have combined to make the modern method of production increasingly attractive. Here ethylene is used as the starting point for the production of both acetylene and acetic acid. Acetylene is made by removing hydrogen from ethylene, while acetic acid is obtained by oxidizing the ethylene to acetaldehyde, which is oxidized further to acetic acid.

B. Polymerization of Vinyl Acetate

Vinyl acetate monomer may be polymerized by many of the conventional polymerization techniques, including mass polymerization, solvent polymerization, and emulsion polymerization. The reaction is usually initiated and controlled by the use of free radical or ionic catalysts, although experimental methods of catalysis, including redox catalysis or activation by light, may be used for specialized products. The polymerization is characterized by three successive stages of reaction: initiation, the growth of the polymer, and termination.

1. *Initiation*

Initiation occurs when a free radical or ion attaches itself to a vinyl acetate molecule. This leads to a rearrangement of the electrons in the double bond, transferring the reactive site to the vinyl acetate monomer.

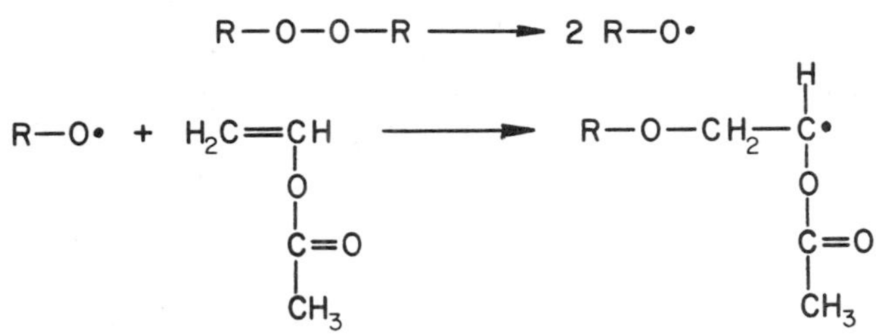

The initiator is usually a free radical derived from a peroxide such as benzoyl, lauroyl, or even hydrogen peroxide, although other initiators, such as persulfates, may also be used.

2. *Polymerization*

This highly reactive initiated molecule reacts with further monomer molecules by the same transfer mechanism, retaining the terminal reactive site for further growth.

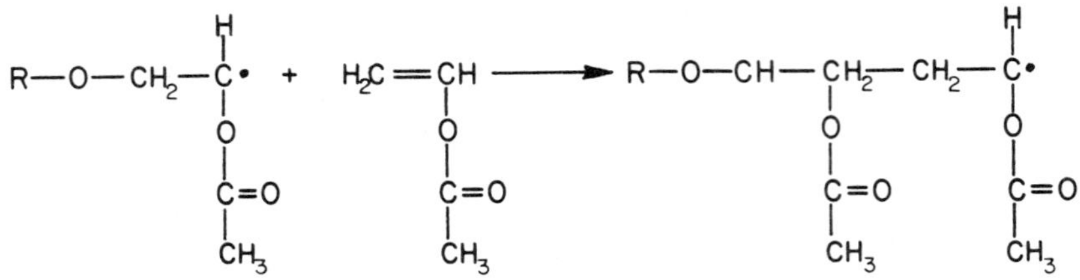

or more generally

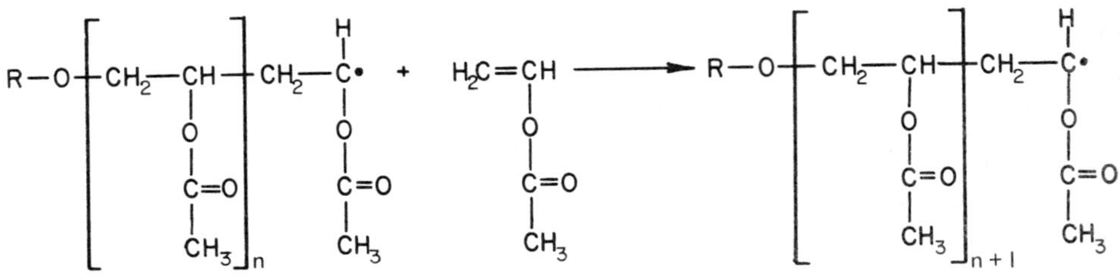

3. *Termination*

Growth of the macromolecule is terminated when the reactive site is removed, either by combination with the reactive site of some other molecule, or by transfer of the reactive site to some other molecule.

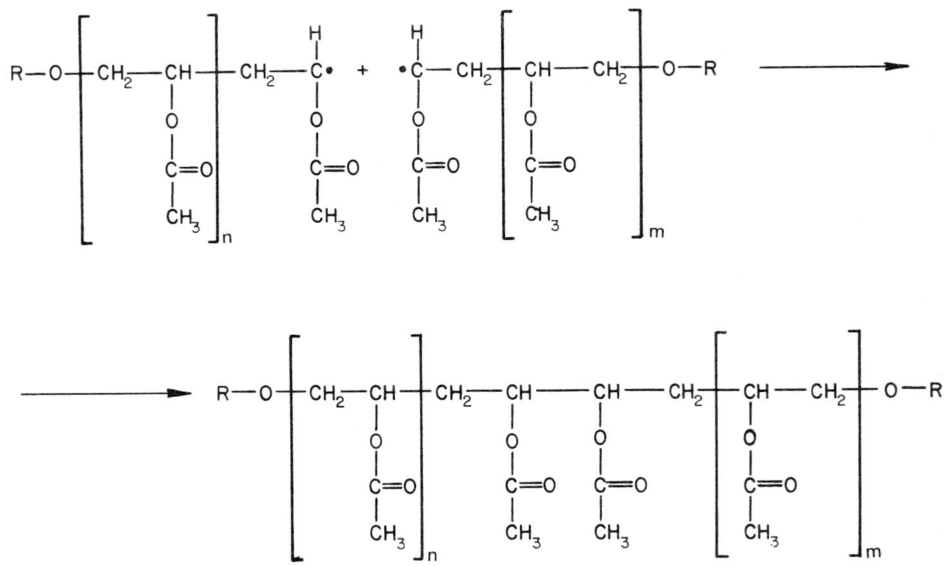

or alternatively,

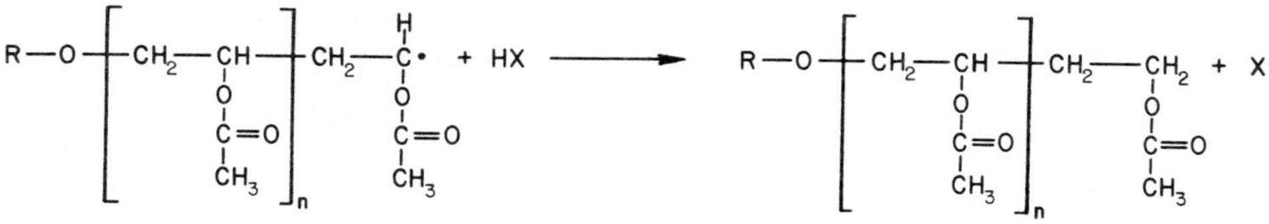

Selection of the initiating catalyst, the ratio of catalyst to monomer, and the reaction conditions allows control over the average molecular weight of the polymer formed, and also of the degree of branching, if any, in the macromolecule.

In the case of polvinyl acetate the polymerization may be carried out by a wide range of techniques, including mass polymerization, solution polymerization, and emulsion polymerization. Most of the polyvinyl acetate produced is made using emulsion polymerization techniques, and this is particularly true of those grades used in the production of wood adhesives. Vinyl acetate is particularly suited to emulsion polymerization, owing to the relatively high solubility of the monomer in water, and the complete solubility of the polymer in the monomer. In this process the vinyl acetate monomer is dispersed by means of relatively high speed stirring in water that contains suitable emulsifiers or protective colloids. A more or less stable suspension of monomer particles will be formed in the water. To this suspension the initiator is added, and typically the mixture will be heated to a temperature which will substantially speed up the rate of reaction, while allowing it to remain controllable. As polyvinyl acetate is soluble in the monomer, the reaction will take place within the individual droplets of the suspension, producing a stable emulsion of polyvinyl acetate. Typical commercial polymer emulsions will contain between 40 and 60 parts by weight of the polymer, a very low residual level of monomer, and a viscosity between 0.1 and 20 Pa-sec. In addition, emulsions intended for use as the basis of formulated wood adhesives will also have a relatively large particle size, usually within the range $0.3-5 \times 10^{-6}$ m.

Other characteristics of the polymer emulsion that will goven its suitability for use in a specific application will include the molecular weight; degree of copolymerization or plasticizing, if any; film strength and film-forming properties at low temperature; and ability of the emulsion to withstand both the mechanical effects of mixing and changes in temperature, especially where freezing may be involved. Manufacturers of PVA emulsions will supply most or all of this information for their products to assist in the selection of the most suitable grade.

IV. FORMULATING A PVA-BASED ADHESIVE

A. General Considerations

When formulating a PVA-based wood adhesive, a number of factors, sometimes conflicting, must be borne in mind. It follows that the final product will often be a compromise in which the conflicting factors have been carefully considered in order to give most weight to those which seem to be the most important in the specific application.

Factors to consider will include the following:

1. *Substrates*

Where the adhesive is to be used for bonding wood to wood, consideration must be given to whether the wood will be hardwood or softwood. For the hardwoods an adhesive of high solid content is usually advantageous, while with softwoods, adhesives of lower solid content may be used. In addition, if the species to be glued is known to be oily, incorporation of a wetting agent or solvent will assist in adhesive penetration.

In many cases, however, wood will be only one of the substrates. The other substrate can vary from concrete in the case of adhesives for parquet or mosaic wood blocks to decorative laminates which may be cellulosic or plastic. Although, in general, PVA adhesives are used principally on cellulosic materials because of their exceptionally good adhesion to such surfaces, special applications may call for wood to be bonded to rubber, to foams, both flexible and rigid, to synthetic or natural fibers, or even to metal or other nonporous surfaces. Each of these will impose restrictions, sometimes severe, on the freedom of the formulator, and corresponding limitations on the applications for which the formulated adhesive may be suitable.

2. *Surface Preparation*

The preparation of the surfaces to be glued will also influence the formulation to a certain extent. In the case of adhesives for gluing wood to wood, inaccurately machined surfaces may necessitate the formulation of gap-filling adhesives in order to produce a satisfactory bond over the entire surface. Adhesives may need to be formulated to give good adhesion to greasy, loose, or dusty surfaces or to seal a very porous substrate. In certain applications the adhesive must be capable of bonding surfaces that have been coated or lacquered.

3. *Application*

The viscosity of the formulated adhesive will largely be governed by the method of application of the adhesive. Application methods will include manual application by brush, roller, smooth or notched trowel or spray, machine application from smooth or embossed rollers, with or without a doctor blade, or by cascade coaters, by nozzle or jet, or even mechanical extruders or sprays. Consideration should always be given to making the adhesive as easy to use as possible, especially where it will be handled by unskilled people unfamiliar with the proper handling of adhesives. This applies particularly to adhesives intended for household use.

4. *Assembly Conditions*

Intricate or multicomponent assemblies will demand an adhesive with a long open assembly period in order to enable all the components to be brought together and placed under pressure before the adhesive has started to dry. At the other extreme, applications such as core composing, where the pressure station is usually very short, require an adhesive that can develop high bond strength very quickly. High production rates will similarly demand a quick-setting adhesive. Application conditions involving high or low temperatures or humidities will also influence the formulation. The formulator must also consider whether or not the glued article requires machining. If this is the case, fillers must be chosen carefully or eliminated in order to minimize damage to cutters. Adhesives containing solvent should only be used in well-ventilated locations.

5. *Service Conditions*

Although PVA adhesives are not, in general, used for joints that are under continuous load or subjected to high temperatures or high humidity, these adhesives can be formulated to give better performance under such conditions. The conditions under which the completed assembly will be expected to operate should always be taken into account when designing the adhesive.

6. *Appearance*

Again, the use to which the completed article is to be put may influence the formulation. The appearance may be marred by unsightly glue lines, in which case it may be necessary to design the adhesive to dry completely transparent, or even to tint the adhesive so that the glue line will be less obtrusive. Tinting or pigmenting the adhesive may be essential where the finished article is to be stained, as squeezed-out adhesive may seal the surface in the vicinity of the glue line and prevent subsequent penetration of stain in that area.

7. *Storage Conditions*

If the adhesive is likely to be stored under adverse conditions, attention must be given to ensuring that it has an adequate freeze-thaw stability. The maximum storage life of the adhesive, where applicable, should also be shown.

8. *Price*

Very often the most serious limitation will be that of formulating to a particular price. In this regard, ease of application, reliability, and spread rate are factors to take into account, as it may be the case that a relatively expensive adhesive will prove more economical in the specific application than an inferior but cheaper product. When comparing prices it is important to take the density of the products into account, especially if they are sold in units of mass.

B. Formulating and Compounding

A number of different components will normally be incorporated into a PVA wood adhesive. Each of these has a specific function in the finished product. To formulate successfully, it is necessary to understand not only the performance criteria, but also the function of these components.

1. The Base Polymer

Since the PVA emulsion provides the major proportion of the adhesive strength, and will in many cases be the only binder in the formulation, it is worth considering its function in some detail. While the mechanism of adhesion is not fully understood, adhesion probably occurs as a result of secondary forces, principally Debye forces and London dispersion forces operating at close range, with some hydrogen bonding with the cellulosic fibers especially where polyvinyl alcohol has been used as the protective colloid. In addition, mechanical bonding occurs as a result of adhesive penetration into the open-cell structure of the wood. All these combine to produce an excellent bond which in a properly formulated adhesive will be stronger than the wood itself.

Since the PVA is in emulsion form, coalescing of the particles must occur in order to produce a continuous film during the drying process. This will happen only if the drying takes place at a temperature above the solidification temperature of the polymer particles. If this condition is not met, the particles will not coalesce properly, leading to a loss of mechanical properties in the dried film. If the temperature at which evaporation of the water takes place is significantly below this minimum film-forming temperature or white point, no coalescing will take place, and a white, chalky film will result with no mechanical strength whatsoever. This white point is thus an important aspect of the base PVA to consider when formulating a wood adhesive.

Viscosity of the emulsion will be determined by the solid content of the emulsion, the particle size distribution, and the emulsifier or protective colloid system used, and may vary between very wide limits. For production of wood adhesives it is most common practice to use grades with a coarse to medium particle size (in the range $0.3-10 \times 10^{-6}$ m) and a solid content between 40 and 60% by weight of the total emulsion. Even with these restrictions viscosity may still be anywhere between 0.05 and 50 Pa-sec, however, and it will usually be necessary to modify this viscosity in the finished product.

The resistance of the dry film from a PVA emulsion to water is mainly dependant on the type and quantity of protective colloid used. Where this is polyvinyl aocohol, the water resistance of the dried film is generally poor. It is possible, however, either by using emulsions protected by cellulose-based colloids, or by incorporating certain additives, to produce PVA adhesives that have a fair resistance to water.

Because polyvinyl acetate is a thermoplastic polymer, it loses cohesive strength as the temperature increases. In general, higher molecular weight polymers lose less strength at elevated temperatures than those of lower molecular weight, but the differences are not great. In addition, thermoplastic polymers are subject to cold creep, which is the tendency for a fully dried film to flow slowly under a sustained load. Plasticized grades are much more subject to cold creep than are unplasticized grades. To a certain limited extent, the tendency to cold creep can be reduced, but in general PVA wood adhesives are not suitable for applications in which the glue line is highly stressed or subject to temperatures above about 50°C, and a combination of these two factors will completely rule out the use of PVA wood adhesives.

Different grades of PVA emulsions will also have different drying times and will therefore offer the possibility of formulating adhesives with long or short open assembly times. Rate of strength development in the bonded structure and the final bond strength attained will similarly vary from grade to grade. Thus selection of the correct grade of polyvinyl acetate emulsion is of major importance in the production of a successful adhesive. All the larger manufacturers issue comprehensive data sheets for their various grades from which it is possible to select, with a good deal of precision, the best grade to use as a starting point.

2. *Other Binders*

In addition to the PVA emulsion it is common practice to incorporate other binders into the formulation of various purposes. Probably the most widely used cobinder is polyvinyl alcohol. Incorporation of polyvinyl alcohol into a formulation increases the cold creep resistance of the dried film, but reduces its water resistance, especially where the polyvinyl alcohol has a low degree of hydrolysis. In addition, polyvinyl alcohol will increase the open assembly time of the adhesive substantially. Use of high molecular weight polyvinyl alcohols is common in low-cost formulations, as these produce relatively high viscosity solutions, allowing incorporation of extra water into the formulation. Adhesives containing polyvinyl alcohol will exhibit good machine stability and running properties and faster initial bond strength development.

The use of starch as an additive is also common practice. Here the major advantage is the cost reduction, which is again achieved at the expense of the water resistance. Because of their affinity for water, starches will also extend significantly the open assembly time of the adhesive. As starches are particularly susceptible to microbial attack, care must be taken to ensure that the formulated adhesive is adequately protected. A wide variety of starch types may be incorporated, including pre-gelatinized, water-soluble, and oxidized starches. Where borax is used to solubilize the starch, the compatibility of the starch solution with the base emulsion must be checked. Borax has the effect of insolubilizing polyvinyl alcohol, and may therefore destroy the protective colloid of the PVA emulsion, thereby destabilizing the emulsion.

While cellulose derivatives such as carboxymethyl cellulose are often added, their function is usually that of viscosity modifier rather than additional binder, and the proportion is invariably small. Thermosetting resins such as phenol, resorcinol, or urea formaldehyde resins are occasionally added to PVA wood adhesives to improve their water resistance. Their use will be more fully discussed subsequently. For specialized application, a range of other binders may be incorprated, especially where wood is to be bonded to some other substrate. Thus all or part of the PVA homopolymer may be replaced by an ethylene-vinyl acetate copolymer for the lamination of polyvinyl chloride film to wood, while vinyl acrylate polymers or copolymers may be added to improve adhesion to nonporous substrates. While dextrins may be added to PVA emulsions used in packaging applications, they are not normally added to PVA wood adhesives.

3. *Plasticizers*

Plasticizers may be regarded as high-boiling solvents with very low vapor pressures at the operating temperature of the adhesive. They will thus remain permanently in the dried film. Plasticizers form a film around the particles of the dispersion, increasing the distance between them and thus lowering the forces between them. In addition to increasing the flexibility of the dried film, they also lower the minimum film-forming temperature of the adhesive. However, they also increase the tendency of the film to creep under load and should thus be used with caution in wood-to-wood adhesives. They should be avoided completely in adhesives for critical applications, especially highly stressed structures, where the use of a solvent to promote film-forming characteristics is preferred.

Where one of the substrates is flexible, they may be used to improve adhesion and match the characteristics of the adhesive film more closely to those of the substrate, especially where this is flexible. Only plasticizers that are compatible with PVA should be used. Those in common use include esters, particularly alkyl phthalates such as dibutyl phthalate, and aromatic phosphates such as tricresyl phosphate, which are chiefly used where flame retardancy is a consideration. Speciality plasticizers are not normally used in wood adhesives. Some physical properties of plasticizers in common use are shown in Table 7.1. As the commonly used plasticizers are more or less immiscible with water, the addition of plasticizer to the base emulsion does not present undue difficulty. It is good practice to add the plasticizer slowly while stirring

Table 7.1. Physical Properties of Some Plasticizers in Common Use[a]

Common name	Chemical name	Molecular weight	Boiling point at 760 mm Hg (°C)	Vapor pressure (mm Hg)
Dibutyl phthalate	Phthalic acid dibutyl ester	278.34	340	10.5 at 200°C
Dioctyl phthalate	Phthalic acid di(2-ethyl hexyl) ester	390.54	385	5 at 230°C
Benzyl butyl phthalate	Phthalic acid benzyl butyl ester	312.35	370	6.5 at 225°C
Tricresyl phosphate	Phosphoric acid tris(methyl phenyl) ester[b]	368.36	430	10 at 265°C
Tributyl phosphate	Tributyl phosphate	266.32	289	5 at 265°C

[a]These figures, where available, are for pure chemicals. Since commercially available raw materials are seldom pure, these figures should only be regarded as a guide.
[b]Avoid phosphoric acid tris(2-methyl phenyl) ester, as this is highly toxic.

Table 7.2. Physical Properties of Some Solvents in Common Use[a]

Common name	Chemical name	Molecular weight	Boiling point at 760 mm Hg (°C)	Evaporation rate[b]	Vapor pressure (mm Hg at 20°C)	Solubility in water at 20°C
Methanol	Methanol	32.04	64.96	6.3	96	Infinite
Ethanol	Ethanol	46.07	78.5	8.3	43	Infinite
Ethyl acetate	Acetic acid ethyl ester	88.12	77.06	2.9	75	7.9%
Butyl acetate	Acetic acid butyl ester	116.16	126.5	12	8	1%
Cellosolve[c]	2-Ethoxyethanol	90.12	135	43	4	Infinite
Acetone	2-Propanone	58.08	56.2	2.1	184	Infinite
Methyl ethyl ketone (MEK)	2-Butanone	72.12	79.6	6	74	26%

Methyl isobutyl ketone (MIBK)	4-methyl 2-Pentanone	100.16	116.85	10	15	2.8%
Ethylene glycol	1,2-Ethandiol	62.07	198	>600	0.1	Infinite
Toluene	Toluene	92.15	110.6	6.1	21	0.05%
Methylene chloride	Dichloromethane	84.93	40	1.8	350	2%
Polysolvan O[d]	Hydroxyacetic acid butyl ester	132.15	183–197	>460	1	8%

[a]These figures, where available, are for pure chemicals. Since commercially available raw materials are seldom pure, these figures should only be regarded as a guide.
[b]Relative to diethyl ether as 1.
[c]Registered trade mark of Union Carbide.
[d]Registered trade mark of Farbwerke Hoechst.

the emulsion vigorously. Once added, the stirring rate may be decreased, but stirring should be continued for at least 30 min to ensure thorough dispersion and to allow the plasticeser to solvate the emulsion particles. Plasticiser will seldom be added at a level above 10% based on polymer solids in formulations for wood adhesives.

4. *Solvents*

Solvents are similar to plasticizers in their influence on the behavior of PVA emulsions with the important difference that the effect they produce is more or less temporary. Thus solvents with a high degree of compatibility with the PVA will lower the solidification temperature of the PVA and thus promote the formation of a homogeneous film, especially at low temperatures. This effect is more marked if the solvent is comparatively slow drying. Since the solvent will ultimately evaporate completely, however, it will not affect the mechanical properties of the dried film. Thus solvents offer a means of lowering the white point of an adhesive without increasing the tendency to cold creep.

Solvents with a high evaporation rate will speed up the drying of the film, especially if they are immiscible with water. They also tend to cause an increase in the viscosity of the emulsion. Solvents which are both miscible with water and compatible with PVA can have a marked destabilizing effect on the emulsion, and special care must be taken when incorporating such solvents. Solvents should be added slowly, dispersing each portion thoroughly before the next portion is added.

Solvents in common use include the lower alcohols, a wide range of esters, ketones, aromatic hydrocarbons, and chlorinated solvents. Aliphatic hydrocarbons are seldom used. A blend of solvents may be used to achieve the desired effects. Solvents will normally be added at a level of 1–5% based on PVA solids. Where plasticizers are also to be added, it is common practice to premix solvent and plasticizer and add the two simultaneously. Certain physical properties of solvents in common use appear in Table 7.2.

5. *Fillers*

Fillers, which are more or less inert, nonadhesive materials are added to reduce cost, increase solid content (and therefore rate of bond strength development), increase viscosity and body, reduce penetration, toughen the film, and improve gap filling properties. In the process they introduce certain disadvantages, in that they raise the white point and increase wear on cutting machines during machining of the bonded articles.

Widely used fillers are gypsum, coated or uncoated calcium carbonate, kaolin or china clay, unmodified starches, and wood flour. Talc or silica may be used provided that no subsequent machining of the bonded article is necessary as they have a drastic blunting effect on cutting tools. Asbestos may be used in special applications as a reinforcing filler.

The organic fillers produce an adhesive with much better machineability, but increase the susceptibility of the adhesive to microbial attack. Where gypsum is used, care should be taken to ensure that it is a fully hydrated grade. Addition of starch to a wood glue produces an improvement in heat resistance of the film.

The level of filler addition has a pronounced influence on adhesive performance. Too high a loading will reduce adhesive strength, and by increasing the specific gravity of the finished product, actually have an adverse effect on the cost of the adhesive. In general, organic fillers are used at a much lower level than inorganic fillers, as they produce a much more marked decrease in tensile strength and increase in viscosity. They will seldom be used at a level above 5–10% based on polymer solids where inorganic fillers may be used at levels as high as 50% of the solid polymer.

Incorporation into the adhesive is straightforward. Nonabsorbent fillers may be added directly as a powder, while absorbent fillers are usually slurried in water with the addition of a suitable wetting agent, prior to incorporation into the adhesive. If this is not done, their high water absorption may cause part or all of the emulsion to coagulate. A formulation containing starch as a filler should not be heated after addition of the starch, as this will tend to solubilize the starch. Some physical properties of fillers in common use appear in Table 7.3.

6. *Other Additives*

A number of other ingredients may be included in a formulated wood adhesive. These include additives to eliminate staining, pigments, dyes or tinters, thickeners, defoamers, preservatives, wetting agents, and in certain applications, tackifiers.

Under acid conditions iron salts form blue dyes with tannin. Since most woods contain varying proportions of tannin and most PVA emulsions are acidic, it is necessary to take precautions to prevent discolorations of this type occuring. The two methods most commonly employed are either to neutralize the emulsion or to add a reagent that forms a colorless complex with iron. Neutralizing the emulsion is most effectively achieved by incorporating 1–2% of an uncoated calcium carbonate into the formulation. This will react with any acid produced during storage and buffer the emulsion at a pH around 7. Alternatively, colorless iron complexes will be produced if the formulation contains about 1% of oxalic acid, phosphoric acid, or a water-soluble polyphosphate salt. If fillers are to be incorporated, a polyphosphate salt will also act as a wetting agent to facilitate wetting out of the fillers.

Pigments or tinters may be added in order to match the color of the glue line to that of the timber to be glued. This will have the effect of rendering the glue line almost invisible and is particularly important where the finished article is to be stained. Squeezed-out adhesive, unless removed promptly, will seal off the surface of the timber and prevent penetration of the stain at that point. Dyes may be included

Table 7.3. Physical Properties of Some Fillers in Common Use[a]

Description	Chemical name	Specific gravity	Oil absorption (%)	Effects on product
Wood flour	–	0.8–1.2	100–300	Marked thickening, good machineability
Starch	–	1.0–1.2	15–30	Good machineability, may cause gradual thickening
Clay or kaolin	Hydrated aluminum silicate	2.6	35–45	Fair machineability
Uncoated whiting	Calcium carbonate	2.7	25–35	Fair machineability; uncoated whiting causes slightly greater thickening
Coated whiting	Calcium carbonate	2.7	20–30	
Talc	Magnesium silicate	2.8	35–45	Bad machineability
Silica	Silicon dioxide	2.65	20–30	Bad machineability
Barytes	Barium sulfate	4.45	6–12	Poor machineability and significant increase in density
Gypsum	Hydrated calcium sulfate	2.4	20–30	May cause thickening of product over first few days if not fully hydrated

[a]The physical properties of these materials may vary widely, depending on source, purity, and method of recovery. These figures should thus be regarded as no more than a guide and do not constitute specifications.

in the formulation in order to produce a colored glue line, either for identification or in order to make it obvious where adhesive has been applied. In either case the nature and quantity of colorant will have to be determined for the specific application. Pigments should be ground to the desired fineness before incorporation into the emulsion. Alternatively, a suitable preground pigment may be obtained from the supplier. Compatibility with the emulsion should be established beforehand, and the smallest acceptable quantity should be used, especially where a paste is used in which the carrier will also have a plasticizing effect on the polyvinyl acetate.

Thickeners may be added in small quantities in order to obtain the desired viscosity. The thickeners in common use are carboxymethyl cellulose, and hydroxyethyl cellulose or sodium polyacrylate. Generally, the grade with the greatest thickening power will be preferred in order to keep the level of addition as low as possible. The modified celluloses may be added predissolved in the smallest possible amount of water, or they may be slurried in a small amount of solvent immediately before addition. Where no solvents or additional water can be tolerated, they may be added very slowly to the emulsion with continuous stirring, which should be maintained for at least 4 hr after the last addition to ensure that the thickener is completely dissolved and dispersed. Where heating is acceptable and available, grades are available which dissolve only once a certain temperature has been reached.

The cellulose-based thickeners lower the water resistance of bonded joints and increase the susceptibility of the adhesive to microbial attack. Hence additions should be kept as low as possible. In this regard it is worth noting that an addition of 1% magnesium carbonate to an emulsion containing both polyvinyl alcohol and cellulose will produce a marked thickening effect.

Fused or pyrogenic silica may also be used to increase the viscosity of an adhesive, but may have undesirable side effects, especially on the rheology of the emulsion. For this reason it is not widely used in wood adhesives. Level of addition will usually be very small—about 0.1–1%. Defoamers may be used wherever foaming of the adhesive is a problem. A wide range of defoamers is available and suitable for incorporation into PVA emulsions. Compatibility tests should always be undertaken and the supplier's recommendation sought as to the most effective level of addition and method of incorporation.

While PVA emulsions have little susceptibility to bacterial attack, once compounded it is usually necessary to add a preservative, especially if starch or cellulose are included in the formulation. In general, mercuric-based compounds are the most effective, but their use is precluded in some countries. Again a number of products may be used, and recommendations should be sought from suppliers to obtain the most effective preservative for any given application. Close attention should be paid to legislation relating to the use of preservatives, also bearing in mind the end use of the product to be glued.

Table 7.4. Illustrative Formulations for Various Types of Wood Adhesives

Low-cost wood glue		*General-purpose wood glue*		*Fast-setting wood glue*		*Long-open-time wood glue*	
Vinac[a] XX-220	65.0	Vinol[a] 523	2.0	Mowilith[e] DH	62.5	Mowilith[e] D50	80.0
Pearl starch	3.9	Vinol[a] 205	1.0	Polysolvan O[e]	1.6	Polysolvan O[e]	2.0
Vinol[a] 540	1.5	Santiciser 160[d]	2.0	Calgon S[f]	0.1	Chalk	2.0
Water	29.4	Ethylene glycol	3.0	Water	0.9	Mowiol[e] 18/88	2.5
Nopco NDW[b]	0.1	Vinac[a] XX210	83.0	150-mesh whiting	30.2	Water	13.5
Acticide MPM[c]	0.1	Water	9.0	MIBK	2.1		100.0
	100.0		100.0	Water	2.6		
					100.0		
Viscosity at 25°C: 6000 cP		Viscosity at 25°C: 5000 cP		Viscosity at 25°C: 12,000 cP		Viscosity at 25°C: 10,000 cP	
Solid content: 41%		Solid content: 52%		Solid content: 65%		Solid content: 44%	

Wood glue with improved Water Resistance		*Woodblock adhesive*		*Adhesive for PVC foil lamination*	
Part A:		Revertex F500[g]	20.8	Elvace 1875[j]	64.0
Mowilith[e] DH	98.0	Revertex F505[g]	20.8	150-mesh whiting	25.0
Polysolvan O[e]	2.0	Barytes	38.8	Gum rosin	3.0
Part B:		Ethyl acetate	0.5	Toluene	3.0
Chromium(III) nitrate	40.0	Cellosize QP 52000[h]	0.8	Reomoll DBP[k]	3.0
		Water	15.0	Calgon S[f]	0.1

Water	60.0	Asbestos powder	0.6	Bevaloid 513[l]	0.1
Mix ratio: 100 parts part A, 5 parts part B		Preventol ON[i]	0.2	Acticide MPM[c]	0.2
		Calgon S[f]	0.1	Water	1.6
		Water	2.4		100.0
			100.0		
Viscosity of mix (25°C): 20,000 cP		Viscosity of 25°C: 40,000 cP		Viscosity at 25°C: 15,000 cP	
Solid content of mix: 49%		Solid content: 60%		Solid content: 66%	
Usable life: within 8 hr of mixing		S.G.: 1.40			

[a]Supplied by Air Products.
[b]Supplied by Nopco Chemical Company.
[c]Supplied by Thor Chemicals.
[d]Supplied by Monsanto Chemical Company.
[e]Supplied by Farbwerke Hoechst AG.
[f]Supplied by I.C.I.
[g]Supplied by Revertex.
[h]Supplied by Union Carbide.
[i]Supplied by Bayer AG.
[j]Supplied by du Pont.
[k]Supplied by Ciba Geigy.
[l]Supplied by Bevaloid.

Wetting agents may occasionally be added for either of two reasons: first, in order to improve the quality of the adhesive by wetting out of fillers prior to their incorporation, and second, to improve adhesive penetration into oily or tight surfaces. For the former purpose soluble polyphosphate salts are preferred, while for the latter, sodium or triethanolamine neutralized long-chain fatty acids are more common. Dosage is usually in the range 0.05–0.15% of total formulation.

Tackifiers are used only where one of the substrates has a surface to which adhesion of the unmodified adhesive will be poor. This applies most often in the case of PVC film, which must often be laminated to wood or chipboard as a decorative finish. Here adhesion can be improved by addition of up to 5% of a resin such as gum rosin or any of its derivatives in admixture with a solvent, a plasticizer, or both. This type of additive is not normally used in adhesives for bonding wood to wood, as it has a markedly deleterious effect on both heat resistance and cold creep resistance.

C. Guide Formulations

A series of formulations is reproduced in Table 7.4. These formulations are intended only as a guide. For the sake of simplicity, trade names have been used. In most instances alternative products will be available from other suppliers.

V. ASPECTS OF APPLICATION

Correct application is as important for the production of a satisfactory bond as is the use of a suitable adhesive. Some of the aspects that are relevant to the correct application of the adhesive are outlined below.

A. Joint Design

In designing an adhesive joint it must at all times be remembered that polyvinyl acetate is a thermoplastic material, subject to phenomena such as cold creep and loss of strength at elevated temperatures. For critical applications, therefore, the joint should be designed to support the adhesive as much as possible. Simple butt joints are unlikely to be suitable for anything but the least demanding applications. Mitre, lap, dado, mortice and tenon, or dowel-supported joints are preferred, the aim being to produce a joint in which the adhesive will be subjected to compression or shear rather than tension or peel stresses. In addition, it should be remembered that the properties of the wood vary with the grain direction, strength being much greater in the direction of the grain than perpendicular to the grain. PVA wood adhesives are unlikely to give satisfactory long-term durability in highly stressed joints or in assemblies that are likely to be subjected to temperatures continuously in excess of 40°C, and in these cases a suitable thermosetting adhesive should be selected.

B. Surface Preparation

The tendency of PVA adhesives to creep under load is increased as the film thickness increases. Thus joints to be glued should be well machined and close fitting. Ideally, surfaces to be bonded should be machined immediately before application of adhesive so that the adhesive has the best opportunity of penetrating the cells ruptured during machining. Surfaces to be glued should be clean and free from sawdust and machine cutting edges should also be clean to prevent contamination of the freshly cut surfaces. The timber will ideally have a moisture content between 8 and 15% by mass, with no great variation between individual components, although under most circumstances it is possible to compensate for higher or lower moisture content in the timber by careful control of the spread rate.

C. Adhesive Preparation

Generally speaking, PVA wood adhesives will be ready for use as supplied. It is generally advisable to stir the contents briefly immediately prior to application to ensure that the adhesive is completely homogeneous. Two component or cross-linked PVA adhesives will require weighing of the components and thorough mixing prior to application. Suitable equipment should be used to ensure that the correct ratio of base to catalyst is maintained, and the mix should be thoroughly stirred to ensure intimate mixing of the components. The recommendations of the adhesive supplier with regard to mix ratio, usable life, and dilution should be obtained and followed at all times. Iron mixing vessels should be avoided, as the presence of iron in the adhesive will increase the likelihood of staining in the timber.

D. Application

Polyvinyl acetate adhesives may be made suitable for application by any of the conventional techniques, including brush, roller, glue spreader, spray, trowel, or glue gun. In consultation with the supplier, the adhesive may be made suitable for application by the most convenient method for the job in hand.

The adhesive may be applied to one or both of the surfaces to be bonded. In general, application to both surfaces is preferred, especially where the time elapsed between application of adhesive and mating of adherends—the open time—is likely to be more than a couple of minutes. Double spreading, as this is called, will markedly improve the penetration of adhesive into both surfaces. Spread rates will vary depending on timber species, closeness of fit in the joint, viscosity and solids content of the adhesive, timber moisture content, and pressing conditions, but will usually be in the region 100–200 g/m^2 where both surfaces are to be coated, and 200–300 g/m^2 where adhesive is applied to one surface only. Again, the recommendations of the supplier should be obtained and followed.

1. *Special Applications*

As well as being well suited to standard joinery practices, PVA adhesives are also used in a number of specialized applications. These include core composing, edge veneering, and high-frequency curing.

For composing of cores for blockboard manufacture, full use is made of polyvinyl acetate's ability to develop bond strength extremely rapidly. Typical core composing machines have a pressure station of about 30–120 sec and the adhesive bond must develop sufficient strength within this period to hold the components of the core during gentle handling. A unique advantage of the PVA adhesives here is that their thermoplasticity will allow the components of the core to realign if necessary during the subsequent heated veneering operation.

In addition to conventional edge veneering techniques, PVA adhesives may also be used for edge veneering by heat reactivation. Here use is again made of the thermoplasticity of the PVA. One or both of the surfaces to be mated will be coated with the adhesive, which is then allowed to dry completely. The bond may be produced at any subsequent time very rapidly by bringing the surfaces together under heat and pressure. Because of the low stresses on the glue line it is usually possible to release pressure while the adhesive is still warm, thus permitting short cycle times.

Because of their rapid bond strength development, PVA adhesives are also well suited for high-frequency gluing. Cycle times of 10–30 sec are possible, again permitting very high production rates.

E. Assembly Conditions

Open and closed assembly periods and clamping conditions must be related to the specific adhesive under consideration. Maximum open assembly time may be made as short as 5 min or as long as 30 min, depending on the formulation. In addition, it will be influenced by external factors, such as the ambient temperature, the relative humidity, the timber species and density, and the moisture content of the timber. High temperatures, low humidity, low-density timber, and low moisture content in the timber will all reduce the allowable open assembly time for any given formulation, while the opposite conditions will increase the open time. Especially where double spreading is employed, an open time of about 25–50% of the maximum open time is usually desirable to allow some of the water to evaporate. This will increase the speed of strength development. Articles to be bonded with PVA emulsions do not generally require a closed assembly period, but may be clamped immediately after mating. Where several components have to be assembled, however, a certain closed assembly period will be inevitable. This is not likely to cause problems as long as the total open and closed assembly time does not exceed the maximum allowable open assembly period for the specific adhesive under the prevailing conditions of use.

The object of clamping is to obtain the thinnest possible glue line without causing either damage to the wood or excessive squeeze out of glue and consequently a starved glue line. Attainment of a thin glue line is particularly important where the glue line is likely to be subject to continuous stress. Clamping should be so arranged that a uniform pressure is applied to the entire glue line, shown by a uniform slight squeeze out along the complete joint.

Excess glue should be removed immediately, especially where the article is to be stained, as dry glue will seal the timber and prevent penetration of the stain at that point. In general, fairly high clamping pressures, of the order of 600–1000 kPa (100–150 psi) will be used for hardwoods, while lower pressures, around 300–600 kPa (50–100 psi) will be sufficient for softwoods.

Pressing times will vary depending on the temperature of the glue line, the adhesive used, its spread rate, the open assembly period used, the timber species, density and moisture content, and the degree of stress in the assembly. With such a range of variables it is not possible to lay down any guidelines and the adhesive supplier should be approached for recommendations in this regard.

F. Influence of Temperature

The glue line temperature has a marked effect on the rate of bond development. High temperatures will produce a much more rapid increase in strength, while temperatures below normal lead to slow strength development and, if allowed to drop below the minimum film-forming temperature, may cause the bond to fail completely. In this regard it is the temperature of the timber rather than the ambient temperature that is the governing factor, as wood has a low thermal conductivity and responds relatively slowly to changes in ambient conditions.

In addition to increasing the rate of strength development, higher temperatures during clamping produce an increase in the ultimate strength developed. This is particularly substantial in the case of two-component adhesives designed to offer improved water resistance. Pressing for 1 hr at 80°C can produce an improvement of up to 10% in ultimate shear strength, up to 30% in shear strength after immersion in boiling water for 3 hr, and up to 50% in shear strength after immersion in cold water for 24 hr compared with results obtained on the same adhesive cured at ambient temperature.

Hot pressing may also lead to a substantial reduction in the pressing time required. As an example, an adhesive requiring a basic pressing time of 1 hr at 20°C or 30 min at 30°C will require a basic pressing time of only $1\frac{1}{2}$ min at 95°C or 1 min at 105°C [8]. These times refer to the temperature at the glue line and will, as a rule, need to be increased substantially depending on the thickness of the wood and consequently the length of time that the heat will need to penetrate to the glue line.

Hot pressing of PVA wood adhesives is not, however, widely employed, as urea-formaldehyde resins will produce equally good bonds equally fast at elevated temperatures at much lower cost. The major advantage of PVA wood glues over urea-formaldehyde resins is their ability to produce bonds rapidly at room temperature, thereby obviating the necessity for expensive heated presses. In addition, as polyvinyl acetate is a thermoplastic material, it is necessary to maintain clamping pressures while the adhesive cools to a point where the film has sufficient strength to hold the assembly together. Thus most of the speed advantage will be lost, especially where the assembly places a stress on the glue line.

VI. PERFORMANCE OF PVA ADHESIVES

Once the adhesive has been chosen correctly and applied correctly, its long-term performance will be determined by the conditions under which the bonded structure is expected to operate. The factors that will affect the durability of the joint include climatic conditions such as temperature, ultraviolet and infrared radiation and humidity, environmental factors such as intermittent or continuous exposure to water or chemicals, the presence or absence of microorganisms, and the existence of stresses wthin the assembly. In a properly designed structure, all these factors will have been taken into consideration before selection of the adhesive to ensure that it will not be operating outside its working limits.

A. Factors Affecting Durability

1. *Temperature*

Because of the thermoplasticity of polyvinyl acetate, wood adhesives based on PVA are inherently unsuited to applications requiring sustained resistance to high temperatures. In general, temperatures continuously in excess of 40°C are likely to prove too much, even for joints that are only mildly stressed. Polyvinyl acetate wood adhesives should thus be viewed as being essentially room temperature adhesives, suitable only for applications in which at worst the glue line temperature will rise only intermittently above normal ambient temperature—and then only by 10–15°C.

Within the framework of this limitation, however, where the adhesive will be required to operate at temperatures above room temperature, the adhesive may be made more suitable by using a high molecular weight emulsion, rigid elimination of anything that might constitute a plasticizer, and incorporation of a high molecular weight polyvinyl alcohol. Such a formulation whould also be buffered by, for example, addition of about 2% of calcium carbonate to prevent any hydrolysis of the polyvinyl acetate, as the acetic acid so produced will tend to plasticize the adhesive film. Incorporation into the formulation of a percentage of thermosetting resin such as a phenol formaldehyde will also improve the heat resistance of the formulated adhesive. As this will, however, require either the addition of a catalyst immediately before application, or heat curing, much of the convenience of this type of adhesive will be lost.

2. *Ultraviolet Light*

Polyvinyl acetate is not normally degraded by ultraviolet light, nor is this likely to cause problems unless other components are present that might break down. In this case addition of a small amount of an ultraviolet light absorber should eliminate the problem.

3. *Humidity*

Emulsion-based adhesives tend to be inherently sensitive to moisture, and polyvinyl acetate is no exception. It will, however, withstand the effect of changes in humidity provided that the moisture sensitivity has not been heavily affected by significant additions of water soluble additives such as methyl cellulose or related products or partially hydrolyzed grades of polyvinyl alcohol.

4. *Water and Solvents*

While polyvinyl acetate adhesives will not usually be adversely affected by changes in humidity, prolonged exposure to water or solvents will generally destroy the bond. Because of the thermoplastic nature of polyvinyl acetate, it will always be vulnerable to solvents, especially those such as ketones, esters, or aromatic hydrocarbons that are true solvents for the base polymer, and there is little that can be done to reduce this susceptibility. PVA wood glues are thus not suitable for applications that involve intermittent or continuous exposure to solvents. It is however, possible to improve the water resistance of the bond to a considerable extent. There are two methods of achieving this.

a. Insolubilizing: Where polyvinyl alcohol is present, either as the original protective colloid or as a subsequent additive, its tendency to redissolve and hence increase the susceptibility of the glue line to water may be reduced by reacting it with a variety of chemicals, including certain trivalent metal salts such as aluminium nitrate or chloride or chromium nitrate, and glyoxal. The most common of these is probably chromium nitrate, which is usually added as a 40% solution of the salt in water. Addition is typically at the level of 5% based on the base emulsion. While mixes of this type do not have a pot life in the normally accepted sense, the water resistance of joints bonded with such a mix will be lower if the adhesive is mixed more than 2–3 days before application, so it is normal practice to supply such products as two-component systems and to mix no more at a time than can be used in a normal working day. It should be noted that the use of chromium nitrate produces a blue glue line and that all three catalysts based on trivalent metal ions lead to a drop in the pH of the glue mix.

The water resistance of bonds produced with two component systems of this type are improved significantly if the assembly is subsequently postcured under heat. Not only is the rate at which water resistance is developed faster, but in addition the ultimate water resistance will be far greater if the glue line is thus postcured.

b. Fortifying: The second method of improving water resistance is by incorporating a fortifying resin into the formulation. This may be based on any of the currently available thermosetting formaldehyde resins, such as urea-formaldehyde, melamine-formaldehyde, phenol-formaldehyde, or resorcinol-formaldehyde. With any of these additives the mix will have a pot life of a few hours, so again the adhesive must be sup- ;lied as a two-component system. It is usual to keep the PVA emulsion in one component and the formaldehyde resin in the other, as there may be storage stability problems if the two are premixed. This means that the PVA component must contain an additive to complete the cross-linking of the other component. This will usually be ammonium chloride in the case of the aminoplast resins and phenol-formaldehyde or paraformaldehyde in the case of resorcinol-based resins. The type and quantity of fortifying resin added will depend on the degree of durability required and the cost but will, in general, be between 5 and 25% of the base polymer. Urea formaldehyde will impart less resistance than melamine- or phenol-formaldehyde, while resorcinol-based resins will give the greatest degree of water resistance. Phenol- and resorcinol-formaldehydes will produce dark glue lines, especially if used at levels above 5%.

In general, the use of fillers is undersirable in water-resistant PVA adhesives, as fillers impair the water resistance. It should also be noted that different grades of PVA emulsion will have varying degrees of water resistance to start with. The recommendations of the supplier should thus be sought for the best grade to use not only with regard to the water resistance of the base emulsion, but also with regard to its compatibility with the fortifying resin.

5. Microbial Attack

The base polymer is not susceptible to attack by microorganisms and should not, therefore, require protection. There is, however, the possibility that certain of the additives may be susceptible and may, therefore, not only increase the vulnerability of the glue line to attack, but may even introduce microorganisms into the product. This may easily be prevented by addition of a suitable preservative as detailed in Sec. IV.B.6. Where long-term durability, especially in a warm and damp environment is required, care must be taken to ensure that the preservative used will not leach out.

B. Specifications

Along with the increasing use of wood adhesives based on PVA, there has been an increase in the number of performance specifications produced by various independant bodies. Organizations such as the American Society for Testing and Materials (ASTM), the British Standards Institute (BSI), and the Deutsche Industrie Normen (DIN), as well as departments of defense, statutory, and private organizations around the world, have drawn up specifications and detailed test methods for PVA-based wood glues, while research bodies have spent years correlating the results of long-term exposure tests in various environments with results obtained by accelerated means, in order to develop short-term tests that will give a close approximation to results obtained in the actual conditions of use. As a result, most of the developed countries have specifications providing the means of predicting the behavior and durability of adhesives under various conditions of use. From this plethora of specifications, some of the better known are summarized here.

1. British Standard BSS 4071 [9]

This standard specifies the minimum resistance the adhesive must possess to alternate cycles of freezing and thawing—the freeze-thaw stability. It also lays down minimum storage life, after which the adhesive must still perform satisfactorily, minimum strength that must be obtained under dry conditions, maximum staining permitted, and maximum degree of cold creep. Cold creep is tested by placing test pieces under a sustained load for 7 days. All strength tests are performed by placing double lap specimens prepared from beechwood in tension. Details of the methods of test are given in BSS 3544 [10].

2. American Military Specification MMM-A-193C [11]

This specification lays down the range in which the viscosity must lie, the minimum freeze-thaw stability, and various minimum shear strengths, tested at both room temperature and elevated temperature on dry specimens prepared from plywood, and on similar pieces subjected to a humid atmosphere, while a minimum block shear strength on test pieces prepared from maple is also specified.

3. Japanese Standard JIS K6804 [12]

This standard provides for two classes of adhesive. Class 1 is for single-component PVA adhesives while Class 2 is for PVA adhesives designed to be mixed with urea-formaldehyde resins. Class 1 is further subdivided into Number 1, adhesives for use at normal temperature, and Number 2, adhesives for use at low temperature. The major difference between the latter two is the much lower minimum film-forming temperature specified for the Number 2 adhesives. Class 2 adhesives must obviously be miscible with urea-formaldehyde resin with and without the hardener added. The standard also specifies pH range, minimum viscosity, solids and strength both before and after immersion in water, maximum ash content, and permissible degree of staining. Strength is tested by performing block shear tests using test pieces machined from maple.

4. German Industrial Standard DIN 68602 [13]

Four classes of adhesive are specified in this standard, depending on the stability of the bond to exposure to low or high humidity conditions or direct weathering. Class B1 is suitable for use in dry interior conditions, Class B2 will withstand exposure to high humidity interior conditions such as those encountered in bathrooms or kitchens, Classes B3 and B4 are able to be used under exterior exposure conditions, with Class B4 being suitable for more severe exposure than Class B3.

C. Testing

Testing of PVA emulsions may be divided into two broad categories: physical testing and performance testing. Physical testing embraces all the directly measurable properties of the base emulsion or formulated product, such as viscosity, pH, volatile content, minimum film-forming temperature, or specific gravity, as well as less easily measured properties such as molecular weight distribution, second-order transition or glass point, particle size distribution, and free monomer content, while performance testing deals with characteristics such as bond strength, rate of strength development, water resistance, and durability which are related to the manner in which the product may be expected to perform in use.

Unfortunately, the result of any of these tests depends to a significant extent on the method of test used, especially in the case of performance tests. The situation is made worse by the fact that many organizations that manufacture or process polyvinyl acetate have developed their own test methods. Again bodies such as the ASTM, JIS, and BSI have drawn up standard test methods so that results obtained from different sources may be compared directly. It is thus essential when quoting results of a test to include all the relevant test conditions: otherwise, the results obtained may be nothing more than a guide. It must also be remembered that seemingly minor deviations from a standard test method may produce disproportionately large variations in results.

While test conditions may usually be tightly controlled in the case of most of the physical tests, performance tests often depend on factors which are not easily controlled. For this reason, performance tests are usually performed at least in quintuplicate in order to reduce as far as possible the effects of these variable factors.

VII. CONCLUSION

While appearing to the casual observer to be highly advanced, polymer chemistry is still relatively speaking, in its infancy. With literally thousands of scientists working in laboratories all over the world, new and exciting discoveries are made daily that extend the scope and applicability of polymers. Polyvinyl acetate is no exception and it seems safe to predict that it will play an increasingly important part in adhesive technology. Improvements to the water and solvent resistance and bond strength at high temperature and reduction of cold creep will be made and will eliminate most of the disadvantages of polyvinyl acetate wood adhesives while retaining the advantages inherent to the resins constituting these adhesives. Even without these improvements, their ease of handling, tolerance to abuse in application, and rapid bond strength development have made them one of the most important wood adhesives today and it is probable that this place will be maintained into the foreseeable future.

ACKNOWLEDGMENTS

The author would like to express his gratitude to the following organizations for information and assistance freely offered and gratefully accepted: Farbwerke Hoechst AG and Hoechst South Africa and their extremely helpful "Mowilith Handbook"; Air Products and Chemicals, Inc., and Air Products South Africa, particularly Mr. F. M. Rosenblum, for permission to quote from his article "White Glues"; African Explosives and Chemical Industries; South African Bureau of Standards; Rubber and Plastics Research Association, United Kingdom; and SIRA Institute, United Kingdom.

REFERENCES

1. F. Klatte, German State Patent No. 271,381 (1912).
2. H. Warson and S. A. Miller, *Ethylene and Its Industrial Derivatives*, BENN, London, 1969.
3. 1979 Census Bureau statistics, *Adhes. Age*, 36 (Sept. 1979).
4. *U.K. Production Business Monitors*, 2nd quarter, 1980 [HMSO, London (PQ276.1).
5. Rubber and Plastics Research Association, private communication, 1980.
6. German State Patent No. 582,544 (1930).
7. German State Patent Nos. 403,784 (1921) and 485,271 (1927).
8. *Borden Technical Data Sheet No. 756e*, Dec. 1979.
9. Specification for polyvinyl acetate emulsion adhesives for wood, British Standard Specification BSS 4071, 1966.
10. Methods of test for polyvinyl acetate adhesives for wood, British Standard Specification BSS 3544, 1962.
11. Adhesive, vinyl acetate resin emulsion, American Military Specification MMM-A-193C, Oct. 1977.
12. Polyvinyl acetate emulsion adhesives for wood, Japanese Industrial Standard Specification JIS K 6804, 1977.
13. Evaluation of adhesives for joining of wood and derived timber products, German Industrial Standard DIN 68602, Apr. 1979.

Author Index

Subject Index